NANOTECHNOLOGY: ENVIRONMENTAL IMPLICATIONS AND SOLUTIONS

NANOTECHNOLOGY: ENVIRONMENTAL IMPLICATIONS AND SOLUTIONS

LOUIS THEODORE
Theodore Tutorials, Consultant

ROBERT G. KUNZ
Environmental Consultant
Hillsborough, NC

Foreword – Rita D'Aquino
Section 1.3 Contributor – Adrian Calderone
Chapter 2 Contributor – Suzanne A. Shelley

WILEY-INTERSCIENCE

A JOHN WILEY & SONS, INC PUBLICATION

Published by John Wiley & Sons, Inc., Hoboken, New Jersey.
Published simultaneously in Canada.

For general information on our other products and services please contact out Customer Care Department
within the U.S. at 877-762-2974, outside the U.S. at 317-572-3993 or fax 317-572-4002.

Wiley also publishes its books in a variety of electronic formats. Some content that appears
in print, however, may not be available in electronic format.

Library of Congress Cataloging-in-Publication Data:

Theodore, Louis.
   Nanotechnology : environmental implications and solutions / Louis Theodore, Robert G. Kunz.
      p. cm.
   Includes bibliographical references and index.
   ISBN 0-471-69976-4 (cloth)
   1. Nanotechnology.   I. Kunz, Robert G. II. Title.

   T174.7.T48 2005
   620'.5--dc22   2004053459

Printed in the United States of America

10 9 8 7 6 5 4 3 2

"Nothing is impossible to the man who doesn't care who gets credit for it."
—Adapted from the comments at President Ronald Reagan's funeral,
June 11th, 2004

CONTENTS

PREFACE

Nanotechnology. Although not to be found in any *Webster's Dictionary*, it is concerned with the world of invisible miniscule particles that are dominated by forces of physics and chemistry that cannot be applied at the macro- or human-scale level. These particles have come to be defined by some as nanomaterials, and these materials possess unusual properties not present in traditional and/or ordinary materials.

Regarding the word *nanotechnology*, it is derived from the words *nano* and *technology*. Nano, typically employed as a prefix, is defined as one-billionth of a quantity or term that is represented mathematically as 1×10^{-9}, or simply as 10^{-9}. Technology generally refers to "the system by which a society provides its members with those things needed or desired." The term nanotechnology has come to be defined as those systems or processes that provide goods and/or services that are obtained from matter at the nanometer level, that is, from sizes in the range of one-billionth of a meter. The new technology thus allows the engineering of matter by systems and/or processes that deal with atoms; or as Drexler (who some view as the godfather of this industry) put it: "... entails the ability to build molecular systems with atom-by-atom precisions, yielding a variety of nanomachines". One of the major problems that remain is the development of nanomachines that can produce other nanomachines in a manner similar to what many routinely describe as mass production.

The classic laws of science are different at the nanoscale. Nanoparticles possess large surface areas and essentially no inner mass, i.e., their surface-to-mass ratio is extremely high. This new "science" is based on the knowledge that particles in the nanometer range, and nanostructures or nanomachines that are developed from these nanoparticles, possess special properties and exhibit unique behavior. These special properties, in conjunction with their unique behavior, can significantly impact physical, chemical, electrical, biological, mechanical, and functional qualities. These new characteristics can be harnessed and exploited by applied scientists to engineer "Industrial Revolution II" processes. Present-day and future applications include chemical products, including plastics, specialty metals, powders, computer chips, computer systems, and miscellaneous parts, pollution prevention areas that can include energy conservation, environmental control, and health/safety issues, plus addressing crime and terrorism concerns. In effect, the sky's the limit regarding efforts in this area, and as far as the environment is concerned, this new technology can terminate pollution as it is known today.

The authors believe that nanotechnology is the second coming of the industrial revolution, or as one of the authors has described it, "Industrial Revolution II." It promises to make that nation (hopefully ours) that seizes the nanotechnology initiative the technology capital of the world. One of the main obstacles to achieving this goal will be to control, reduce, and ultimately eliminate environmental and environmentally related problems associated with this technology; the success or failure of this new use may well depend on the ability to address these environmental issues.

Only time will provide answers to three key environmental questions:

1. What are the potential environmental concerns associated with this new technology?
2. Can industries and society expect toxic/hazardous material to be released into the environment during either the manufacture or use of nanoproducts?
3. Could nanoapplications lead to environmental degradation, particularly from bioaccumulation of nanoproducts in living tissue?

Regarding these three questions, the environmental health and hazard risks associated with both nanoparticles and the applications of nanotechnology for industrial uses are not known. Some early studies indicate that nanoparticles can serve as environmental poisons that accumulate in organs. Although these risks may prove to be either minor, avoidable, or both, the engineer and scientist is duty bound to determine if there are in fact any health, safety, and environmental impacts associated with nanotechnology.

Regarding education, it is becoming more and more apparent that engineering and science curricula must include courses that are concerned with nanotechnology material that the student will need and use both professionally and socially later in life. It is no secret that the teaching of nanotechnology and nano-related courses will soon be required in most engineering curricula. It is also generally accepted as one of the key state-of-the-art courses in applied science. The need to develop an understanding of this general subject matter for the practicing engineer and scientist of the future cannot be questioned.

Regarding students, the Accreditation Board for Engineering and Technology (ABET) requires that engineering graduates understand the engineer's responsibilities both to protect the environment and address occupational health and safety concerns. Traditionally, engineering schools have done a superb job of educating students on the fundamental laws of nature governing their fields and on the application of these laws to real-world situations. Unfortunately, they have yet to be successful in conveying to the students the importance of not only nanotechnology itself but also the accompanying environmental, health, and safety issues in the design and application of this new technology. These concerns served as the driving force for the writing of this book.

This project was a unique undertaking. Rather than prepare a textbook on nano-technology, the authors considered writing a book that was concerned with the environmental implications of nanotechnology. Because of the dynamic nature of this emerging field, it was also decided to write an overview of this subject rather than to provide a comprehensive treatise. One of the key features of this book is that it could serve both academia (students) and industry. Thus, this book offers material not only to individuals with limited technical background but also to those with extensive industrial experience. As such, it can be used as a text in either a general engineering or environmental engineering/science course and (perhaps primarily) as a training tool for industry.

As suggested above, this book is intended primarily for environmental regulatory officials, company administrators, engineers, scientists, and both undergraduate and first-year graduate engineering and science students. It is assumed that the reader has taken basic courses in physics and chemistry; only a minimum, background in mathematics is required (though calculus is desired). The authors' aim is to offer the reader the fundamentals of the subject title. The reader is encouraged to use the works cited in the references to continue development beyond the scope of this book.

In the final analysis, the problem of what to include and what to omit was particularly difficult. However, every attempt was made to offer course materials to individuals at a level that should enable them better to cope with some of the nanoenvironmental problems they will later encounter in practice. As such, the book (as indicated above) was not solely written for the student; rather, it was primary written for these individuals who plan to work in the nanotechnology field as engineers and/or scientists solving future real-world problems.

The book contains 10 chapters. Chapter 1 provides an introduction to background issues surrounding nanotechnology and the environmental implications associated with this new technology. Chapter 2 covers nanofundamentals. Chapters 3, 4, and 5 address air, water, and land (solid waste) issues, respectively, particularly as they relate to control and monitoring. The general subject of multimedia approaches is reviewed in Chapter 6. Chapters 7 and 8 address risk concerns—health in the former and hazard in the latter chapter. A brief introduction to ethics can be found in Chapter 9, and the text concludes with a short chapter (10) on future trends.

The authors wish to acknowledge those individuals who served as either contributors or partial contributors to earlier works that are referenced in the body of the text. The name of these engineers—along with the section that contains some of their contributed work—follows: David Gouveia, Section 1.5; Christine Hellwege, Section 1.6; Jeanmarie Spillaine, Section 1.7; Christopher Reda, Section 3.5; Michael Reid, Section 4.3; Kristina Neuser, Section 4.4; Peter Damore, Section 5.1; Romeo Fuentebella, Section 5.2; Dorothy Caraher, RN, Section 5.4; James Merlin, Section 5.5; Christine Jolly, Section 5.7; Julie Shanahan, Section 7.6; Patricia Brady, Section 7.7; Eleanor Capasso, Section 8.7; Ruth Richardson, Section 9.1; and Abdool Jabar, Section 10.9. We gratefully acknowledge their contributions.

Finally, every effort has been made to acknowledge material drawn from the literature. The authors' trust that their apology will be accepted for any errors, or omissions, and changes will be included in a later printing or edition.

Regarding acknowledgments, the authors would be remiss if they did not note the major contributions of the following three colleagues: Adrian Calderone—J.D., Rita D'Aquino—*Chemical Engineering Progress*, and Suzanne Shelly—*Chemical Engineering* magazine.

LOUIS THEODORE
ROBERT G. KUNZ

FOREWORD

RITA D'AQUINO, BSChE, MSChE
Senior Editor, *Chemical Engineering Progress*

When Lou Theodore and Bob Kunz approached me about writing a foreword to their book on environmental nanotechnology, I was both surprised and honored. As an editor with *Chemical Engineering Progress* magazine, the flagship publication of the American Institute of Chemical Engineers (New York, New York; *www.aiche.org*), I have followed the topic closely, bearing witness to its ability solve environmental problems, monitor cellular metabolism, deliver drugs to inaccessible parts of the body, strengthen polymers, render inert materials violently reactive, and get once-impossible chemical reactions to occur. Equally powerful is its potential to act as poison, accumulating in the environment and in animal organs.

Therein lies the seed of dissent. While some can't wait to seize the nanotechnology initiative, others won't touch it with a 40-ft pole. Activists have begun to organize against the science, calling for a moratorium on nanotechnology products until the social and environmental risks are better understood and regulations are put into place.

Indeed, if ignorance is the sickness, then knowledge must be the cure. That is why it is with such great pleasure that I introduce *Nanotechnology: Environmental Implications and Solutions* to the scientific and engineering audience. Drawing on their extensive knowledge and experience in the field of environmental science and chemical engineering, Theodore and Kunz touch upon all the key environmental issues of which anyone involved in nanotechnology should be aware. Given their collective background in the environmental arena, the book is certain to be a benchmark in the field.

"The concerns raised by those involved with the environmental implications of nanoparticle emissions from nanoapplications appear not be to justified," writes Lou Theodore in the April 24, 2004, issue of, "As I See It," his monthly column in the *Williston Times*, a local Long Island, New York, newspaper. In their current work, the authors apply the classical works of Cunningham (1906) and Einstein (1910) to demonstrate that submicron particles are easier (for recovery and/or control purposes) than their microplus counterparts, a hypothesis for which there is limited experimental evidence. Breaking new ground, the authors present data generated with high-effciency control devices (e.g., baghouses, electrostatic precipitators, and venturi scrubbers) in which particles in the submicron regime were collected with 100 percent efficiency.

The laws of chemistry and physics work differently when particles reach the nanoscale, as the powers of hydrogen bonding, quantum energy, and van der Waals' forces endow some nanomaterials with unusual properties. Carbon nanotubes, for instance, discovered in the sooty residue of vaporized carbon rods, defy standard physics. Stronger and more flexible than steel, yet measuring about 10,000 times smaller than the diameter of a human hair, these cylindrical sheets of carbon atoms are useful as coatings on computers and other electrical devices despite their ability to conduct heat and electricity. Nanoparticles, another manifestation of nanotechnology, are known to foster stubborn reactions because they have enormous surface area relative to their volume.

What I have learned after months of exploring the applications and implications of nanotechnology is simply that these developments are not only changing our lives every day, but they are moving so fast we have not yet grasped their tremendous impact. I commend Theodore and Kunz for embracing the challenges that come with exploring the environmental opportunities of nanotechnology and ways to pursue and mitigate the potential dangers ahead. If there's one point the authors drive home, it's that nanotechnology will revolutionize the way industry operates, creating chemical processes and products that are more efficient and less expensive. A primary obstacle to achieving this goal will be to control, reduce, and ultimately eliminate the environmental and related problems associated with nanotechnology, or dilemmas that may develop through its misuse. Nanotechnology enables control over properties of materials and structures at an atomic scale, so that engineering of materials and devices specific to a need—even a malevolent misuse (e.g., the release of nanoproducts and devices that consume water or flesh; emissions of deadly virus-like organisms; ill-suited alterations of DNA, cells, and biological beings)—is simple and inexpensive. The radical age sparked by taking a nanoscale approach to technology, or what Theodore has dubbed, "the second Industrial Revolution," is, in his own words, "so powerful that it promises to make the country that conquers nanotechnology the capital of the world."

CHAPTER 1

NANOTECHNOLOGY/ ENVIRONMENTAL OVERVIEW

1.1 INTRODUCTION

Nanotechnology is concerned with the world of invisible miniscule particles that are dominated by forces of physics and chemistry that cannot be applied at the macro- or human-scale level. These particles have come to be defined by some as nanomaterials, and these materials possess unusual properties not present in traditional and/or ordinary materials.

Regarding the word *nanotechnology*, it is derived from the words *nano* and *technology*. Nano, typically employed as a prefix, is defined as one billionth of a quantity or term that is represented mathematically as 1×10^{-9}, or simply as 10^{-9}. Technology generally refers to "the system by which a society provides its members with those things needed or desired." The term *nanotechnology* has come to be defined as those systems or processes that provide goods and/or services that are obtained from matter at the nanometer level, that is, from sizes at or below one billionth of a meter. The new technology thus allows the engineering of matter by systems and/or processes that deal with atoms; or as Drexler (whom some view as the godfather of this industry) put it: "... entails the ability to build molecular systems with atom-by-atom precision yielding a variety of nanomachines." One of the major problems that remain is the development of nanomachines that can produce other nanomachines in a manner similar to what many routinely describe as mass production.

The classic laws of science are different at the nanoscale. Nanoparticles possess large surface areas and essentially no inner mass, that is, their surface-to-mass ratio is extremely high. This new "science" is based on the knowledge

Nanotechnology: Environmental Implications and Solutions, by L. Theodore and R. G. Kunz
ISBN 0-471-69976-4   Copyright © 2005 John Wiley & Sons, Inc.

that particles in the nanometer range, and nanostructures or nanomachines that are developed from these nanoparticles, possess special properties and exhibit unique behavior. These special properties, in conjunction with their unique behavior, can significantly impact physical, chemical, electrical, biological, mechanical, and functional qualities. These new characteristics can be harnessed and exploited by applied scientists to engineer "Industrial Revolution II" processes. Present-day and future applications (to be discussed in more detail in the next section) include chemical products—to be parts, plastics, specialty metals, and powders; computer chips, computer systems, and miscellaneous parts; and pollution prevention areas that can include energy conservation, environmental control, and health/safety issues; plus crime and terrorism concerns. In effect, the sky's the limit regarding efforts in this area, and as far as the environment is concerned, this new technology can terminate pollution as it is known today.

The authors believe that nanotechnology is the second coming of the industrial revolution, or as one of the authors[1] has described it, "Industrial Revolution II." It promises to make that nation (hopefully ours) that seizes the nanotechnology initiative the technology capital of the world. One of the main obstacles to achieving this goal will be to control, reduce, and ultimately eliminate environmental and environmental-related problems associated with this technology; the success or failure of this new use may well depend on the ability to address these environmental issues. Only time will provide answers to three key environmental questions:

1. What are the potential environmental concerns associated with this new technology?
2. Can industries and society expect toxic/hazardous material to be released into the environment during either the manufacture or use of nanoproducts?
3. Could nanoapplications lead to environmental degradation, particularly from bioaccumulation of nanoparticles in living tissue?

Regarding these three questions, it is important to note that the environmental health and hazard risks associated with both nanoparticles and the applications of nanotechnology for industrial uses are at present not fully known. Some early studies indicate that nanoparticles can serve as environmental poisons that accumulate in organs. Although these risks may prove to be either minor, avoidable, or both, the engineer and scientist is duty bound to determine if there are in fact any health, safety, and environmental impacts associated with nanotechnology. These concerns served as the driving force for the writing of this book.

Note that Sections 1.2 to 1.4 discuss nanotechnology while Sections 1.5 to 1.8 address environmental issues.

1.2 SURVEY OF NANOTECHNOLOGY APPLICATIONS

The extreme surface-to-volume ratio of nanoparticles is a key attribute that accounts for their range of superior performance characteristics. As the functional advantages of ultra-small particles continue to be deciphered, and processes are perfected to make and manipulate then, there seems to be no limit to what nanomaterials can do.

Once an academic curiosity, nanotechnology has sweeping implications for many electronic, optical, magnetic, catalytic, and medical-therapeutic applications. Nanomaterials are being used to produce composite materials with improved electroconductivity and catalytic activity, hardness, scratch resistance, and self-cleaning capabilities. And, they are being exploited to improve the performance of gas sensors and other devices, the way drugs reach targets in the human body, and the aesthetic appeal and efficiency of consumer products.

A diverse array of ultra-small-scale materials, including metal oxides, ceramics and polymeric materials, and wide-ranging processing methods including techniques that employ 'self-assembly' on a molecular scale, are either in use today or are being groomed for commercial-scale use.[2]

Examples of nanotechnology in actual commercial use, under serious investigation, or on the verge of commercialization include:

- Semiconductor chips and other microelectronics applications
- High surface-to-volume catalysts, which promote chemical reactions more efficiently and selectively
- Ceramics, lighter-weight alloys, metal oxides, and other metallic compounds
- Coatings, paints, plastics, fillers, and food-packaging applications
- Polymer–composite materials, including tires, with improved mechanical properties
- Transparent composite materials, such as sunscreens containing nanosize titanium dioxide and zinc oxide particles
- Use in fuel cells, battery electrodes, communications applications, photographic film developing, and gas sensors
- Nanobarcodes
- Tips for scanning probe microscopes
- Purification of pharmaceuticals and enzymes

Promising medical applications encompass diagnostic and drug delivery systems, including specific targeting of cancer cells. However, these appear to be many years away from commercialization because of lengthy approval procedures by the Food and Drug Administration (FDA) in the United States and its counterpart agencies overseas.[3]

Other examples, in various stages of development, focused on pollution prevention and treatment are listed below:

- Sensing of pollutants, pH, and chemical warfare agents
- Ultraviolet light (UV)-activated catalysts for treatment of environmental contaminants
- Removal of environmental contaminants from various media, including in situ remediation of pesticides, polychlorinated biphenyls (PCBs), and chlorinated organic solvents, such as trichloroethylene (TCE)

- Posttreatment of contaminated soils, sediments, and solid wastes
- Sorption of contaminants for air and water pollution control, in a manner said to be vastly superior to activated carbon
- Chelating agents for polymer-supported ultrafiltration
- Oil–water separation
- Destruction of bacteria (including anthrax)
- Purification of drinking water, without the need for chlorination

Further details can be found in the references cited below. The reader is encouraged to stay abreast of the latest developments in this rapidly changing field.[3]

1. Baum, R. M., editor-in-chief, "Biotech and Nanotech," *Chemical & Engineering News (C&EN)*, **82**(15), 3 (April 12, 2004).
2. Dagani, R., "Nanotech Hoopla," *Chemical & Engineering News (C&EN)*, **82**(15), 31 (April 12, 2004).
3. Masciangioli, T. and W. Zhang, "Environmental Technologies at the Nanoscale," *Environmental Science & Technology*, **37**(5), 102A–108A (Mar. 1, 2003).
4. Shelley, S., "Carbon Nanotubes: A Small-Scale Wonder," *Chemical Engineering*, **110**(1), 27–29 (Jan. 2003).
5. Various Authors, "Nanotechnology," *Chemical Engineering Progress*, **99**(11), 34S–48S (Nov. 2003).
6. Zhang, W., "Nanoscale Iron Particles for Environmental Remediation: An Overview," *Journal of Nanoparticle Research*, **5**, 323–332 (2003).

In addition to the publications cited above, possible sources of additional information include (1) NanoFocus page of C&EN *Online*, (2) *Small*, a Wiley-VCH journal scheduled for start-up on January 2005, and (3) The American Chemical Society's *Nano Letters*.

It has been reported that market opportunities for nanoparticles as materials are limited; the present real value of the nanos are in their future application. In time, these applications will increase, probably at an exponential rate. Once these opportunities have been identified by the engineer and applied scientist, the appropriate nanotechnology fit will be made. Normally, this will involve successfully engineering the interface between the aforementioned nanoparticles and other materials. In the meantime, there are the areas briefly described above to which industry is currently applying nanoparticles and nanotechnology to real-world applications. Shelley,[2] in an outstanding review article provides additional information on some of the above applications. These are detailed in the next chapter.

Two current and specific nanotechnology projects follow. The first is an industrial application with academic ties. The second involves a government-sponsored project. These are detailed in the following paragraphs.

Project (1): NanoScale Materials, Inc., has developed scaled-up production processes for FAST-ACT (First Applied Sorbent Treatment Against Chemical Threats), an advanced nanoengineered family of products designed to provide first responders, Hazmat teams, and other emergency personnel with a single technology to counteract a variety of chemical warfare agents and toxic industrial chemicals. Nontoxic, noncorrosive, and nonflammable, FAST-ACT is particularly useful when response personnel are confronted with a chemical spill whose exact nature is unknown. While substances such as activated carbon only physically absorb toxic substances, FAST-ACT neutralizes, destroys, and renders them harmless. The material's large surface area gives it the ability to capture and destroy toxic chemicals. Just 25 g (a little less than an ounce) has the surface area of almost three football fields. Independent testing by chemical warfare experts showed that FAST-ACT removed more than 99 percent of such agents as VX, soman, and mustard gas from surfaces in less than 90 s. The initial research that led to FAST-ACT was conducted by the Kansas State University laboratory of Kenneth Klabunde. The National Science Foundation (NSF) Small Business Innovation Research (SBIR) program supported NanoScale's research to make the production processes commercially viable. This scaling-up required dramatic process changes, development of quality control standards and testing to confirm the safety and efficacy of FAST-ACT.

Project (2): Mercury pollution is widely recognized as a growing risk to both the environment and public health. It is estimated that coal-burning power plants contribute about 48 tons of mercury to the U.S. environment each year. The Centers for Disease Control and Prevention (CDC) estimate that one in eight women have mercury concentrations in their bodies that exceed safety limits. The U.S. Environmental Protection Agency (EPA) is currently reconsidering proposed rules on the release of mercury from coal-burning power plant effluents and may impose greater restrictions. Mercury found in liquid effluents comes from water-based processes the facilities used to scrub, capture, and collect the toxic material.

Scientists at the Department of Energy's Pacific Northwest National Laboratory (PNNL) have developed a novel material that can remove mercury and other toxic substances from coal-burning power plant wastewater. PNNL's synthetic material features a nanoporous ceramic substance with a specifically tailored pore size and a very high surface area. The surface area of one teaspoon of this substance is equivalent to that of a football field. This material has proven to be effective for absorbing mercury. Pore sizes can be tailored for specific tasks. The material relies on technology previously developed at PNNL—self-assembled monolayers on mesoporous support, or SAMMS. SAMMS integrates a nanoporous silica-based substrate with an innovative method for attaching monolayers, or single layers of densely packed molecules that can be designed to attract mercury or other toxic substances. In recent tests at PNNL, a customized version of SAMMS with an affinity of mercury, referred to as thiol-SAMMS, was developed. According to Shas Mattigod, lead chemist at PNNL, test results revealed mercury-absorbing capabilities that surpassed the devel-opers' expectations. After three successive treatments, 99.9 percent of the mercury in the simulated wastewater was captured, reducing levels from

145.8 to 0.04 parts per million (ppm). This is below the EPA's discharge limit of 0.2 ppm. The mercury-laden SAMMS also passed Washington State's dangerous waste regulatory limit of 0.2 ppm, allowing for safe disposal of the test solution directly to the sewer. Tests have shown that the mercury-laden SAMMS also passed EPA requirements for land disposal. This technology may result in huge savings to users who are faced with costly disposal of mercury in the waste stream. It appears that SAMMS technology can be easily adapted to target other toxins such as lead, chromium, and radionuclides.

With regard to economics and finance, the latest reports indicate that worldwide research and development (R&D) spending is now up to approximately $12 billion, with the biggest increases occurring in defense and security projects. Expenditures/investments also continue to increase in:

1. Information technology
2. Life sciences
3. Food
4. Energy
5. Water

The nanotechnology market is expected to grow by 30 percent annually in this decade. This growth will be further fueled as individuals become more aware of the impact nanotechnology will have on society. Additional details are provided in the next chapter.

1.3 LEGAL CONSIDERATIONS FOR NANOTECHNOLOGY

SECTION AUTHOR: ADRIAN CALDERONE, BChE, ME, J.D.

A new area of technological research and development is rapidly emerging. It is called *nanotechnology* because it pertains to technology operating at the scale of nanometers. For the present, the full implications of the meaning of the term must remain speculative. Nanotechnology is being introduced into a society with an existing legal framework. Nevertheless, because of the potential of this technology to introduce changes as to what human beings can do and expect in the realm of materials, mechanics, information systems, and biological systems, the legal system may have a difficult time catching up with the vast new potentials. Not the least of the reasons for this difficulty is that technological development begets more technological development. More technological development has occurred in the past 30 years than in the previous century. Twenty years ago it was relatively uncommon for a household to have a personal computer. Now, most households in the United States have personal computers. In the early 1990s the Internet was just becoming popular and there was conflict over its commercialization. Now, even grade-school children communicate over the Internet and "surf the net" for research or amusement.

Technology affects almost every area of human activity in one way or another. And so one can expect that the legal relations between people will have to be taken into account. Even though anticipated developments in nanotechnology may yet be in the realm of speculation, those involved with nanotechnology and those involved with law should consider how nanotechnology and law might interact.

One area of law with which the nanotechnologist must be concerned is intellectual property law. Technological development is all about ideas. Ideas have commercial value only if they can be protected by excluding others from exploiting those ideas. Typically, the way to protect ideas is through intellectual property rights such as patents, trademarks, copyrights, trade secrets, and maskworks. Patents can be used to protect useful inventions, ornamental designs, and even botanical plants. The patent allows the owner of the patent the right to prevent anyone else from making, using, or selling the invention covered by the claims of the patent. Trademarks are distinctive marks associated with a product or service (these are usually referred to as service marks), which the owner of the mark can use exclusively to identify himself as the source of the product or service. Copyrights protect the expression of an idea, rather than the idea itself, and are typically used to protect literary works and visual and performing arts, such as books, photographs, paintings and drawings, sculpture, movies, songs, and the like. Trade secret law protects technical or business information that a company uses to gain a competitive business advantage by virtue of the secret being unknown to others. Customer or client lists, secret formulations, or methods of manufacture are typical business secrets. Maskwork protection is for original circuit designs in a chip layout.

Of these intellectual property rights, the most pertinent for nanotech inventions are patents. (Recent patent activity is discussed in the next section.) A patent can protect, for example, a composition of matter, an article of manufacture, or a method of doing something.

It is incumbent upon those engaged in any area of technological development to acquire a basic understanding of patent law because the patent portfolio of a company, particularly one focused on research and development, may represent its most valuable asset(s). Certain activities, such as premature sale or public disclosure, can jeopardize one's right to obtain a patent. Patents are creatures of the national law of the issuing country and are enforceable only in that country. Thus, a U.S. patent is enforceable only in the United States. To protect one's invention in foreign countries, one must apply in the countries in which protection is sought. One can obtain a general idea of the developmental progress of a new technological field by monitoring the number of patents issued in that field.

Patent rights are private property rights. Infringement of a patent is a civil offense, not criminal. The patent owner must come to his or her own defense through litigation, if necessary. And this is a very expensive undertaking. Lawsuits costing more than a million dollars are not unusual. But at stake can be exclusive rights to technology worth a hundred times as much.

Another legal area that will be relevant to nanotechnology is contract law. Any time two or more parties agree upon something, the principles of contract law come into play. The essential components of a contract are parties competent to enter into a contractual agreement, subject matter (what the contract is about), legal consideration (the inducement to contract such as the promises or payment exchanged, or some other benefit or loss or responsibility incurred by the parties), mutuality of agreement, and mutuality of contract. While oral contracts can be legally binding, in the event of a dispute it may be difficult to establish in court who said what. It is far better to memorialize the agreement in the form of a written contract.

One of the basic principles of contract law is that the parties should have a meeting of minds. That is, they should have a common understanding of what the terms of the contract mean. Sometimes it is not so clear what particular terms mean, or the meaning or its implications may change in time. What, for example, qualifies as "nanotechnology?" Not only is nanotechnology not well defined now, in the future it may encompass things that are not even imagined today.

Generally, contracts are employed with the sale and licensing of exclusive rights to a technology. Also, there are agreements to fund technological research and development.

Who are the entities engaged in the contract? Typically, these are business entities. So one must also consider whether there may be some peripheral issues of corporation or partnership law.

Another area of concern is government regulations. Environmental regulations, in particular, have become a central concern for technological development. In fact, in many technologies it is the increasingly stringent clean air and water regulations that drive new technological development. This is particularly true, for example, in the automotive industry and in heavy industries where emissions are released into the atmosphere or water. What effects will nanotechnology have on the environment? This is yet unknown because the field is still, relatively speaking, in its infancy. One can only speculate. But suppose, for example, nanoproducts are developed, products that are so small they can be ingested or inhaled, and that act upon the interior organs of the human body. Suppose now that such products are released into the environment. They might be carried by the air within the state or across state lines. The state and federal environmental regulatory agencies would understandably be taking a great interest in the environmental impact of such a release.

Also, one can imagine that nanotechnology intended for medical applications will arouse the interest of the Food and Drug Administration. Perhaps there may be a court case in the future in which a question to be decided is whether a dose of nanomachines qualifies as medical instrumentation or a drug.

Statutory regulations are prospective. They deal with the future. Vague regulations are not enforceable. And, generally speaking, one cannot be accused of violating a law or regulation that was not in effect when the act in question was performed. Hence, it is usually (but not always) the case that regulations and laws are put into effect only after some disaster has occurred.

But there is a branch of law that can address retrospectively certain situations in which property or people are harmed. That is tort law. A tort is a civil wrong, other than a breach of contract, for which the law provides a remedy. One can recover damages under tort law if a legal duty has been breached that causes foreseeable harm. These duties are created by law other than duties under criminal law, governmental regulations, or those agreed to under a contract. Tort law can be very encompassing.

Nanotechnologists have to consider the possibilities of reasonably foreseeable harm arising from their developments and take prudent precautions to avert such harm. In the event that a technology is inherently dangerous, nanotechnologists may be held to a standard of strict liability for any harm caused by the technology regardless of whether an accident was foreseeable.

Sometimes a very unexpected series of events can arise that cause harm. In the *Palsgraf v. Long Island Railroad* case the plaintiff, Mrs. Palsgraf, was standing on the platform of the station after buying a ticket to go to Rockaway Beach. A train bound for another destination stopped at the station and a man carrying a package under his arm jumped aboard the car. A railroad employee, a guard who was holding the door, reached out to help the man in. But the package was dislodged and fell onto the tracks. Unknown to the guard, the package happened to contain fireworks. Upon hitting the ground the fireworks exploded, treating the bystanders to a dazzling, if violent, pyrotechnic display. Unfortunately, the shock of the explosion knocked over a scale that fell upon the hapless Mrs. Palsgraf. She was injured and sued the Long Island Railroad. The court dealt with issues of causation and foreseeability, and ultimately decided in favor of the defendant. Nevertheless, the fact that the defendant in this case was the Long Island Railroad and not the gentleman with the fireworks leads to the conclusion that, insofar as the plaintiff is concerned, causal linkage between an act and an injury is not unrelated to the depth of pockets of a prospective defendant.

Most nanotechnologists are not at all interested in deliberately causing harm. But some will be. As described earlier, nanotechnology can have military applications, and governments may be interested in developing nanoweapons. To take the example given earlier, let us suppose that nanoproducts, perhaps nanomachines, are developed that can invade the human body and do harm. Is a cloud of such nanomachines to be considered a poison gas? Or is it a collection of antipersonnel mechanical implements, like shrapnel? How will nanomachines as weapons be treated under the Geneva convention? And suppose such a cloud of nanoweapons drifts over, or is released over, a civilian population? The devastating effect of land mines, which remain lethal long after hostilities are ended and which wreak havoc upon unsuspecting civilians wandering into mine fields, has been amply documented. Will nanoweapons remain harmful years after their deployment? What responsibilities do governments have morally, and under international law?

But suppose these nanoproducts are not designed to cause harm but simply to obtain and transmit information. For example, suppose such nanoproducts, if ingested, provide information about bodily functioning. Or suppose they enable a person to be tracked wherever he or she goes. Larger devices are already known that enable a person to be tracked by global positioning satellites. These devices are worn voluntarily. There are also implantable devices. But nanomachines would be undetectable and could very well be implanted in a person without his or her knowledge or consent. Under what circumstances should such invasions of privacy be allowed or forbidden?

Then there is the matter of the criminal use of nanotechnology. The past decade has already seen the growth of a new area of crime: computer crime. For example, in addition to conventional theft, law enforcement agencies must now become technologically proficient to handle computer fraud, identity theft, theft of information, embezzlement, copyright violation, computer vandalism, and like activities, all accomplished over the computer network under conditions such that not only is the criminal hard to trace but even the crime may go undetected. The computer criminals are technologically very savvy and willing to exploit the potentials of any new technology. About the only thing one can expect is that if nanotechnology provides great new potentials, someone will use those potentials for criminal purposes, and the laws will again be forced to play catch-up in response to the crimes, after the harm has occurred.

These speculations presume the feasibility of such futuristic devices. And one comes back to the original point that it is not known what will be feasible in 10 or 20 years, not only with respect to nanotechnology but with respect to the convergence of nanotechnology with other technologies such as biotechnology, information technology, wireless technology, materials science, and quantum physics.

It would be prudent for those involved in nanotechnology and those involved in law to interact constructively, and to share knowledge and expertise to avoid blindly traveling along a road whose features and direction are yet unknown. As always, the question is whether human beings will control technology for their own constructive uses, and that depends upon whether people can control themselves. The alternative is to be mastered by one's own creations.

1.4 RECENT PATENT ACTIVITY[2]

The publication of patent applications by the U.S. Patent and Trademark Office (USPTO) has given researchers in both academia and industry a means of following new developments in their particular fields of interest. This is particularly true in nanotechnology. The following survey, developed with the assistance of Examiner Vivek Koppikar of the U.S. Patent and Trademark Office, a chemical engineer and member of the American Institute of Chemical Engineers (AIChE), shows some of the new directions in the field while suggesting opportunities for further exploration and future patent protection.

In terms of issued patents alone, the USPTO reports that those involving nanotechnology have increased by over 600 percent in the last 5 years, from 370 in 1997 to 2650 in 2002. While these patents made up only 2 percent of all patents issued in 2002, this compares to a figure of 0.3 percent in 1997. New filings of nanotechnology-related patent applications are evenly split between process inventions and product inventions, as is typical for all patent applications. Most of these applications (approximately 90 percent) come from private corporations, with

Hewlett-Packard, Texas Instruments, 3M, and Motorola filing the largest number. Universities are filing approximately 7 percent, with the University of California and Stanford University in the lead. About 3 percent are being filed by agencies of the U.S. government and collaboration research centers such as the Department of Energy, the Department of Defense, and Sandia National Laboratories. Most of the inventions in these applications are refinements to known technology, but a significant number can be considered "revolutionary" or pioneering in nature. The following is a sampling of what has appeared in published patent applications, showing the wide diversity of chemical engineering disciplines utilizing nanotechnology.

In heat transfer and reaction engineering, the extremely high surface area of nanoparticles has been known to raise reaction rate coefficients and ultimately reduce the energy requirements for a reactor. Patent application no. 2002/0100578 now discloses that nanoparticles are also useful in heat transfer fluids where the high surface area of the particles raises the convective heat transfer coefficients and heat capacities of the fluids and produces a greater degree of collision-induced motion, all of which produce significant energy and cost savings for heat transfer and exchanger systems.

In fiber chemistry, nanoparticle-containing fibers formed by melt-spinning resins containing nanoparticle dispersions are improved upon by the invention of patent application no. 2003/0083401. In this invention, the fibers are formed by nanoparticles that are surface-derivatized with functional groups that allow the nanoparticles themselves to link together in a "polymerization" reaction. The resulting fiber is claimed to benefit from reduced agglomeration of the particles and greater control of their size, both of which enhance product uniformity.

Drug delivery has also benefited from nanotechnology, including nanoparticles, nanotubes, and nanowires. Among the new patent applications are those claiming nanowire compositions for specific treatments and therapies. Orally administered nanowires (termed *nanorobots*), for example, are disclosed for delivering antibodies to specific cells in the body and for probing and manipulating biomolecules at the cellular and subcellular level. In application no. 2002/0187504, nanowires are made to act as antibodies themselves by functionalizing one end of a nanowire with a ligand to bind to a specific target molecule and the other end with a ligand that attracts immune cells.

Nanoparticles are also used for controlled-release drug delivery. In application no. 2003/0095928, the drug itself is formed into nanoparticles with the assistance of a surface stabilizer. The drug in this application is insulin, and the resulting nanoparticle offers both a rapid onset of insulin activity and prolonged activity. Other applications disclose the use of nanotechnology to deliver Taxol, antigens, and antibiotics such as penicillin.

In soaps, detergents, and industrial cleaners, nanoparticles have been used to provide controlled release of fragrances, biocides, and antifungals on textiles by adhering to the textile fibers. Application no. 2003/0013369, discloses "textile-reactive" nanoparticles that bind to the textile fiber through a covalent bond. Nanoscale polymers have been disclosed for use in hair and skin care cleansing preparations to increase the stability and consistency of the preparations. Recent improvements are disclosed in patent applications nos. 2003/0086894, 2003/0059385, and 2003/0003070, in which nanoscale polymers with a mean diameter of 10 to 500 nm provide increased dermatological compatibility and improved sensory properties on skin and hair due to the high compatibility of these polymers with keratin.

In coatings technology, nanoparticle hardcoats have been applied to the surfaces of glass, quartz, wood, and metal, as well as thermoset and thermoplastic materials such as acrylics and polycarbonates, providing protection of the substrate in each case. Application no. 2003/0068486 discloses that smudge resistance can be increased by using nanoparticles with exposed reactive groups that covalently attach to smudge-resistant moieties.

The use of nanoparticles in separation technology has seen numerous improvements as well. In application no. 2003/0038083, for example, superparamagnetic nanoparticles are placed in contact with a liquid or gaseous medium to bind to specified targets in the medium. A magnetic field then separates the target-laden nanoparticles from the medium. The advantage of nanoparticles is their lack of retention of magnetism after the magnetic field is removed, since retained magnetism causes the laden nanoparticles to aggregate, which inhibits separation.

In composite materials, recent patent applications disclose the incorporation of nanoparticles to add novel properties to macrosized materials without altering other, favorable properties of the materials. One example is application no. 2003/0099834 in which silver nanoparticles are embedded in glass to make the glass photochromic, that is, capable of light-induced, reversible darkening. The nanoparticles are incorporated into the glass at room temperature without the need for annealing or other processing steps at higher temperatures. Another example is application no. 2003/0037443, which discloses the incorporation of metallic nanoparticles in plastic forks and spoons to add luster.

As with any invention, the qualities that make nanotechnology-related inventions patentable are novelty, nonobviousness, and utility. While many unique properties of nanotechnology are already known, more are being discovered as are new ways of exploiting known properties, all of which can lead to patentable inventions. The growth of patents in nanotechnology is a clear indication of the industry's recognition of the potential in this field.

1.5 ENVIRONMENTAL IMPLICATIONS[5,6]

Any technology can have various and imposing effects on the environment and society. Nanotechnology is no exception, and the results will be determined by the extent to which the technical community manages this technology. This is an area that has, unfortunately, been seized upon by a variety of environmental groups.

There are two thoughts regarding the environmental implications of nanotechnology: One is positive and the other is potentially negative. Some of the positive features of this new technology are discussed in this section. Much of this has been drawn from an excellent report that recently appeared in the literature.[7] Additional details are provided in Section 2.7.

The other implication of nanotechnology has been dubbed by many in this diminutive field as "potentially negative." The reason for this label is as simple as it is obvious: The technical community is dealing with a significant number of unforeseen effects that could have disturbingly disastrous impacts on society. Fortunately, it appears that the probability of such dire consequences actually occurring is near zero . . . but *not* zero. This finite, but differentially small, probability is one of the reasons this book was written; and it is the key topic that is addressed in the pages that follow. Air, water, and land (solid waste) concerns with emissions from nanotechnology operations in the future, as well companion health and hazard risks, receive extensive treatment. All of these issues arose earlier with the Industrial Revolution, the development/testing/use of the atomic bomb, the arrival of the Internet, Y2K, and so forth; and all were successfully (relatively speaking) resolved by the engineers and scientists of their period. An example of how the U.S. EPA chose to address the impacts of toxic air pollution (TAP) concerns of the 1980s is presented in the next paragraph.

Basic air pollutant toxicology, or the science that treats the origins of toxics, must be considered in terms of entering the body through inhalation. This makes the respiratory tract the first site of attack. Among the primary air pollutants, only lead and carbon monoxide exert their major effects beyond the lung. The more reactive a compound, the less likely it is to penetrate the lung. However, many individuals breathe a mixture of air contaminants, and many of the TAP compounds are known to cause cancer. The total nationwide cancer incidence due to outdoor concentrations of air toxics in the United States was estimated to range from approximately 1700 to 2700 excess cancer cases per year. This is roughly equivalent to between 7 and 11 annual cancer cases per million population (data obtained from a 1986 population of 240 million). The U.S. EPA initiated a broad "scoping" study with a goal of gaining a better understanding of the size and causes of the health problems caused by outdoor exposure to air toxics. This broad scoping study was referred to as the Six-Month Study. The objective was to assess the magnitude and nature of the air toxics problem by developing quantitative estimates of the cancer risks posed by selected air pollutants and their sources from a national and regional perspective. The main conclusion of the Six-Month Study was that the air toxics problem is widely thought to be related to the elevated cancer mortality. Table 1.1 provides a summary of the estimated annual cancer cases by pollutant.[8] The EPA classifications used in this report were: A = proven human carcinogen; B = probable human carcinogen (B1 indicates limited evidence from human studies and sufficient evidence from animal studies; B2 indicates sufficient evidence from animal studies, but inadequate evidence from human studies); C = possible human carcinogen.

To the authors knowledge there are no documented nano human health hazards. Statements in the literature refer to *potential* human health problems. A typical recent study follows.

Günter Oberdörster, professor of Toxicology in Environmental Medicine at the University of Rochester (UR) and director of the university's EPA Particulate Matter Center, has already completed one study showing that inhaled nanosized

TABLE 1.1 **Summary of Estimated Annual Cancer Cases by Pollutant**

Pollutant	EPA Classification[a]	Estimated Annual Cancer Cases
1. Acrylonitrile	B	113
2. Arsenic	A	68
3. Asbestos	A	88
4. Benzene	A	181
5. 1,3-Butadiene	B	2266
6. Cadmium	B	110
7. Carbon tetrachloride	B	241
8. Chloroform	B	2115
9. Chromium (hexavalent)	A	147–265
10. Coke oven emissions	A	7
11. Dioxin	B	22–125
12. Ethylene dibromide	B	268
13. Ethyl dichloride	B	245
14. Ethlene oxide	B	1–26
15. Formaldehyde	B	1124
16. Gasoline vapors	B	219–276
17. Hexachlorobutadiene	C	9
18. Hydrazine	B	26
19. Methylene chloride	B	25
20. Perchloroethylene	B	26
21. PIC[b]		438–1120
22. Radionuclides	A	3
23. Radon[c]	A	2
24. Trichloroethylene	B	27
25. Vinyl chloride	A	25
26. Vinylidene chloride	C	10
27. Miscellaneous[d]		15
Totals		1726–2706

[a]For a discussion of how EPA evaluates suspect carcinogens and more information on these classifications, refer to "Guidelines for Carcinogen Risk Assessment" (51 *Federal Register* 33992).
[b]EPA has not developed a classification for the group of pollutants that compose products of incomplete combustion (PICs), although EPA has developed a classification for some components, such as benzo(*a*)pyrene (BaP), which is a B2 pollutant.
[c]From sources emitting significant amounts of radionuclides (and radon) to outdoor air. Does not include exposure to indoor concentrations of radon due to radon from soil gases entering homes through foundations and cellars.
[d]Includes approximately 68 other individual pollutants, primarily from the TSDF study and the Sewage Sludge Incinerator study.

particles accumulate in the nasal cavities, lungs, and brains of rats. Scientists speculate this buildup could lead to harmful inflammation and the risk of brain damage or central nervous system disorders. "I'm not advocating that we stop using nanotechnology, but I do believe we should continue to look for adverse health effects," says

Oberdörster, who also leads the UR Division of Respiratory Biology and Toxicology. "Sixty years ago scientists showed that in primates, nanosized particles traveled along nerves from the nose and settled into the brain. But this has mostly been forgotten. The difference today is that more nanoparticles exist, and the technology is moving forward to find additional uses for them—and yet we do not have answers to important questions of the possible health impact." Oberdörster is leading a 5-year study employing a multidisciplinary team from 10 departments at three universities (UR, University of Minnesota, and University of Washington at St. Louis). They plan to test a hypothesis that the chemical characteristics of nanoparticles determine how they will ultimately interact with human or animal cells. A negative cellular response may indicate impaired function of the central nervous system. In previous studies, Oberdörster showed that nanosized particles depositing in the nose of rats traveled into the olfactory bulb. At this point the team is not entirely opposed to nanotechnology, Oberdörster explains. In fact, researchers hope to work with industry, as well as with the American and Canadian governments, to seek solutions if problems arise. Another goal is to develop an educational program so that future engineers, and scientists will understand the health consequences of nanotechnology. For decades, Oberdörster has studied how the body interacts with ambient ultrafine particles, including automotive and power plant emissions and dust from the World Trade Center disaster. What's different about nanotechnology is that these particles appear to seep all the way into the mitochondria, or energy source, of living cells. "We must consider many different issues before we come to a judgment on risk," he says. "Foremost is an assessment of potential human and environmental exposure by different routes: inhalation, ingestion, dermal. Then, what is their fate in the organism? And what are the risks of cumulative effects, given that these particles are being mass produced? At this point we're trying to balance the tremendous opportunity that nanotechnology presents with any potential harm."

It should be noted that nanoenvironmental concerns are starting to be taken seriously around the globe. There are a variety of studies going on into the health and environmental impacts of many applications of nanotechnology. It is in everyone's interest to ensure that any new compound is fully characterized and the long-term implications studied before it is commercialized. Class action suits in the United States against both tobacco companies and engineering companies, coupled with a new era of corporate responsibility, have ensured that most companies are well aware of this need. Now that potential risks that may have been overlooked are becoming widely known, these companies are more inclined to be proactive than they have been with risks in the past.[7]

Returning to the positive features of this new activity, nanotechnology will be one of the key technologies used in the quest to improve the global environment in the 21st century. While there will be some direct effects, much of the technology's influence on the environment will be through indirect applications of nanotechnology. Although any technology, whether nanotechnology or a box of matches, can always be put to both positive and negative uses, there are many areas in which the positive aspects of nanotechnology look promising. These extend from pollution reduction through environmental remediation to sustainable development.

There has already been a considerable shift in both public and corporate attitudes to the environment. Major scandals such as Enron and WorldCom have led not only to tighter corporate governance but also to calls for greater corporate responsibility. The end result of this shift will be to make companies focus on the environment, and look to leveraging nanotechnology as a way of not only improving efficiency and lowering costs, but doing this by reducing energy consumption and minimizing waste. A typical example would be in the use of nanoparticle catalysts that are not only more efficient, owing to more of the active catalyst being exposed, but also require less precious metal (thus reducing cost), are more tightly bound to the support (increasing the lifetime of the catalyst) and may also increase selectivity, that is, produce more of the desired reaction product, rather than by-products.

Though nanotechnology could have some significant effects on environmental technologies, environmental considerations have not historically been given anywhere near the priority in new developments that commercial considerations are given, and this balance, though swinging gradually more toward environmental considerations, still largely dominates. Many of the direct applications of nanotechnology relate to the removal of some element or compound from the environment, through, for example, the use of nanofiltration, nanoporous sorbents (absorbents and adsorbents), catalysts in cleanup operations, and filtering, separating, and destroying environmental contaminants in processing waste products. Most effects, as with other technologies, are likely to be indirect.

The application of nanotechnology to the environment is already being hailed by some as the "killer application," offering companies a chance to enhance their green credentials without hurting their balance sheets and also creating some huge new markets. Improved efficiency of energy production and supply has both commercial and environmental advantages. One is likely to see the biggest impacts in this area, in savings through lighter composite materials, growth of the use of alternative energy (e.g., through improved economic viability of solar and wind energy generation), and the advent of commercially viable fuel cells in a number of applications. Such technologies certainly have the ability to help considerably in the reduction of global carbon emissions, and other emissions, but probably the most dramatic effects, from the point of view of the daily lives of many individuals in both developed and developing countries, will come with clean water and a reduction in health costs. Further, air in cities around the world could become as unpolluted as the air in the country. Some, including the authors of this work, have concluded that the major environmental nanotechnology breakthroughs will occur naturally through pollution prevention principles.[10]

Pollution prevention in the EPA's definition involves the reduction, to the extent feasible, of generated waste. It includes any source reduction or closed-loop recycling activity undertaken by a generator that results in either: (1) the reduction of the total volume or quantity of waste or (2) the reduction of toxicity of the waste, or both, so long as such reduction is consistent with the goal of minimizing present and future threats to human health and the environment.[11] Source reduction is defined as any activity that reduces or eliminates the generation of waste at the source, usually within a process. A material is "recycled" if it is used, reused, or reclaimed (40 CFR 261.1 [c][7]). A material is "used or reused" if it is either: (1) employed as an ingredient to make a product, including its use as an intermediate (however, a material will not satisfy this condition if distinct components of the material are recovered as separate end products, as when metals are recovered from metal containing secondary materials), or (2) employed in a particular function as an effective substitute for a commercial product (40 CFR 261.1 [c][5]). A material is "reclaimed" if it is processed to recover a useful product or if it is regenerated. Examples include the recovery of lead from spent batteries and the regeneration of spent solvents (40 CFR 261.1 [c][4]). (The reader should note that the working definition of pollution prevention in this text also includes energy conservation and health, safety, and accident management considerations. All these effects are considered in more detail later in this book.) Because of the very nature of nanotechnology, applications will lead to and produce less waste/pollution because this new technology involves exact and/or perfect manufacturing.

There are other areas of environmental advances currently in progress. Areas include:

1. Sensors
2. Treatment
3. Remediation

Some details of these topics are provided below.[12]

Sensors—Novel Sensing Technologies or Devices for Pollutant and Microbial Detection Protection of human health and ecosystems requires rapid, precise sensors capable of detecting pollutants at the molecular level. Major improvement in process control, compliance monitoring, and environmental decision making could be achieved if more accurate, less costly, more sensitive techniques were available. Examples of research in sensors include the development of nanosensors for efficient and rapid in situ biochemical detection of pollutants and specific pathogens in the environment; sensors capable of continuous measurement over large areas, including those connected to nanochips for real-time continuous monitoring; and sensors that utilize lab-on-a-chip technology. Research also may involve sensors that can be used in monitoring or process control to detect or minimize pollutants or their impact on the environment.

Treatment—Technologies to Effectively Treat Environmental Pollutants Cost-effective treatment poses a challenge for the EPA and others in the development of effective risk management strategies. Pollutants that are highly toxic, persistent, and difficult to treat, present particular challenges. EPA supports research that addresses new treatment approaches that are more effective in reducing contaminant levels and more cost effective than currently available techniques. For example, nanotechnology research that results in improved treatment options might include removal of the finest contaminants from water (under 300 nm) and air (under 50 nm) and "smart" materials or reactive surface coatings that destroy or immobilize toxic compounds.

Remediation—Technologies to Effectively Remediate Environmental Pollutants   Cost-effective remediation techniques also pose a major challenge for the EPA in the development of adequate remedial techniques that protect the public and safeguard the environment. EPA supports research that addresses new remediation approaches that are more effective in removing contamination in a more cost-effective manner than currently available techniques. Substances of significant concern in remediation of soils, sediment, and groundwater, because of both their cancer and noncancer hazards, include heavy metals (e.g., mercury, lead, cadmium) and organic compounds (e.g., benzene, chlorinated solvents, creosote, toluene). Reducing releases to the air and water, providing safe drinking water, and reducing quantities and exposure to hazardous wastes also are areas of interest.

1.6   CURRENT ENVIRONMENTAL REGULATIONS[5,6,13]

Many environmental concerns are addressed by existing health and safety legislation; most countries require a health and safety assessment for any new chemical before it can be marketed, with the European Union (EU) recently introducing the world's most stringent labeling system. Prior experience with materials such as PCBs and asbestos, and a variety of unintended effects of drugs such as thalidomide, mean that both companies and governments have an incentive to keep a close watch on potential negative health and environmental effects.[12]

It is very difficult to predict future nanoregulations. In the past, regulations have been both a moving target and confusing. What can be said for certain is that there will be regulations, and the probability is high that they will be contradictory and confusing. Past and current regulations provide a measure of what can be expected. And, it is for this reason that this section is included in this book.

Introductory Comments

Environmental regulations are not simply a collection of laws on environmental topics. They are an organized system of statutes, regulations, and guidelines that minimize, prevent, and punish the consequences of damage to the environment. This system requires each individual—whether an engineer, field chemist, attorney, or consumer—to be familiar with its concepts and case-specific interpretations. Environmental regulations deal with the problems of human activities and the environment, and the uncertainties the law associates with them.

With the onset of the Industrial Revolution, environmental law has increased in popularity due to an interest in public health and safety and the environment.

Companies are concerned with exposing themselves to future potentially disastrous liabilities by polluting the environment where their operations are located. Businesses are now taking a proactive approach to environmental compliance. The recent popularity of environmental issues with public and private interest groups has brought about changes in legislation and subsequent advances in technology.

The EPA's authority is increasingly broadening. It is the largest administrative agency in the federal government and accounts for nearly one-seventh of the federal budget. The agency has nearly 20,000 employees and churns out more pages of regulation than any other administrative agency. The EPA's comprehensive environmental programs encompass regulations for air pollution, hazardous waste management, solid waste management, drinking water standards, emergency management, and permitting requirements for discharges to the air, land, or waters of the United States and its territories and districts.

Most major environmental statutes stem from dramatic and highly publicized incidents such as Love Canal, New York, where hazardous substances from an abandoned dump site polluted a nearby community in 1970; Union Carbide Corporation's Bhopal, India, methyl isocyanate release from a chemical plant in 1984; and the *Exxon Valdez* oil spill off the coast of Alaska in 1988. As a result, the focus on environmental issues has been magnified as official government policies on the environment can be the swing vote in an election year.

Legislation, such as the Clean Air Act Amendments of 1990, is requiring members of society to familiarize themselves with its myriad provisions or become lost in the proverbial shuffle. Who would have thought the federal government would be telling citizens not to drive our cars to work, wash cars in one's yard, or to retrofit automobiles with air conditioning units that use a nonchlorofluorocarbon coolant? Even sellers of new homes are required to identify whether there is radon, lead, asbestos, drinking water pollution, or a leaking underground oil storage tank.

This is by far the longest section in the book ... and for good reason. Environmental regulations usually play a key role in any environmental management issue. For this reason, this section attempts to explain some of the major topics in environmental regulation (along with its seemingly never-ending list of acronyms!). With society moving toward almost a state of total litigiousness it may pay, by saving exorbitant attorney fees, to examine in more detail than presented here, those environmental regulations that most affect each individual or organization. As a general rule, based on the authors' experience as environmental professionals, using common sense (unfortunately) in environmental regulation does not always render the appropriate choice. Much of the material to follow as well as the writing style has been drawn from the *Federal Register*.

Air Pollution

The American Cancer Society says it is a matter of life and breath. On November 15, 1990, President George Bush signed into law the Clean Air Act Amendments

(CAAA) of 1990. This section will key on the 1990 amendments. Details regarding the entire amendments are available in 40 Code of Federal Regulations (CFR) 50 through 99. It is the major piece of legislation that today applies to the air. The CAAA build upon the regulatory framework of the earlier Clean Air Act's programs and their amendments and expands their coverage to many more industrial and commercial facilities. The CAAA established a new permit program, substantially tightened requirements for air pollution emission controls, and dramatically increased the potential civil and criminal liability for noncompliance for individuals and companies. A brief overview of the 10 titles of the 1990 CAAA is presented below.

Title I: Provisions for Attainment and Maintenance of National Ambient Air Quality Standards This provision of the CAAAs provides a new strategy for controls of urban air pollution problems of tropospheric ozone, carbon monoxide (CO), and particulate matter (PM-10). (The troposphere extends upward from the surface of Earth to approximately 12 km and is the air one breathes.) The new law mandates that the federal government, which in turn empowers the states, address the problem of urban air pollution by designating geographical locations according to the extent of their contamination of ozone, CO, and NO_X (oxides of nitrogen) as "nonattainment" areas, that is, areas that cannot meet specified air quality standards.

The CAAA require states to revise their State Impementation Plans (SIPs) to include more sources of ozone, both mobile and stationary. For the pollutant ozone, nonattainment areas are further categorized as to the severity of the ozone contamination. Each nonattainment area is classified, as shown in Table 1.2 by the area's "ozone design value," as either extreme, severe, serious, moderate, or marginal.

States with nonattainment areas have to enact various control measures to reduce these emissions, depending on the classification of the areas. An area that is classified as marginal nonattainment will require less control measures than an area classified as extreme to bring it into compliance.

Carbon monoxide nonattainment areas are designated as either "moderate" or "serious." Moderate areas are those that have a CO concentration of 9.1 to 16.4 ppm. Serious areas are those that have CO concentrations of 16.5 ppm or greater. Moderate areas were to meet milestones for CO reductions by 1995 and

TABLE 1.2 Ozone Nonattainment Classifications Specified by the Act

Area Class	Ozone Design Value (ppm)	Number of Metropolitan Areas	Attainment Date
Marginal	0.121–0.138	41	11/15/93
Moderate	0.138–0.160	32	11/15/96
Serious	0.160–0.180	18	11/15/99
Severe	0.180–0.280	8	11/15/2005[a]
Extreme	>0.280	1	11/15/2010

[a]11/15/2007 for Chicago, Houston, and New York.

serious areas had until 2000 to attain the standard. The statute specifically requires controls for CO by enhanced vehicle inspection and maintenance and clean fuels for fleet vehicles, both mobile source controls.

The PM-10 nonattainment areas (PM-10 is particulate matter with an aerodynamic diameter less than or equal to 10 μm) are designated as either "moderate" or "serious." Moderate areas had until 1994, while serious areas had until 2001 to meet the standard. States must develop attainment plans that will require implementation of control measures to attain the standard. Depending upon their classifications, PM-10 areas will have to implement reasonably available control measures (RACMs) or best available control measures (BACMs), among other requirements.

Title I also contains provisions for emissions offset requirements at new and modified sources for ozone, PM-10, and CO. Sources will be required to use the lowest achievable control technology (LEAR). Any resulting increase in emissions must be offset by equivalent or greater emissions reductions elsewhere (P.L. 101–549, 1990; 40 CFR, December 10, 1993).

Title II: Provisions Relating to Mobile Sources   Cars and trucks in many nonattainment areas account for over 50 percent of the precursors to ground-level ozone (different from stratospheric ozone), volatile organic carbons (VOCs), NO_X, and over 90 percent of CO emissions. It has been speculated that this was brought about by an unexpected growth in motor vehicle emissions. A summary of the major requirements is presented below.

The CAAAs establish emissions standards applicable to heavy- and light-duty trucks and conventional motor vehicles. These standards will help reduce tailpipe emissions of hydrocarbons, CO, NO_X, and PM-10. The EPA will also regulate evaporative emissions from all gasoline-fueled vehicles both during operation and during periods of nonuse. CO emissions from light-duty vehicles may not exceed 10.0 g per mile when operated at 20°F.

Title II also contains provisions to establish rules for reformulated gasoline in specific nonattainment areas. Methyl tertiary butyl ether (MTBE) is used in some cities. The clean fuel fleet program requires that states revise their SIPs to include a clean fuel program for fleets. The program impacts anyone owning a fleet that is capable of being centrally fueled. Fleet owners and operators who purchase additional new vehicles had to do so according to the earlier schedule in Table 1.3.

Each of these provisions is aimed at decreasing pollution in nonattainment areas that are in excess of the guideline values. The degree of control placed on industrial sources of CO, VOCs, and NO_X will depend on reductions of these pollutants achieved through Title II programs (P.L. 101–549, 1990).

Title III: Air Toxics   In its prime, the federal air toxics program, known as the National Emissions Standards for Hazardous Air Pollutants (NESHAP), only encompassed seven toxic air pollutants. Prior to the CAAAs, NESHAP regulated emissions for arsenic, asbestos, beryllium, mercury, radionuclides, benzene, and vinyl chloride. Title III was revamped to give the EPA the authority to establish

TABLE 1.3 Clean Fuel Fleet Program: Clean Fuel Vehicle Phase-in

Model Year	Light-Duty Vehicles	Heavy-Duty Vehicles
1988	30%	50%
1999	50%	50%
2000	70%	50%

an elaborate program to regulate emissions of air toxics from sources that emit hazardous air pollutants (HAPs).

The EPA has identified, via an industry-specific list, major sources of HAPs. These industries are categorized by the amount of pollutants they emit. A "major source" is defined as a source that emits 10 tons or more per year of any one of the list of 189 HAPs, or 25 tons per year or more of any combination of HAPs. The regulations also empower the federal government and states to regulate sources that emit less than the aforementioned levels if they are found to pose a "threat of adverse effects to health and the environment." Dry cleaners using perchloroethylene as a cleaning solvent would fall under this category.

The EPA requires each of these affected sources to comply with maximum achievable control technology (MACT) standards. These technology-based emissions standards are enacted to consider not only emissions control technology that removes pollutants at the point of discharge but also "measures, processes, methods, systems or techniques which:"

1. Reduce or eliminate emissions through process changes, materials substitution, or other changes
2. Enclose processes to monitor emissions
3. Are design, equipment, work practice, or operational standards, including operator training and certification

Title III also establishes a new and comprehensive accidental release program modeled in part after New Jersey's Toxic Catastrophic Prevention Act. The Process Safety Management Regulation and the Accidental Release and Emergency Preparedness Regulation, respectively, are regulations required to be promulgated by the EPA and the Occupational Safety and Health Administration (OSHA) under Title III. These new regulations provide for accidental release prevention and define consequences within and external to a facility that has in excess of a threshold quantity of 100 substances that are known to cause, or may be reasonably anticipated to cause, death, injury, or serious adverse health effects to human health and the environment (P.L. 101–549, 1990; 40 CFR, October 20, 1993).

Title IV: Acid Deposition Control Title IV mandates controls to reduce the deposition of acidic sulfates and nitrates by reducing emissions of sulfur dioxide (SO_2) and nitrogen oxides (NO_X), which are precursors of acid rain. Sulfur emissions are attributed largely to coal burning at electric utilities that contain sulfur as an impurity.

Congress set a two-phased goal for utility emissions reduction. The first phase requires 110 fossil-fuel-fired public power plants to reduce their emissions to a level equivalent to the product of an emissions rate of 2.5 lb of SO_2/MMBtu times the average of their 1985 to 1987 fuel usage. The second phase requires approximately 2000 utilities to reduce their emissions to a level equivalent to the product of an emissions rate of 1.2 lb of SO_2/MMBtu times the average of their 1985 to 1987 fuel use.

Title IV also mandates a program in which allowances are created as market shares, which utilities are allowed to buy and sell in order to meet reductions of emissions. These shares currently have a value of approximately $2000. The regulation also contains requirements for reduction of NO_x from tangentially fired boilers and dry bottom wall-fired boilers by 1995 and all other utility boilers by 1997.

Affected sources will be required to perform continuous emissions monitoring, permit their sources, and keep records of their activities. Continuous emissions monitoring devices must be installed on all affected sources. The permit application must include a compliance plan for the source to comply with the requirements of Title IV. Affected sources are also required to pay an annual fee of $2000 per ton for emissions in excess of allowance levels (P.L. 101–549. 1990).

Title V: Permits   Under Title V, the EPA has attempted to model the permit requirements from the Clean Water Act (CWA). (See a later section on Water Pollution for additional details on the CWA.) The CWA permitting system is known as the National Pollution Discharge Elimination System (NPDES). The permitting program is enacted to ensure compliance with various portions of the program. States were required to submit a proposed program to the EPA by November 1993 for issuing permits to certain significant sources of air emissions.

The Title V operating permit program requires all affected sources under the CAAAs to submit a permit application to the appropriate permitting authority within one year of the effective date of the state program. The operating permit program applies to the following major sources:

1. Air toxic sources as defined under Title III, with the potential to emit 10 tons per year (tpy) or more of a single HAP or 25 tpy or more of the aggregate of HAPs or a lesser quantity if the administrator so specifies.

2. Sources emitting more than 100, 50, 25, or 10 tons per year depending on their nonattainment designation; that is, marginal, moderate, severe, or extreme.

3. Sources of air pollutants with the potential to emit 100 tpy or more of any pollutant.

4. Any other source, including an area source, subject to a hazardous air pollutant standard under Title III.

5. Any affected source under the acid rain program under Title IV.

6. Any source required to have a preconstruction permit pursuant to the requirements of the prevention of significant deterioration (PSD) program under Title I, or the nonattainment new source review (NSR) program under Title I.

7. Any other stationary source in a category the EPA designates in whole or in part by regulation after notice and comment.

Once subject to the operating permit program for one pollutant, a major source must submit a permit application including all emissions of all regulated pollutants from all emissions units located at the plant, except that only a generalized list needs to be included for insignificant events or emissions levels. The program applies to all geographic areas within each state, regardless of their attainment status (P.L. 101–549, 1990).

Title VI: Stratospheric Ozone and Global Climate Protection   Title VI requires the EPA to promulgate controls pertaining to the protection of stratospheric ozone. In response to growing evidence that chlorine and bromine could destroy stratospheric ozone on a global basis, many members of the international community concluded that an international agreement to reduce the global production of ozone-depleting substances was needed. Because releases of chlorofluorocarbons (CFCs) from all areas mix in the atmosphere to affect stratospheric ozone globally, efforts to reduce emissions from specific products by only a few nations could quickly be offset by increases in emissions from other nations, leaving the risk to the ozone layer unchanged. In September 1987 the United States and 22 other countries signed the Montreal Protocol on Substances that Deplete the Ozone Layer. The Montreal Protocol called for a freeze in the production and consumption of certain CFCs. The Montreal Protocol was the main thrust for Title VI, although in response to recent scientific evidence indicating a hole in the ozone layer, the EPA has adopted a phase-out schedule more aggressive than the protocol.

The CAAAs established a phase-out schedule as fast as, and in some instances faster than, the 1990 Amendments to the Montreal Protocol. Specifically this Title required the United States to phase out production and consumption of class I substances, that is, CFCs, Halons, carbon tetrachloride, and methyl chloroform by 1996, 1994, 1996, and 1996, respectively. In addition, the CAAAs require a freeze in the production of class II chemicals—hydrochlorofluorocarbons (HCFCs)—by the year 2015, and a phaseout by 2030. The phaseout includes exemptions for certain uses that the administrator of the EPA deems essential, such as CFC as a propellant in metered-dose inhalers (an asthma prescription pharmaceutical). As of January 1, 1994, methyl bromide and hydrobromofluorocarbons (HBFCs) were added to the list of class I chemicals. Recently the EPA has accelerated the phase out of HCFC-22, HCFC-141b, and HCFC-142b, three relatively heavily weighted ozone depleters. Production and consumption will be frozen in 2010 with a complete phaseout by 2020.

The Safe New Alternatives Program (SNAP) evaluates the overall effects on human health and the environment of the potential substitutes for ozone-depleting substances. SNAP will render it unlawful to replace an ozone-depleting substance with a substitute chemical or technology that may present adverse effects to human health and the environment if the Administrator determines that some other alternative is commercially available and that this alternative poses a lower overall threat to human health and the environment. The SNAP program is a powerful tool to assure that safe alternatives are developed.

The refrigerant recycling regulations promulgated May 14, 1993, require recycling, emissions reduction, and disposal for ozone-depleting refrigerants. The regulations require technicians servicing and disposing of air-conditioning and refrigeration equipment to observe certain service practices, to be certified by an EPA-approved organization for technicians servicing equipment. They also establish reclaimer certification programs.

Other Title VI provisions require the EPA to promulgate additional controls pertaining to the protection of stratospheric ozone for banning nonessential products, such as party streamers and noise horns, and mandating warning labels for products manufactured with containers of, and products containing, specific ozone-depleting substances (P.L. 101–549, 1990; 40 CFR, December 10, 1993 and February 11, 1993; "Complying with the Refrigerant Recycling Rule," June 1993).

Title VII: Provisions Relating to Enforcement The 1990 provisions bring with them penalties unlike other environmental laws. Congress has taken several measures in order to make air quality violations felonies or crimes, punishable by jail sentences, where previously they were misdemeanors that were punishable by fines. Under the CAAAs there are now four classes of criminal offenses: negligent; knowing; knowing endangerment; falsification, failure to make required reports, and tampering with monitors. Each of these offences is detailed below.

1. Negligent: Anyone who negligently releases any of the 189 hazardous air pollutants designated under section 112, as well as the 360 or so extremely hazardous substances listed under 40 CFR Part 350 into the ambient air is subject to criminal prosecution. In order to be a criminal offense, the negligent release must place another person in imminent danger of death or serious bodily injury. Violators may be imprisoned for up to one year.

2. Knowing Violations: Any knowing violations of a state implementation plan, a nonattainment provision, an air toxics law, and a standard associated with the permits, acid rain, or stratospheric ozone titles has been upgraded to a felony. These violations are now punishable by a fine and 5 years imprisonment. The CAAAs also increase the maximum fine associated with criminal knowing violations to $250,000 for individuals and $500,000 for organizations. It is also a criminal offense to knowingly fail to pay any fee owed to the United States under Titles I to VI. Violators may be imprisoned for up to one year and fined.

3. Knowing Endangerment: Anyone who knowingly releases any of the 189 hazardous air pollutants designated under section 112, as well as the 360 or so extremely hazardous substances listed under 40 CFR Part 350 is subject to criminal

prosecution. In order to be a criminal offense, the knowing release must place another person in imminent danger of death or serious bodily injury. This is a felony offense punishable by a fine and/or 15 years in prison. The statute authorizes a $1 million maximum fine for organizations and companies.

4. Falsifications, Failures to Report, and Tampering: Knowing actions to falsify reports, failure to keep necessary monitoring records, and material omissions from such reports and records as well as failures to report or notify as required by the CAAAs or failure to properly install the monitoring equipment are criminal violations. The CAAAs also increase the maximum fine associated with criminal knowing violations to $250,000 for individuals and $500,000 for organizations.

Citizens will be allowed to sue violators for penalties that will go to the U.S. Treasury fund for the EPA in compliance and enforcement activities. Administrative enforcement now includes provisions for issuance of field citations for violations observed in the field (this is like receiving a speeding ticket). These citations may not exceed $5000 for each day that the violation continues. Administrative penalties are capped at $200,000. The CAAAs authorize the EPA to issue administrative compliance orders with compliance schedules of up to one year. The EPA may act in emergency situations to protect public health and the environment. Companies can pay fines of up to $25,000 per day for failure to comply with an emergency order. Knowing violations of an emergency order are punishable by a fine and 5 years in prison.

The act also includes other titles that are not included here. It is suggested that the interested reader contact his or her specific EPA region for information and literature, call the stratospheric ozone hot line for specific questions, or refer directly to the CAAA (P.L. 101–549, 1990).

Hazardous and Solid Waste

Resource Conservation and Recovery Act   Defining what constitutes "solid waste" and "hazardous waste" requires consideration of both legal and scientific factors. The basic definition is derived from the Resource Conservation and Recovery Act (RCRA) of 1976 and the subsequent Hazardous and Solid Waste Amendments (HSWA) of 1984. Hazardous substances are regulated under the Comprehensive Environmental Response Compensation and Liability Act (CERCLA) of 1980 and the subsequent Superfund Amendments and Reauthorization Act (SARA) of 1986. These two governing bodies regulate management of currently generated hazardous waste and remediation of hazardous waste sites, respectively.

The first comprehensive federal effort to deal with the solid waste problem in general, and hazardous waste specifically, came with the passage of RCRA. The act provides for the development of federal and state programs for otherwise unregulated disposal of waste materials and for the development of resource recovery programs. It regulates anyone engaged in the creation, transportation, treatment, and disposal of "hazardous wastes." It also regulates facilities for the disposal of all solid waste and prohibits the use of open (essentially uncontrolled) dumps for solid wastes in favor of sanitary (essentially controlled) landfills.

The hazardous waste management program identifies specific hazardous wastes either by listing them or identifying characteristics that render them hazardous. Under RCRA, a solid waste that is not excluded as a hazardous waste is a hazardous waste if it exhibits any of the characteristics of reactivity, corrosivity, ignitability, or toxicity under 40 Code of Federal Regulations (CFR) 261.21–261.24, or it is listed hazardous waste under 40 CFR Parts 261.31–261.33. The EPA has established three lists:

1. "K" Listed Wastes—These hazardous wastes are from specific sources, that is, those wastes generated in a specific process that is specific to an industry group. Examples include wastewater treatment sludge from the production of chrome yellow and orange pigments and still bottoms from the distillation of benzyl chloride.

2. "F" Listed Wastes—These hazardous wastes are from nonspecific sources, that is, those wastes that are generated by a nonspecific industry that are generated from a standard operation that is part of a particular manufacturing process. Examples include spent solvent mixtures and blends used in degreasing containing, before use, a total of 10 percent or more (by volume) of various solvents.

3. "P" and "U" Listed Wastes—The third list has been broken into two distinct subsets. The "U" list contains chemicals that are deemed toxic and the "P" list contains chemicals that are deemed acutely hazardous. These hazardous wastes are discarded commercial chemical products, off-specification products, container residues, and spill residues. Examples include beryllium, fluorine, and methyl isocyanate.

As mentioned above, those wastes that are not listed in either the F, K, P, U lists may still be a hazardous waste if they exhibit one or more of the four following characteristics: reactivity, ignitability, corrosivity, or toxicity. These are listed below.

1. Reactivity—The waste will react violently with or release toxic gases or fumes when mixed with water, or is susceptible to explosions or detonations

2. Ignitability—A solid liquid or gas that is easily ignitable

3. Corrosivity—Alkaline or acidic material normally having a pH in the range of less than 2 and greater than 12.5

4. Toxicity—Wastes that have the ability to bioaccumulate in various aquatic species

Facilities generating these wastes are required to notify the EPA of such activity, as well as comply with the standards authorized by RCRA under Subtitle C. Facilities that generate hazardous waste are classified as either large-quantity generators, small-quantity generators, or conditionally exempt small-quantity generators. Large-quantity generators receive this classification if they generate more than 1000 kg in a calendar month. Small-quantity generators receive this classification if they generate less than 1000 kg but more than 100 kg in any calendar month. Both large- and small-quantity generators are subject to all EPA hazardous waste regulations, except small-quantity generators are not required to have formal training programs for employees; an RCRA contingency plan and accumulation time is determined differently. Conditionally exempt small-quantity generators, generating no more than 100 kg of hazardous waste in a calendar month, are not subject to EPA regulations, except that all waste(s) generated must be disposed of at a site approved by state or federal authority.

Generators are allowed to accumulate waste on site in two related circumstances. First, the generator is allowed to accumulate up to 55 gallons of a hazardous waste in a satellite accumulation area, provided the waste container is properly marked and compatible with the waste, and is removed to a storage area within three days of reaching the 55-gal limit. Second, generators may store waste on site for 90 days or less in storage areas and containers that adhere to strict requirements. Small-quantity generators may store waste on site for up to 180 days (270 days if the waste must be transferred 200 miles or more). Conditionally exempt small-quantity generators are conditionally exempt under RCRA but must still meet certain minimum requirements (40 CFR, Part 261, July 1, 1992).

Comprehensive Environmental Response, Compensation, and Liability Act
Congress enacted the Comprehensive Environmental Response, Compensation, and Liability Act of 1980 (CERCLA) commonly known as "Superfund." Superfund establishes two related funds to be used for the immediate removal of hazardous substances released into the environment. Superfund is intended to establish a mechanism of response for the immediate cleanup of hazardous waste contamination from accidental spills and from chronic environmental damage such as associated with abandoned hazardous waste disposal sites. CERCLA works in concert with RCRA to provide full coverage of present and past hazardous waste activities.

A hazardous substance under CERCLA is any substance the EPA has designated for special consideration under the CAA, CWA, Toxic Substance Control Act (TSCA), as well as any hazardous waste under RCRA. Furthermore, the EPA must designate additional substances as hazardous that may present substantial danger to health and the environment. A list of these hazardous substances is found in 40 CFR Part 302.

Under CERCLA, the EPA is empowered to undertake removal and/or remedial action where the pollutant may present an imminent and substantial danger. Potentially responsible parties (PRPs) are those private parties that are encouraged by the EPA to clean up sites on the National Priorities List (NPL). Only NPL sites are eligible for fund-financed remedial action. Sites make the list by undergoing evaluation and scoring under the hazard ranking system (HRS), which estimates the degree of risk each suspected site poses to human health and the environment. Factors such as waste volume, toxicity, and potential pathways are evaluated and combined to form a single numbered HRS score. If this HRS score is over 28.5 the site will be included on the NPL.

Spill reporting under CERCLA covers releases to all environmental media: air, surface water, groundwater, and soil. The EPA assigns reportable quantities, that is, the exceeding of the assigned value of a hazardous substance released within any 24-h period for which the CERCLA reporting requirements are triggered, to all hazardous substances. Immediately after a person in charge of the facility has knowledge of a release in excess of a reportable quantity, that person must immediately contact the National Response Center (NRC) to relay telephone information on the details of the release. Written follow-up information is required for establishing the details of cleanup mitigation efforts.

In response in part to the Bhopal, India, tragedy, the Emergency Planning and Community Right to Know Act (EPCRA) was enacted under the Superfund Amendments and Reauthorization Act of 1986. Under EPCRA, state and local governments are required to develop emergency response plans for unanticipated releases of a number of acutely toxic materials known as extremely hazardous substances. EPCRA mandates the formation of State Emergency Response Commissions (SERC) and Local Emergency Planning Committees (LEPC), which are in charge of developing and implementing emergency response plans.

The EPCRA also establishes reporting provisions for informing the public of hazardous substances being used, as indicated below.

1. Material Safety Data Sheets (MSDS) Reporting—Facilities for which hazardous chemicals are present in excess of a threshold quantity are required to submit MSDSs to the SERC, LEPC, and local fire departments.
2. Tier I or Tier II Reporting—Facilities for which hazardous substances are present are required to provide information on the annual and daily inventory information on the quantities of those materials and their locations on site. Tier I reports provide the required information on hazardous chemicals grouped by hazard category and Tier II reports provide the information on individual hazardous chemicals. Tier II may be submitted in lieu of Tier I reports.
3. Toxic Release Inventory (TRI)—Facilities that manufacture, process, or otherwise use a toxic chemical in excess of a threshold quantity are required to provide annual reports on the quantities released (40 CFR, Parts 300–372, July 1, 1992).

Solid Waste   Under RCRA, 40 CFR 261.4(a), a solid waste is defined as any material that is discarded, abandoned, recycled, or inherently wastelike. To gain further insight into the meaning of solid waste, the RCRA Section 1004(27) definition is:

A solid waste is defined as any garbage, refuse, sludge from a waste treatment plant, water supply treatment plant, or air pollution control facility and other discarded material, including solid, liquid, semisolid, or contained gaseous material resulting from industrial, commercial, mining, and agricultural operations and community activities, but does not include solid or dissolved material in domestic sewage (P.L. 80–272, 1965).

A material can be considered a solid waste if it is recycled in a manner constituting disposal. This type of recycling includes materials used to produce products that are applied to the land, burned for energy recovery, or accumulated speculatively before recycling. A material that fits the definition of a solid waste may be regulated as a hazardous waste if it poses a threat to human health or the environment. The definitions of solid waste and hazardous waste interlock, which results in EPA regulating a plethora of materials that may not be commonly thought of as wastes for certain industries.

The following materials are excluded from the definition of solid waste:

1. Domestic sewage
2. Industrial wastewater discharges that are point-source discharges subject to regulation under section 402 of the Clean Water Act (This exclusion applies only to point-source discharges. It does not exclude industrial wastewaters while they are being collected, stored, or treated before discharge, nor does it exclude sludges that are generated by industrial wastewater treatment.)
3. Irrigation return flows
4. Source, special nuclear, or by-product material as defined by the Atomic Energy Act
5. Materials subjected to in situ mining techniques that are not removed from the ground as part of the extraction process
6. Pulping liquors (i.e., black liquor) that are reclaimed in a pulping liquor recovery furnace and then reused in the pulping process, unless it is accumulated speculatively
7. Spent sulfuric acid used to produce virgin sulfuric acid, unless it is accumulated speculatively
8. Secondary materials that are reclaimed and returned to the original process or processes in which they were generated where they are reused in the production process provided (This exemption has limitations stipulating closed processes, reclamation not involving controlled flame combustion, accumulation, use as fuel, and not used in a manner constituting disposal.)

Obviously, solid waste is far less regulated, which, when translated by hazardous waste generators, means a savings on disposal costs. Certain materials are not regulated as a solid waste when recycled. These include materials shown to be recycled by being:

1. Used or reused as ingredients in an industrial process to make a product, provided the materials are not being reclaimed
2. Used or reused as effective substitutes for commercial products

3. Returned to the original process from which they are being generated, without first being reclaimed. The material must be returned as a substance for raw material feedstock, and the process must use raw materials as principal feedstock.

Certain materials may still be considered solid wastes, that is, RCRA-regulated, even if the recycling involves use, reuse, or return to the original process. This includes:

1. Materials used in a manner constituting disposal, or used to produce products that are applied to the land; or materials burned for energy recovery, used to produce a fuel, or contained in fuels
2. Materials accumulated speculatively
3. Hazardous waste numbers F020, F021 (unless used as an ingredient to make a product at the site of generation), F022, F023, F026, and F028.

It is important to examine the manner of recycling and the material being recycled in determining whether the waste is considered a solid waste.

Once a material is found to be a solid waste, the next question is whether it is a hazardous waste. The EPA automatically exempts certain solid waste from being hazardous waste. A few examples are household waste, including household waste that has been collected, transported, stored, treated, disposed, recovered (e.g., refuse-derived fuel), or reused; solid waste generated in the growing and harvesting of agricultural crops that is returned to the soil as fertilizer; fly ash waste; bottom ash waste, slag waste; flue gas emissions control waste generated primarily from the combustion of fossil fuels; and cement kiln dust. (Cement kiln dust is under attack by private interest groups to be regulated as a hazardous waste.)

The EPA may also grant a variance from classification as a solid waste on a case-by-case basis. Eligible materials include those that are reclaimed and then reused within the original primary production process in which they were generated, reclaimed partially, but require further processing by being completely recovered, and accumulating speculatively with less than 75 percent of the volume having the potential for recycling (40 CFR, Part 261, July 1, 1992).

Water Pollution Control

Congress put the framework together for water pollution control by enacting the Federal Water Pollution Control Act (FWPCA), the Marine Protection, Research and Sanctuaries Act (MPRSA), the Safe Drinking Water Act (SDWA), and the Oil Pollution Control Act (OPA). Each statute provides a variety of tools that can be used to meet the challenges and complexities of reducing water pollution in the nation. These are discussed below.

Federal Water Pollution Control Act   In 1977, Congress renamed the FWPCA the Clean Water Act (CWA) and substantially revamped and revised the control of toxic water pollutants. The CWA has two basic components: a statement of goals and objectives and a system of regulatory mechanisms calculated to achieve these goals and mechanisms. The objective of the program as outlined in section 101 of the CWA is to "restore and maintain the chemical, physical and biological integrity of the nation's waters." To achieve this, the CWA provides water quality for protection of fish, shellfish, and wildlife for recreational use and eliminates the discharge of pollutants into the waters of the United States.

The system of achieving these goals and objectives has six basic elements:

1. A two-stage system of technology-based effluent limits establishing base level or minimum treatment required to prevent industries and publicly owned treatment works (POTW) from discharging pollutants.
2. A program for imposing more stringent limits in permits where such limits are necessary to achieve water quality standards or objectives.
3. A permit program known as the National Pollution Discharge Elimination Program (NPDES). The NPDES permit requires the dischargers to disclose the volume and nature of their dischargers as well as monitor and report the results to the authorizing agency. The NPDES permit program authorizes the EPA and citizens enforcement in the case of noncompliance.
4. A set of specific deadlines for compliance or noncompliance with the limitations, with attached enforcement provisions for the EPA and citizens.
5. A set of provisions applicable to certain toxic and other pollutant discharges of particular concern, e.g., stormwater and oil spills.
6. A loan program to help fund POTW attainment of the applicable requirements.

Effluent limitations have been established for various categories of point sources that include but are not limited to the chemical, pharmaceutical, paper manufacturing, and pesticide manufacturing industries. The CWA is far more than the six-part framework indicates. The reader is referred to 40 CFR Subchapter N, "Effluent Guidelines and Standards" for more information on how a specific industry is regulated (40 CFR, Part 401, July 1, 1992).

Safe Drinking Water Act   The Safe Drinking Water Act (SDWA) was originally passed in 1974 to ensure that public water supplies are maintained at high quality by setting national standards for levels of contaminants in drinking water, by regulating underground injection-wells, and by protecting sole source aquifers.

The SDWA requires the EPA to establish maximum contamination level goals (MCLGs) and national primary drinking water regulations (NPDWR) for contaminants that, in the judgment of the EPA Administrator, may cause any adverse effect on the health of persons and that are known or anticipated to occur in public water systems. The NPDWRs are to include maximum contamination levels (MCLs) and "criteria and procedures to assure a supply of drinking water that dependably complies" with such MCLs. If it is not feasible to ascertain the level of a contaminant in drinking water, the NPDWRs may require the use of a treatment technique instead of an MCL. The EPA is mandated to establish MCLGs and promulgate NPDWRs for 83 contaminants in public water systems. (The SDWA was amended in 1986 by establishing a list of 83 contaminants for which EPA is to develop MCLGs and NPDWRs). MCLGs and MCLs were to be promulgated simultaneously.

The MCLGs do not constitute regulatory requirements that impose any obligation on public water systems. Rather, MCLGs are health goals that are based solely upon consideration of protecting the public from adverse health effects of drinking water contamination. The MCLGs reflect the aspirational health goals of the SDWA that the enforceable requirements of NPDWRs seek to attain. MCLGs are to be set at a level where "no known or anticipated adverse effects on the health of persons occur and which allows an adequate margin of safety."

The House Report on the bill that eventually became the SDWA of 1974 provides congressional guidance on developing MCLGs:

> [T]he recommended maximum contamination level [renamed maximum contamination level goal in the 1986 amendments to the SDWA] must be set to prevent the occurrence of any known or anticipated adverse effect. It must include an adequate margin of safety, unless there is no safe threshold for a contaminant. In such a case, the recommended maximum contamination level would be set at the zero level (40 CFR, Parts 141 and 142, June 7, 1991).

The NPDWRs include either MCLs or treatment technique requirements as well as compliance monitoring requirements. The MCL for a contaminant must be set as close to the MCLG as feasible. Feasible means "feasible with the use of the treatment techniques and other means which the Administrator of the EPA finds, after examination for efficacy under field conditions and not solely under laboratory conditions, are available (taking cost into consideration)." A treatment technique must "prevent known or anticipated adverse effects on the health of a person to the extent feasible." A treatment technique requirement can be set only if the EPA administrator makes a finding that "it is not economically or technically feasible to ascertain the level of the contaminant." Also the SDWA requires the EPA to identify the best available technology (BAT) for meeting the MCL for each contaminant.

The EPA sets national secondary drinking water regulations (NSDWRs) to control water color, odor, appearance, and other characteristics affecting consumer acceptance of water. The secondary regulations are not federally enforceable but are considered guidelines for the states (40 CFR, Parts 141 and 142, June 7, 1991).

Oil Pollution Control Act   The Oil Pollution Control Act (OPA) of 1990 was enacted to expand prevention and preparedness activities, improve response capabilities, ensure that shippers and oil companies pay the costs of spills that do occur, and establish and expand research and development programs. This was all in response to the *Exxon Valdez* oil spill in Prince William Sound in 1989.

The OPA establishes a new Oil Spill Liability Trust Fund, administered by the U.S. Coast Guard. This fund replaces the fund established under the CWA and other oil pollution funds. The new act mandates prompt and adequate compensation for those harmed by oil spills and an effective and consistent system of assigning liability. The act also strengthens requirements for the proper handling, storage, and transportation of oil and for the full and prompt response in the event discharges occur. The act does so in part by amending section 311 of the CWA.

There are eight titles codified under the act, details of which are available in the literature (40 CFR, Part 112, July 1, 1992; 40 CFR, February 17, 1993).

Occupational Safety and Health Act

The Occupational Safety and Health Act (OSH Act) was enacted by Congress in 1970 and established the Occupational Safety and Health Administration (OSHA), which addressed safety in the workplace; at the same time, EPA was created. Both EPA and OSHA are mandated to reduce the exposure of hazardous substances over land, sea, and air. The OSH Act is limited to conditions that exist in the workplace, where its jurisdiction covers both safety and health. Frequently, both agencies regulate the same substances but in a different manner. In effect, they are overlapping environmental organizations.

Congress intended that OSHA be enforced through specific standards. Employers would follow these standards in an effort to achieve a safe and healthful working environment. A "general duty clause" was added to attempt to cover those obvious situations that were admitted by all concerned but for which no specific standard existed. The OSHA standards are an extensive compilation of regulations, some that apply to all employers—such as eye and face protection—and some that apply to workers who are engaged in a specific type of work, such as welding or crane operation. Employers are obligated to familiarize themselves with the standards and comply with them at all times.

Health issues, most importantly, contaminants in the workplace, have become OSHA's primary concern. Health hazards are complex and difficult to define. Because of this, OSHA has been slow to implement health standards. To be complete, each standard requires medical surveillance, record keeping, monitoring, and physical reviews. On the other side of the ledger, safety hazards are aspects of the work environment that are expected to cause death or serious physical harm immediately or before the imminence of such danger can be eliminated.

Probably one of the most important safety and health standards ever adopted is the OSHA hazard communication standard, more popularly known as the "right-to-know" laws. The hazard communication standard requires employers to communicate information to the employee on hazardous chemicals that exist within the workplace. The program requires employers to craft a written hazard communication program, keep material safety data sheets (MSDSs) for all hazardous chemicals at the workplace and provide employees with training on those hazardous chemicals, and assure that proper warning labels are in place.

The Hazardous Waste Operations and Emergency Response Regulation enacted in 1989 by OSHA addresses the safety and health of employees involved in cleanup operations at uncontrolled hazardous waste sites being cleaned up under government mandate, and in certain hazardous waste treatment, storage, and disposal operations conducted under RCRA. The standard provides for employee protection during initial site characterization and analysis, monitoring activities, training, and emergency response.

Four major areas are under the scope of the regulation:

1. Cleanup operations at uncontrolled hazardous waste sites that have been identified for cleanup by a government health or environmental agency
2. Routine operations at hazardous waste TSD (Transportation, Storage, and Disposal) facilities or those portions of any facility regulated by 40 CFR Parts 264 and 265
3. Emergency response operations at sites where hazardous substances have or may be released
4. Corrective actions at RCRA sites

The regulation addresses three specific populations of workers at the above operations. First, it regulates hazardous substance response operations under CERCLA, including initial investigations at CERCLA sites before the presence or absence of hazardous substance has been ascertained; corrective actions taken in cleanup operations under RCRA; and those hazardous waste operations at sites that have been designated for cleanup by state or local government authorities. The second worker population to be covered is those employees engaged in operations involving hazardous waste TSD facilities. The third employee population to be covered is those employees engaged in emergency response operations for releases or substantial threat of releases of hazardous substances, and postemergency response operations to such facilities (29 CFR, March 6, 1989; 29 CFR, February 24, 1992).

Toxic Substance Control Act

The Toxic Substance Control Act (TSCA) of 1976 provides EPA with the authority to control the risks of thousands of chemical substances, both new and old, that are not regulated as drugs, food additives, cosmetics, or pesticides. TSCA essentially mandates testing of chemical substances to regulate their uses in industrial, commercial, and consumer products. TSCA fills in the gaps and supplements other laws regulating toxic substances, such as the Clean Air Act, the Occupational Safety and Health Act, and the Federal Water Pollution Control Act. TSCA allows EPA to tailor its regulation to specific sources of risk.

The TSCA essentially contains two sections: requirements for information on the substance to identify risks to health and the environment from chemical substances [a premanufacturing notification (PMN)]; and regulations on the production and distribution of new chemicals and regulations on the manufacturing, processing, distribution, and use of existing chemicals (recordkeeping and reporting requirements).

TSCA Inventory No person may manufacture a new chemical substance, or manufacture or process an existing chemical substance for a significant new use, without EPA approval. (Note: This will apply to nanochemicals and products.) There are two kinds of chemical substances—"new chemical substances" and "existing chemical substances." A chemical substance that is not on the TSCA inventory (an inventory of existing chemical substances) is a "new chemical substance." Notification is required before a chemical substance can be put to a significant new use.

The TSCA inventory was initially compiled in 1977 but is updated by EPA to include new chemical substances for which manufacturers have filed Notices of Commencement. Every 4 years, beginning in 1986, manufacturers (and importers) of certain chemicals must submit updated information including chemical identity, plant site, whether the substance is manufactured or imported, whether the substance is distributed offsite for commercial purposes, and production volume on their TSCA-regulated chemicals. Manufacturers of polymers, microorganisms, naturally occurring substances, and inorganics are exempt from inventory updating. Any company that manufactures (or imports) more than 10,000 lb of any chemical substances (except those excluded) in the latest complete fiscal year preceding a reporting period must submit updated information.

Premanufacturing Notification Any company that wishes to manufacture (or import) a chemical substance must first determine whether the chemical substance is on the TSCA inventory. If it is not on the inventory, a manufacturer must file a PMN with the EPA. The PMN must be filed 90 days prior to manufacture of the chemical substance. If a PMN is not necessary, EPA will notify the submitter that submissions are not necessary. Companies may request a 90-day extension if EPA determines there is good cause to extend the notice.

If the EPA takes no action by the end of the review period, the submitter may begin to manufacture. Within 30 days after the first day of manufacturing, the manufacturer must submit a Notice of Commencement (NOC). The NOC essentially places the chemical substance on the TSCA inventory. Companies are exempt from the PMN reporting requirements for the following reasons: The chemical is not a "chemical substance," e.g., a nuclear material, tobacco, foods, and drugs; any chemical substance that is manufactured or imported in small quantities solely for research and development, provided certain conditions are met; any chemical substances that will be manufactured or imported solely for test-marketing purposes under an exemption granted pursuant to 40 CFR 720.38; any new chemical substance manufactured solely for export provided certain requirements are met; any new chemical substance that is manufactured or imported under the terms of a rule promulgated under section 5(h) (4) of TSCA, e.g., certain chemicals used for instant photographic and peel-apart film articles; any by-product if its only commercial purpose is for use by a public or private organization that burns it as fuel, disposes of it as waste, or extracts chemical substances from it for commercial purposes; certain chemical substances described in 40 CFR 720(h), e.g., any impurity or by-product, provided it is not used or manufactured for commercial purposes; and any chemical substance that is manufactured solely for noncommercial research and development purposes.

Recordkeeping and Reporting Requirements Section 8 of TSCA contains a variety of reporting requirements to fulfill the statute's information-gathering objectives. The scope of chemicals and regulated entities changes between sections. Sections 8(a) and (d) apply only to chemicals listed by regulation. Sections 8(c) and (e) cover the full spectrum of TSCA chemicals. Section 8(a) reporting requirements consist of two reports: the Comprehensive Assessment Information Rule (CAIR) and the Preliminary Assessment Information Rule (PAIR). Manufacturers, importers, and processors must report on each substance such that the EPA can formulate risk assessments and develop regulatory strategies. CAIR requires companies to report on listed substances under 40 CFR 704.200 Subparts C and D. PAIR requires manufacturers and importers to report on each listed substance during the reporting period for that substance as given in the rule (see 40 CFR Part 712). Ultimately, PAIR is to be replaced by CAIR.

Section 8(c) requires affected companies to maintain records of allegations of significant adverse reactions to health or the environment caused by the substances. Section 8(d) sets forth requirements for the submission of health studies on chemical substances and mixtures selected for priority consideration testing rules under Section 4(a) of TSCA and on other substances on which EPA requires health and safety studies. Section 8(e) requires affected companies to inform the administrator of the EPA or any chemical substance that presents a substantial risk of injury to health or the environment as soon as this information is discovered.

The TSCA also specifically regulates PCBs, CFCs, and asbestos (40 CFR, Parts 700–766, July 1, 1992).

Nanotechnology Environmental Regulations Overview

Completely new legislation and regulatory rulemaking will almost certainly be necessary for environmental control of nanotechnology. However, in the meantime, one may speculate on how the existing regulatory framework might be applied to the nanotechnology area as this emerging field develops over the next several years. One experienced Washington, D.C., attorney has done just that, as summarized below.[14–16] The reader is encouraged to consult the cited references as well as the text of the laws that are mentioned and the applicable regulations derived from them.

As indicated above, commercial applications of nanotechnology are likely to be regulated under TSCA, which authorizes EPA to review and establish limits on the manufacture, processing, distribution, use, and/or disposal of new materials that EPA determines to pose "an unreasonable risk of injury to human health or the environment." The term *chemicals* is defined broadly by TSCA. Unless qualifying for an exemption under the law [R&D (a statutory exemption requiring no further approval by EPA), low-volume production, low environmental releases along with low volume, or plans for limited test marketing], a prospective manufacturer is subject to the full-blown PMN procedure. This requires submittal of said notice, along with toxicity and other data to EPA at least 90 days before commencing production of the chemical substance.

Approval then involves recordkeeping, reporting, and other requirements under the statute. Requirements will differ, depending on whether EPA determines that a particular application constitutes a "significant new use" or a "new chemical substance." EPA can impose limits on production, including an outright ban when it is deemed necessary for adequate protection against "an unreasonable risk of injury to health or the environment." EPA may revisit a chemical's status under TSCA and change the degree or type of regulation when new health/environmental data warrant. EPA is expected to be issuing several new TSCA test rules in 2004.[17] If the experience with genetically engineered organisms is any indication, there will be a push for EPA to update regulations in the future to reflect changes, advances, and trends in nanotechnology.[18]

Workplace exposure to chemical substances and the potential for pulmonary toxicity is subject to regulation by the OSHA under the OSH Act, including the requirement that potential hazards be disclosed on MSDS. (An interesting question arises as to whether carbon nanotubes, chemically carbon but with different properties because of their small size and structure, are indeed to be considered the same as or different from carbon black for MSDS purposes.) Both governmental and private agencies can be expected to develop the requisite threshold limit values (TLVs) for workplace exposure. Also, EPA may once again utilize TSCA to assert its own jurisdiction, appropriate or not, to minimize exposure in the workplace.

Another likely source of regulation would fall under the provisions of the Clean Air Act (CAA) for particulate matter less than 2.5 μm ($PM_{2.5}$). Additionally, an installation manufacturing nanomaterials may ultimately become subject as a "major source" to the CAA's Section 112 governing hazardous air pollutants (HAPs).

A waste from a commercial-scale nanotechnology facility would be captured under RCRA, provided that it meets the criteria for a RCRA waste. RCRA requirements could be triggered by a listed manufacturing process or the RCRA's specified hazardous waste characteristics. The type and extent of regulation would depend on how much hazardous waste is generated and whether the wastes generated are treated, stored, or disposed of onsite.

Finally, opponents of nanotechnology, especially, may be able to use the National Environmental Policy Act (NEPA) to impede nanotechnology research funded by the U.S. government. A "major Federal action significantly affecting the quality of the human environment" is subject to the environmental impact provision under NEPA. (Various states also have environmental impact assessment requirements that could delay or put a stop to construction of nanotechnology facilities.) Time will tell.

1.7 CLASSIFICATION AND SOURCES OF POLLUTANTS[5,6,19]

It is relatively safe to say that there will be two classifications of nanoemissions: particulates and gases. Additional details cannot be provided at this time since many of these new processes, and their corresponding emissions, have yet to be formulated. Nonetheless, it seems reasonable to conclude that many of these emissions will be similar in classifications to what presently exits. The classification and sources of pollutants of necessity have to be emitted to "traditional" contaminants. At the time of this writing, very little could be said about those pollutants generated from nanoapplications.

Not long ago, the nation's natural resources were exploited indiscriminately. Waterways served as industrial pollution sinks, skies dispersed smoke from factories and power plants, and the land proved to be a cheap and convenient place to dump industrial and urban wastes. However, society is now more aware of the environment and the need to protect it. The American people have been involved in a great social movement known broadly as "environmentalism." Society has been concerned with the quality of the air one breathes, the water one drinks, and the land on which one lives and works. While economic growth and prosperity are still important goals, opinion polls show overwhelming public support for pollution controls and a pronounced willingness to pay for them. This section presents the reader with information on pollutants and categorizes their sources by the media they threaten.

Air Pollutants

Since the Clean Air Act was passed in 1970, the United States has made impressive strides in improving and protecting air quality. As directed by this Act, the EPA set National Ambient Air Quality Standards (NAAQS) for those pollutants commonly found throughout the country that posed the greatest overall threats to air quality. These pollutants, termed *criteria pollutants* under the act, include: ozone, carbon monoxide, airborne particulates, sulfur dioxide, lead, and nitrogen oxide. Although the EPA has made considerable progress in controlling air pollution, all of the six criteria except lead and nitrogen oxide are currently a major concern in a number of areas in the country. The following subsections focus on a number of the most significant air quality challenges: ozone and carbon monoxide, airborne particulates, airborne toxics, sulfur dioxide, acid deposition, and indoor air pollutants.

Ozone and Carbon Monoxide   Ozone is one of the most intractable and widespread environmental problems. Chemically, ozone is a form of oxygen with three oxygen atoms instead of the two found in regular oxygen. This makes it very reactive, so that it combines with practically every material with which it comes in contact. In the upper atmosphere, where ozone is needed to protect people from ultraviolet radiation, the ozone is being destroyed by manmade chemicals, but at ground level, ozone can be a harmful pollutant.

Ozone is produced in the atmosphere when sunlight triggers chemical reactions between naturally occurring atmospheric gases and pollutants such as volatile organic compounds (VOCs) and nitrogen oxides. The main source of VOCs and nitrogen oxides is combustion sources such as motor vehicle traffic.

Carbon monoxide is an invisible, odorless product of incomplete fuel combustion. As with ozone, motor vehicles are the main contributor to carbon monoxide formation. Other sources include wood-burning stoves, incinerators, and industrial processes. Since auto travel and the number of small sources of VOCs are expected to increase, even strenuous efforts may not sufficiently reduce emissions of ozone and carbon monoxide.

Airborne Particulates   Particulates in the air traditionally include dust, smoke, metals, and aerosols. Major sources include steel mills, power plants, cotton gins, cement plants, smelters, and diesel engines. Other sources are grain storage elevators, industrial haul roads, construction work, and demolition. Wood-burning stoves and fireplaces can also be significant sources of particulates. Urban areas are likely to have wind-blown dust from roads, parking lots, and construction work.

Airborne Toxics   Toxic pollutants are one of today's most serious emerging problems and found in all media. Many sources emit toxic chemicals into the atmosphere: industrial and manufacturing processes, solvent use, sewage treatment plants, hazardous waste handling and disposal sites, municipal waste sites, incinerators, and motor vehicles. Smelters, metal refiners, manufacturing processes, and stationary fuel combustion sources emit such toxic metals as cadmium, lead, arsenic, chromium, mercury, and beryllium. Toxic organics, such as vinyl chloride and benzene, are released by a variety of sources, such as plastics and chemical manufacturing plants, and gas stations. Chlorinated dioxins and furans are emitted by some chemical processes and the high-temperature burning of plastics in incinerators.

Sulfur Dioxide   Sulfur dioxide can be transported long distances in the atmosphere because of its ability to bond to particulates. After traveling, sulfur dioxide combines with water vapor to form acid rain. Sulfur dioxide is released into the air primarily through the burning of coal and fuel oils. Today, two-thirds of all national sulfur dioxide emissions come from electric power plants. Other sources of sulfur dioxide include refiners, pulp and paper mills, smelters, steel and chemical plants, and energy facilities related to oil shale, syn (synthetic) fuels, and oil and gas production. Home furnaces and coal-burning stoves are sources that directly affect residential neighborhoods.

Acid Deposition   Acid deposition is a serious environmental concern in many parts of the country. The process of acid deposition begins with the emissions of sulfur dioxide (primarily from coal-burning power plants) and nitrogen oxides (primarily from motor vehicles and coal-burning power plants). As described in the previous subsection, these pollutants interact with sunlight and water vapor in the upper atmosphere to form acidic compounds. During a storm, these compounds fall to Earth as acid rain or snow; the compounds may also join dust or other dry airborne particles and fall as "dry deposition."

Indoor Air Pollutants

Indoor air pollution is rapidly becoming a major health issue in the United States. Indoor pollutant levels are quite often higher than outdoors, particularly where buildings are tightly constructed to save energy. Since most people spend 90 percent of their time indoors, exposure to unhealthy concentrations of indoor air pollutants is often inevitable. The degree of risk associated with exposure to indoor pollutants depends on how well buildings are ventilated and the type, mixture, and amounts of pollutants in the building. Indoor air pollutants of special concern are described below.

Radon Radon is a unique environmental problem because it occurs naturally. Radon results from the radioactive decay of radium-226, found in many types of rocks and soils. Most indoor radon comes from the rock and soil around a building and enters structures through cracks or openings in the foundation or basement. Secondary sources of indoor radon are well water and building materials.

Environmental Tobacco Smoke Environmental tobacco smoke is smoke that nonsmokers are exposed to from smokers. This smoke has been judged by the Surgeon General, the National Research Council, and the International Agency for Research on Cancer to pose a risk of lung cancer to nonsmokers. Tobacco smoke contains a number of pollutants, including inorganic gases, heavy metals, particulates, VOCs, and products of incomplete combustion (PICs), such as polynuclear aromatic hydrocarbons.

Asbestos Asbestos has been used in the past in a variety of building materials, including many types of insulation, fireproofing, wallboard, ceiling tiles, and floor tiles. The remodeling or demolition of buildings with asbestos-containing materials frees tiny asbestos fibers in clumps or clouds of dust. Even with normal aging, materials may deteriorate and release asbestos fibers. Once released, these asbestos fibers can be inhaled into the lungs and can accumulate.

Formaldehyde and Other Volatile Organic Compounds The EPA has found formaldehyde to be a probable human carcinogen. The use of formaldehyde in furniture, foam insulation, and pressed wood products, such as some plywood, particle board, and fiberboard, makes formaldehyde a major indoor air pollutant.

The VOCs commonly found indoors include benzene from tobacco smoke and perchlorethylene emitted by dry-cleaned clothes. Paints and stored chemicals, including certain cleaning compounds, are also major sources of VOCs. VOCs can also be emitted from drinking water; 20 percent of water supply systems have detectable amounts of VOCs.

Pesticides Indoor and outdoor use of pesticides, including termiticides and wood preservatives, are another cause of concern. Even when used as directed, pesticides may release VOCs. In addition, there are about 1200 inert ingredients added to pesticide products for a variety of purposes. While not "active" in attacking the particular pest, some inert ingredients are chemically or biologically active and may cause health problems. EPA researchers are presently investigating whether indoor use of insecticides and subsurface soil injection of termiticides can lead to hazardous exposure.

Water Pollutants

The EPA, in partnership with state and local governments, is responsible for improving and maintaining water quality. These efforts are organized around three themes. The first is maintaining the quality of drinking water. This is addressed by monitoring and treating drinking water prior to consumption and by minimizing the contamination of the surface water and protecting against contamination of groundwater needed for human consumption. The second is preventing the degradation and destruction of critical aquatic habitats, including wetlands, nearshore coastal waters, oceans, and lakes. The third is reducing the pollution of free-flowing surface waters and protecting their uses. The following is a discussion of various pollutants categorized by these themes.

Drinking Water Pollutants The most severe and acute public health effects from contaminated drinking water, such as cholera and typhoid, have been eliminated in America. However, some less acute and immediate hazards remain in the nation's tap water. These hazards are associated with a number of specific contaminants in drinking water. Contaminants of special concern to the EPA are lead, radionuclides, microbiological contaminants, and disinfection by-products.

The primary source of lead in drinking water is corrosion of plumbing materials, such as lead service lines and lead solders, in water distribution systems and in houses and larger buildings. Virtually all public water systems serve households with lead solders of varying ages, and most faucets are made of materials that can contribute some lead to drinking water.

Radionuclides are radioactive isotopes that emit radiation as they decay. The most significant radionuclides in drinking water are radium, uranium, and radon, all of which occur in the natural environment. While radium and uranium enter the body by ingestion, radon is usually inhaled after being released into the air during showers, baths, and other activities, such as washing clothes or dishes. Radionuclides in drinking water occur primarily in those systems that use groundwater. Naturally occurring radionuclides seldom are found in surface waters (such as rivers, lakes, and streams).

Water contains many microbes—bacteria, viruses, and protozoa. Although some organisms are harmless, others can cause disease. The Centers for Disease Control reported 112 waterborne disease outbreaks from 1981 to 1983. Microbiological contamination continues to be a national concern because contaminated drinking water systems can rapidly spread disease.

Disinfection by-products are produced during water treatment by the chemical reactions of disinfectants with naturally occurring or synthetic organic materials present in untreated water. Since these disinfectants are essential to safe drinking water, the EPA is presently looking at ways to minimize the risks from by-products.

Critical Aquatic Habitat Pollutants Critical aquatic habitats that need special management attention include the nation's wetlands, near coastal waters, oceans, and lakes. In recent years, the EPA has been focusing on addressing the special problems of these areas. The following is a discussion of pollutants categorized by the habitats they affect.

Wetlands in urban areas frequently represent the last large tracts of open space and are often a final haven for wildlife. Not surprisingly, as suitable upland development sites become exhausted, urban wetlands are under increasing pressure for residential housing, industry, and commercial facilities.

Increasing evidence exists that our nation's wetlands, in addition to being destroyed by physical threats, also are being degraded by chemical contamination. The problem of wetland contamination received national attention in 1985 due to reports of waterfowl deaths and deformities caused by selenium contamination. Selenium is a trace element that occurs naturally in soil and is needed in small amounts to sustain life. However, for years it was being leached out of the soil and carried in agricultural drainwater used to flood the wildlife refuge's wetlands, where it accumulated to dangerously high levels.

Coastal water environments are particularly susceptible to contamination because they act as sinks for the large quantities of pollution discharged from municipal sewage treatment plants, industrial facilities, and hazardous waste disposal sites. In many coastal areas, non-point-source runoff from agricultural lands, suburban developments, city streets, and combined sewer and stormwater overflows poses an even more significant problem than point sources. This is due to the difficulty of identifying and then controlling the source of the pollution.

Physical and hydrological modifications from such activities as dredging channels, draining and filling wetlands, constructing dams, and building shorefront houses may further degrade near coastal environments. In addition, growing population pressures will continue to subject these sensitive coastal ecosystems to further stress.

The Great Lakes provide an inestimable resource to the 45 million people living in the surrounding basin. A 1970 study by the International Joint Commission identified nutrients and toxic problems in the lakes. They suffered from eutrophication problems caused by excessive nutrient inputs. Since then the United States and Canada have made joint efforts to reduce nutrient loadings, particularly phosphorus. However, contamination of the water and fish by toxics from pesticide runoff, landfill leachates, and in-place sediments remains a major problem.

Ocean dumping of dredged material, sewage sludge, and industrial wastes is a major source of ocean pollution. Sediments dredged from industrialized urban harbors are often highly contaminated with heavy metals and toxic synthetic organic chemicals like PCBs and petroleum hydrocarbons. Although ocean dumping of dredged material, sludge, and industrial wastes is now less of a threat, persistent disposal of plastics from land and ships at sea has become a serious problem. Debris on beaches from sewer and storm drain overflows or mismanagement of trash poses public safety and aesthetic concerns.

Surface Water Pollutants   Pollutants in waterways come from industries or treatment plants discharging wastewater into streams or from waters running across urban and agricultural areas, carrying the surface pollution with them (nonpoint sources). The following is a discussion of surface water pollutants categorized by their main sources.

Raw or insufficiently treated wastewater from municipal and industrial treatment plants still threatens water resources in many parts of the country. In addition to harmful nutrients, poorly treated wastewater may contain bacteria and chemicals.

Sludge, the residue left from wastewater treatment plants, is a growing problem. Although some sludges are relatively "clean," or free from toxic substances, other sludges may contain organic, inorganic, or toxic pollutants and pathogens.

An important source of toxic pollution is industrial wastewater discharged directly into waterways or indirectly through municipal wastewater treatment plants. Industrial wastes discharged indirectly are treated to remove toxic pollutants. It is important that those wastes be treated because toxics may end up in sludge, making them harder to dispose of safely.

Nonpoint sources present continuing problems for achieving national water quality in many parts of the country. Sediment and nutrients are the two largest contributors to non-point-source problems. Nonpoint sources are also a major source of toxics, among them pesticide runoff from agricultural areas, metals from active or abandoned mines, gasoline, and asbestos from urban areas. In addition, the atmosphere is a source of toxics since many toxics can attach themselves to dust, later to be deposited in surface waters hundreds of miles away through precipitation.

Land Pollutants

Historically, land has been used as the dumping ground for wastes, including those removed from the air and water. Early environmental protection efforts focused on cleaning up air and water pollution. It was not until the 1970s that there was much public concern about pollution of the land. It is now recognized that contamination of the land threatens not only future uses of the land itself, but also the quality of the surrounding air, surface water, and groundwater. There are five different forms of land pollutants. These include:

1. Industrial hazardous wastes
2. Municipal wastes
3. Mining wastes
4. Radioactive wastes
5. Underground storage tanks

A short description of each is provided next.

Industrial Hazardous Wastes The chemical, petroleum, and transportation industries are major producers of hazardous industrial waste. Ninety-nine percent of the hazardous waste is produced by facilities that generate large quantities (more than 2200 lb) of hazardous waste each month.

A much smaller amount of hazardous waste, about one million tons per year, comes from small quantity generators (between 220 and 2200 lb of waste each month). These include automotive repair shops, construction firms, laundromats, dry cleaners, printing operations, and equipment repair shops. Over 60 percent of the these wastes are derived from lead batteries. The remainder includes acids, solvents, photographic wastes, and dry cleaning residue.

Municipal Wastes   Municipal wastes include household and commercial wastes, demolition materials, and sewage sludge. Solvents and other harmful household and commercial wastes are generally so intermingled with other materials that specific control of each is virtually impossible.

Sewage sludge is the solid, semisolid, or liquid residue produced from treating municipal wastewater. Some sewage sludges contain high levels of disease-carrying microorganisms, toxic metals, or toxic organic chemicals. Because of the large quantities generated, sewage sludge is a major waste management problem in a number of municipalities.

Mining Wastes   A large volume of all waste generated in the United States is from mining coal, phosphates, copper, iron, uranium, other minerals, and from ore processing and milling. These wastes consist primarily of overburden, the soil and rock cleared away before mining, and tailings, the material discarded during ore processing. Runoff from these wastes increases the acidity of streams and pollutes them with toxic metals.

Radioactive Wastes   Radioactive materials are used in a wide variety of applications, from generating electricity to medical research. The United States has produced large quantities of radioactive wastes that can pose environmental and health problems for many generations.

Pollutants from Underground Storage Tanks   Leaking underground storage tanks are another source of land contamination that can contribute to groundwater contamination. The majority of these tanks do not store waste but instead store petroleum products and some hazardous substances. Most of the tanks are bare steel and subject to corrosion. Many are old and near the end of their useful lives. Hundreds of thousands of these tanks are presently thought to be leaking, with more expected to develop leaks in the next few years.

Hazardous Pollutants

Before the early 1970s, the nation paid little attention to industrial production and the disposal of the waste it generated, particularly hazardous waste. As a result, billions of dollars must now be spent to clean up disposal sites neglected through years of mismanagement. The EPA often identifies a waste as hazardous if it poses a fire hazard (ignitable), dissolves materials or is acidic (corrosive), is explosive (reactive), or otherwise poses danger to human health or the environment (toxic). Most hazardous waste results from the production of widely used goods such as polyester and other synthetic fibers, kitchen appliances, and plastic milk jugs. A small percentage of hazardous waste (less than 1 percent) is comprised of the used commercial products themselves, including household cleaning fluids or battery acid.

Definitions of hazardous substances are not as straightforward as they appear. For purposes of regulation, Congress and the EPA have defined terms to describe wastes and other substances that fall under regulation. The definitions below show the complexity of the EPA's regulatory task.

1. Hazardous Substances [Comprehensive Environmental Response, Compensation and Liability Act (CERCLA), or "Superfund"]—Any substance that, when released into the environment, may cause substantial danger to public health, welfare, or the environment. Designation as a hazardous substance grows out of the statutory definitions in several environmental laws: CERCLA, RCRA, CWA, CAA, and TSCA. Currently there are 717 CERCLA hazardous substances.

2. Extremely Hazardous Substance (CERCLA as amended)—Substances that could cause serious, irreversible health effects from a single exposure. For purposes of chemical emergency planning, EPA has designated 366 substances extremely hazardous. If not already so designated, these also will be listed as hazardous substances.

3. Solid Waste (RCRA)—Any garbage, refuse, sludge, or other discarded material. All solid waste is not solid; it can be liquid, semisolid, or contained gaseous material. Solid waste results from industrial, commercial, mining, and agricultural operations from community activities. Solid waste can be either hazardous or nonhazardous. However, it does not include solid or dissolved material in domestic sewage, certain nuclear material, or certain agricultural wastes.

4. Hazardous Waste (RCRA)—Solid waste, or combinations of solid waste, that because of its quantity, concentration, or physical, chemical, or infectious characteristics, may pose a hazard to human health or the environment.

5. Nonhazardous Waste (RCRA)—Solid waste, including municipal wastes, household hazardous waste, municipal sludge, and industrial and commercial wastes that are not hazardous.

Toxic Pollutants

Today's high standard of living would not be possible without the thousands of different chemicals produced. Most of these chemicals are not harmful if used properly. Others can be extremely harmful if people are exposed to them, even in minute amounts. The following is a discussion of four toxic chemicals under control of the Toxic Substance Control Act of 1976.

Polychlorinated biphenyls (PCBs) provide an example of the problems that toxic substances can present. PCBs were used in many commercial activities, especially in heat transfer fluids in electrical transformers and capacitors. They also were used in hydraulic fluids, lubricants, and dye carriers in carbonless copy paper, and in paints, inks, and dyes. Over time, PCBs accumulated in the environment, either from leaking electrical equipment or from other materials such as inks.

Like PCBs, *asbestos* was widely used for many purposes, such as fireproofing and pipe and boiler insulation in schools and other buildings. Asbestos was often mixed with a cementlike material and sprayed or plastered on ceilings and other surfaces. Now these materials are deteriorating, releasing the asbestos.

Dioxins refer to a family of chemicals with similar structure, although it is common to refer to the most toxic of these—2,3,7,8-tetrachlorodinitro-*p*-dioxin, or TCDD—as dioxin. Dioxin is an inadvertent contaminant of the chlorinated herbicides 2,4,5-T and silvex, which were used until recently in agriculture, forest management, and lawn care. It is also a contaminant of certain wood preservatives and the defoliant Agent Orange used in Vietnam. Dioxins and the related chemicals known as furans also are formed during the combustion of PCBs.

Several other sources of dioxin contamination have been identified in recent years. These include pulp and paper production and the burning of municipal wastes containing certain plastics or wood preserved by certain chlorinated chemicals.

In 1978, the use of *chlorofluorocarbons (CFCs)* as a propellent in aerosol cans and other nonessential uses were prohibited by the EPA. The EPA took this action as a result of evidence that CFCs caused a decrease in stratopsheric ozone.

1.8 EFFECTS OF POLLUTANTS[5,6]

As with the previous section, it is difficult to provide specific information on this topic since technical individuals are essentially dealing with unknown nanoemissions for unknown processes/sources. The material that follows, therefore, primarily addresses "traditional" pollutants. Detailing the effects of emissions from nanoapplications that would be described as pollutants is an order of magnitude more difficult than classifying then, as discussed in the previous section. However, the future looks bright for those individuals involved with toxicology and epidemiology studies of nanotechnology. This topic will be revisited in earnest in Chapter 7—Health Risk Assessment.

Pollutants are various noxious chemicals and refuse materials that impair the purity of the water, soil, and the atmosphere. The area most affected by pollutants is the atmosphere or air. Air pollution occurs when wastes pollute the air. Artificially or synthetically created wastes are the main sources of air pollution. They can be in the form of gases or particulates, which result from the burning of fuel to power motor vehicles and to heat buildings. More air pollution can be found in densely populated areas. The air over largely populated cities often becomes so filled with pollutants that it not only harms the health of humans, plants, and animals but also adversely affects materials of construction.

Water pollution occurs when wastes are dumped into the water. This polluted water can spread typhoid fever and other diseases. In the United States, water supplies are disinfected to kill disease-causing germs. The disinfection, in some instances, does not remove all the chemicals and metals that may cause health problems in the distant future.

Wastes that are dumped into the soil are a form of land pollution, which damages the thin layer of fertile soil that is essential for agriculture. In nature, cycles work to keep soil fertile. Wastes, including dead plants and wastes from animals, form a substance in the soil called humus. Bacteria then decays the humus and breaks it down into nitrates, phosphates, and other nutrients that feed growing plants.

This section will review the effects of air pollutants, water pollutants, and land (solid waste) pollutants on:

1. Humans
2. Plants
3. Animals
4. Materials of construction

For obvious reasons, the material will key on the effects on humans. It will also focus primarily on air pollutants since this has emerged as the leading environmental issue with the passage of the Clean Air Act Amendments of 1990.

Air Pollution

Humans   Humans are in constant contact with pollutants, whether they are indoors or outdoors. The pollutants, primarily air pollutants, may have negative effects on human health. In some instances humans adapt and do not realize that they are being affected. For example, people living in smog-covered cities know that smog is bad for their health but just consider it "normal." There are still some who do not think that there is anything that can be done about it.

Indoors, which includes the home and the workplace, a definite correlation seems to exist between some of the most important indoor activities and the resulting pollutants that are generated. Some examples of these are smoking, the use of personal products, cleaning, cooking, heating, maintenance of hair and facial care, hobbies, and electrical appliances such as washing machines and dryers.[20,21] Fumes from these activities can get trapped in the home or workplace, and the buildup of these over time will cause health problems in the short- and long-term future.

When people go outside, they usually say they are going to "get some fresh air." This "fresh air" to them usually means breathing in the air from a different location. Although the common term for the air outside is "fresh air." the air may not necessarily be very "fresh." The outside air can be full of air pollutants that can cause negative effects on the health of humans.

The influence of air pollution on human productivity has not been firmly established. In addition, a number of authorities suspect (and some are convinced) that air pollution is associated with an increasing incidence of lung and respiratory ailments and heart disease.[21] Table 1.4 shows some of the health effects of the regulated air pollutants.

"Air toxics" is the term generally used to describe cancer-causing chemicals, radioactive materials, and other toxic chemicals not covered by the National

TABLE 1.4 Health Effects of the Regulated Air Pollutants

Pollutant	Health Concerns
Criteria Pollutants	
Ozone	Respiratory tract problems such as difficult breathing and reduced lung function. Asthma, eye irritation, nasal congestion, reduced resistance to infection, and possibly premature aging of lung tissue
Particulate matter	Eye and throat irritation, bronchitis, lung damage, and impaired visibility
Carbon monoxide	Ability of blood to carry oxygen impaired; cardiovascular, nervous and pulmonary systems affected
Sulfur dioxide	Respiratory tract problems, permanent harm to lung tissue
Lead	Retardation and brain damage, especially in children
Nitrogen dioxide	Respiratory illness and lung damage
Hazardous Air Pollutants	
Asbestos	A variety of lung diseases, particularly lung cancer
Beryllium	Primary lung disease, although also affects liver, spleen, kidneys, and lymph glands
Mercury	Several areas of the brain as well as the kidneys and bowels affected
Vinyl chloride	Lung and liver cancer
Arsenic	Causes cancer
Radionuclides	Cause cancer
Benzene	Leukemia

Ambient Air Quality Standards for conventional pollutants. Air toxics result from many activities of modem society, including driving a car, burning fossil fuel, and producing and using industrial chemicals or radioactive materials. The latter is one of the highest health risk problems with which the EPA is wrestling.[22]

Some major contributors to pollution that affect human health are: sulfur dioxide, carbon monoxide, nitrogen oxides, ozone, carcinogens, fluorides, aeroallergens, radon, smoking, asbestos, and noise. These are treated in separate paragraphs below.

Sulfur dioxide (SO_2) is a source of serious discomfort and in excessive amounts is a health hazard, especially to people with respiratory ailments. In the United States alone, the estimated amount of SO_2, emitted into the atmosphere is 23 million tons per year. SO_2 causes irritation of the respiratory tract; it damages lung tissue and promotes respiratory diseases. The taste threshold limit is 0.3 ppm and SO_2 produces an unpleasant smell at 0.5 ppm concentration. In fact, sulfur dioxides in general have been considered as prime candidates for an air pollution index. Such an index would be a measure reflecting the presence and action of harmful environmental conditions. This would aid in rendering meaningful analyses of the effect of air pollutants on human health, especially since health effects are most probably due to the complementing action of pollutants and meteorological variables. SO_2 is more harmful in a dusty atmosphere. This effect may be explained as follows: The respiratory tract is lined with hairlike cilia, which by means of regular sweeping action force out foreign substances entering the respiratory tract through the mouth. SO_2 and H_2SO_4 (sulfuric acid) molecules paralyze the cilia, rendering it ineffective in rejecting these particulates, causing them to penetrate deeper into the lungs. Alone, these molecules are too small to remain in the lungs; but, some SO_2 molecules are absorbed on larger particles, which penetrate into the lungs and settle there, bringing concentrated amounts of the irritant SO_2 into prolonged contact with the fine lung tissues. SO_2 and the other sulfur dioxide–particulate combinations are serious irritants of the respiratory tract. In high-pollution intervals they can cause death. Their action of severely irritating the respiratory tract may cause heart failure due to the excessive laboring of the heart in its pumping action to circulate oxygen through the body.

Carbon monoxide (CO) levels have declined in most parts of the United States since 1970, but the standards are still exceeded in many cities throughout the country. Carbon monoxide pollution is the basic concern in most large cities of the world where traffic is usually congested and heavy. CO cannot be detected by smell or sight, and this adds to its danger. It forms a complex with hemoglobin, called carboxy-hemoglobin (COHb). The formation of this complex reduces the capability of the bloodstream to carry oxygen by interfering with the release of the oxygen carried by remaining hemoglobin. Also, since the affinity of human hemoglobin is 210 times higher for CO than it is for oxygen, a small concentration of CO markedly reduces the capacity of the blood to act as an oxygen carrier. The threshold limit value (TLV), or maximum allowable concentration (MAC), of CO for industrial exposure is 50 ppm; concentrations of CO as low as 10 ppm produce effects on the nervous system and give an equilibrium level of COHb larger than 2 percent. A concentration of 30 ppm produces a level greater than 5 percent COHb, which affects the nervous system and causes impairment of visual acuity, brightness discrimination, and other psychomotor functions. Carbon monoxide concentrations of 50 to 100 ppm are commonly encountered in the atmosphere of crowded cities, especially at heavy-traffic rush hours. Such high concentrations adversely affect driving ability and cause accidents. In addition, an estimate of the average concentration of CO inhaled into the lungs from cigarette smoking is 400 ppm.[23]

Two major pollutants among *nitrogen oxides* (NO_X) are nitric oxide (NO), and nitrogen dioxide (NO_2). Emissions from stationary sources are estimated to be 16 million tons of NO_X per year. Mobile sources of NO_X pollution are automobiles emitting an estimated average of 10.7 million tons per year. NO is colorless, but it is photochemically converted to nitrogen dioxide, which is one of the components of smog. Nitrogen dioxide also contributes to the formation of aldehydes and ketones through the photochemical reaction with hydrocarbons of the atmosphere. Nitrogen dioxide is an irritant; it damages lung tissues, especially through the formation of nitric acid. Breathing nitrogen dioxide at 25 ppm for 8 hours could cause spoilage of lung tissues, while breathing it for one-half hour at 100 to 150 ppm could produce serious pulmonary edema, or swelling of lung tissues. A few breaths at 200 to 700 ppm may cause fatal pulmonary edema.

Ozone (O_3) is produced from the activation of sunlight on nitrogen dioxide, pollutants such as volatile organic compounds (VOCs), and atmospheric gases such as oxygen. It is an irritant to the eyes and lungs, penetrating deeper into the lungs than sulfur dioxide. In air it forms complex organic compounds; dominant among these are aldehydes and peroxyacetyl nitrate (PAN), which also causes eye and lung irritation. Rural areas have concentrations of 2 to 5 parts per hundred million (pphm) of ozone, which is distinguished by an odor of electrical shorting. A few good smells and the individual's sensitivity for this odor disappears. At 5 to 10 pphm, the odor is unpleasant and pungent. Exposure to ozone for 30 min at 10 to 15 pphm, which is normally encountered in large cities, causes serious irritation of the mucous membranes and reduces their ability to fight infection. At 20 to 30 pphm it affects vision, and exposure to concentrations of 30 pphm for a few minutes brings a marked respiratory distress with severe fatigue, coughing, and choking. When volunteers were exposed intermittently for 2 weeks to a 30-pphm ozone atmosphere, they experienced severe headaches, fatigue, wheezing, chest pains, and difficulty in breathing. It reduces the activity of individuals, especially those with previous heart conditions. Even young athletes tire on smoggy days.

Carcinogens, which are often polycyclic hydrocarbons inducing cancer in susceptible individuals, are present in the exhaust emissions of the internal combustion engine, be it diesel or gasoline. Two major carcinogens are benzopyrene, which is a strong cancer-inducing agent, and benzanthracene, which is a weak one. They are essentially nonvolatile organic compounds associated with solids or polymeric substances in the air. These compounds are not very stable, and they are destroyed at varying rates by other air pollutants and by sunlight. However, as a result of industrialization and urbanization, these substances are discharged into the atmosphere in significant quantities, reportedly causing a steady increase in the frequency of human lung cancer in the world.

Aeroallergens are airborne substances causing allergies. These are predominantly of natural origin, but some are industrial. Allergic reactions in sensitive persons are caused by allergens such as pollens, spores, and rusts. A large percentage of the population is affected by hay fever and asthma each year; ragweed pollen may be the worst offender—it is about 20 μm in diameter, and under normal conditions, nearly all of it will be deposited near the source. Organic allergens come from plants, yeasts, molds, and animal hair, fur, or feathers. Fine industrial materials in the air cause allergies; for example, the powdered material given off in the extraction of oil from castor beans causes bronchial asthma in people living near the factory.[23]

Radon (as described earlier) is a radioactive, colorless, odorless, naturally occurring gas that is found everywhere at very low levels. It seeps through the soil and collects in homes. Radon problems have been identified in every state, and millions of homes throughout the country have elevated radon levels. Radon in high concentrations has been determined to cause lung cancer in humans.

Smoking can be categorized as voluntary pollution. The smokers not only create a health hazard for themselves but also for the nonsmokers in their company. Cigarette smoke causes lung cancer, and in pregnant women it may cause premature birth and low birthweight in newborns.

Asbestos is a mineral fiber that has been used commonly in a variety of building construction materials for insulation as a fire retardant. The EPA and other organizations have banned several asbestos products. Manufacturers have also voluntarily limited the use of asbestos. Today asbestos is most commonly found in older homes in pipe and furnace insulation materials, asbestos shingles, millboard, textured paints, and floor tiles. The most dangerous asbestos fibers are too small to see. After the fibers are inhaled, they can remain and accumulate in the lungs. Asbestos can cause lung cancer, cancer of the chest and abdominal linings, and asbestosis (irreversible lung scarring that can be fatal). Symptoms of these diseases do not show up until many years after exposure. Most people with asbestos-related disease were exposed to elevated concentrations on the job, and some developed disease from clothing and equipment brought home from job sites.

Noise pollution is not usually placed among the top environmental problems facing the nation; however, it is one of the more frequently encountered sources of pollution in everyday life. Recent scientific evidence shows that relatively continuous exposures to sound exceeding 70 dB can be harmful to hearing. Noise can also cause stress reactions that include: (1) increases in heart rate, blood pressure, and blood cholesterol levels, and (2) negative effects on the digestive and respiratory systems. With persistent, unrelenting noise exposure, it is possible that these reactions will become chronic stress diseases such as high blood pressure or ulcers.

Plants Pollutants, especially in the air, cover a wide spectrum of particulate and gaseous matter, damaging and effecting the growth of many types of vegetation.[20] Whether particulate matter is harmful to vegetation depends upon the type of particulate matter predominating, upon the concentration of particulate matter versus time, the type of vegetation under consideration, climatic conditions, the duration of exposure, and similar factors.[21]

Different types of plants are affected differently by pollutants. The three major types of plants are trees, vegetative plants (crops), and flowers. Only a few kinds of trees can live in the polluted air of a big city. Sycamores and Norway maples seem to resist air pollution best. That is why those trees are planted among most city streets. However, air pollution can kill even sycamores and Norway maples. The danger to the trees is greatest at street corners. That is where cars and buses may have to stop and wait for traffic lights to change. While they are waiting, exhaust pours out of their tailpipes, resulting in tree kills. Pine trees do not resist air pollution as well as sycamores and Norway maples. Air pollutants, even in small amounts, are very harmful to pine trees. For example, the San Bernadino forest was a beautiful forest about 60 miles east of Los Angeles. Most of the trees in the forest were pines. Winds usually blow from west to east. The winds carried polluted air from the streets of L.A. to the San Bernadino forest and harmed the pine trees.[24] Crops and flowers cannot be planted within many miles of industry because they will not grow due to the pollution emitted from the factories. The major contributors to plant pollution are sulfur dioxide, ethylene, acid deposition, smog, ozone, and fluoride.

Metallurgical smelting processes emit substantial quantities of *sulfur dioxide*, and they are in, general associated with a good degree of defoliation. Serious damage from sulfur dioxide is usually characterized by loss of chlorophyll and suppression of growth. Leaf and needle tissues are damaged, and die as the time of exposure increases. The attack starts at the edges, moving progressively toward the main body of the leaf or needle. A concentration as low as 2 pphm could suppress growth. Cereal crops, especially barley, are readily damaged at concentrations less than 50 pphm. The presence of soot particles in the air can increase the damage because sulfur dioxide and sulfuric acid mist are enriched at the surface of particles. It has been determined that pine trees cannot survive the damage when the mean annual concentrations of sulfur dioxide exceed 0.07 to 0.08 ppm.[23]

Ethylene in the air causes injury to many flowers, whether they are orchids, lilacs, tulips, or roses. The first symptom of ethylene damage is the drying of the sepals, which are leaflike formations located at the bottom of the flower bloom. This attack destroys the beauty of the flowers and contributes to extensive economic losses to growers. In addition, accidental escape of ethylene from a polyethylene plant caused 100 percent damage to cotton fields a mile away.[23]

The process of *acid deposition* begins with emissions of SO_2 and NO_X. These pollutants interact with sunlight and water vapor in the upper atmosphere to form acidic compounds. When it rains (or snows), these compounds fall to the Earth. Forests and agriculture may be vulnerable because acid deposition can leach nutrients from the ground, killing nitrogen-fixing microorganisms that nourish plants and release toxic metals.

Smog damage to vegetation is serious, especially in locations such as Los Angeles; the lower leaf surfaces of petunias and spinach become silvery or bronze in color. The most toxic substance of the Los Angeles air has been identified as peroxyacetyl nitrate (PAN), formed by photochemical reactions of hydrocarbons and nitrogen oxides emanating mostly from automobile exhausts.[23]

Ozone is a major component of the Los Angeles smog; it is phytotoxic at concentrations of 0.2 ppm, even when exposure time is only a few hours. Its effect on spinach is strong and destructive, causing whitening or bleaching of the leaves. Certain tobaccos are damaged by concentrations as small as 5 to 6 ppm. Ozone hinders plant growth even if bleaching or other distinctive marks are not found.

Fluorides are given off by factories that make aluminum, iron, and fertilizer. Due to these factories, growers have complained about the damage to fruits and leaves of peach, plum, apple, fig, and apricot trees. Fluorides also damage grapes, cherries, and citrus. It has been observed that the average yield of fruit per tree decreases 27 percent for every increase of 50 ppm of fluoride in the leaves.[23]

Animals   Animals are also affected by pollutants in the air. There are many similarities between the effects on humans and the effects on animals. For example, animals in zoos suffer the same effects of air pollution as humans. They also are beset with lung disease, cancer, and heart disease. Their babies have more birth defects than those of wild animals.

Air that is polluted with *fluorides* can be deadly for sheep, cows, and some other animals. However, inhaling the polluted air is not what causes the damage. Some plants that are eaten by the animals store up the fluoride that they have taken from the air, and after a while contain dangerous amounts of fluoride. Animals become ill and even die after eating these plants.[23]

The bald eagle is the U.S. national bird, but it is being killed mainly with *insecticides.* The eagles are not killed by breathing the polluted air but are dying because they cannot reproduce. When an insecticide is sprayed on plants, some of it misses the plants and gets into the air as a pollutant. Certain kinds of insecticides do not change into harmless substances; they are referred to as "persistent" insecticides because they remain harmful for years. Rain washes these insecticides out of the air and into the water. Small animals bioaccumulate the insecticides, and these animals are often eaten by larger animals, who also have absorbed the insecticide, thereby doubling their insecticide intake. Bald eagles eat large animals (fish), and they may store enough insecticide to kill them. Even if they do not die, the insecticide prevents the bird from reproducing. The eggs that the females lay either have very thin shells or no shells at all, causing the inability of baby eagles to hatch.[24]

Materials of Construction Air pollution has long been a significant source of economic loss in urban areas. Damage to nonliving materials may be exhibited in many ways, such as corrosion of metal, rubber cracking, soiling and eroding of building surfaces, deterioration of works of art, and fading of dyed materials and paints.

An example of the deterioration of works of art is Cleopatra's Needle standing in New York City's Central Park. It has deteriorated more in 80 years in the park than in 3000 years in Egypt. Another example is the Statue of Liberty located on Liberty Island in New York Harbor. When the statue arrived from France in 1884 it was copper, and 100 years later it has turned a greenish color. It was so deteriorated that the internal and external structures had to be renovated. The steady deterioration of the Acropolis in Athens, Greece, is yet another example.

Water Pollution

Pollution in waterways impairs or destroys aquatic life, threatens human health, and simply fouls the water such that recreational and aesthetic potential are lost. There are several different types of water pollution and there are several different ways in which water can be polluted. This section will focus on:

1. Drinking water and its sources (groundwater and tap water)
2. Critical aquatic habitats (wetlands, near coastal waters, the Great Lakes, and oceans)
3. Surface water (municipal wastes, industrial discharge, and nonpoint sources)

Drinking Water   Half of all Americans and 95 percent of rural Americans use groundwater for drinking water. Through testing water in different areas and at different times, pollutants were found in the drinking water. Several public water supplies using groundwater exceeded EPA's drinking water standards for inorganic substances (fluorides and nitrates). Major problems were reported from toxic organics in some wells in almost all states east of the Mississippi River. Trichloroethylene, a suspected carcinogen, was the most frequent contaminant found. The EPA's Ground Water Supply Survey showed that 20 percent of all public water supply wells and 29 percent in urban areas had detectable levels of at least one VOC. At least 13 organic chemicals that are confirmed animal or human carcinogens have been detected in drinking water wells.

The most severe and acute public health effects from contaminated drinking water from the tap, such as cholera and typhoid, have been eliminated in America. However, some less acute and immediate hazards still remain in the nation's tap water. Contaminants of special concern to the EPA are lead, radionuclides, microbiological contaminants, and disinfection by-products. Each of these is discussed below.

Lead in drinking water is due primarily to the corrosion of plumbing materials. The health effects related to the ingestion of too much lead are very serious and can lead to impaired blood formation, brain damage, increased blood pressure, premature birth, low birth weight, and nervous system disorders. Young children are especially at high risk.

Radionuclides are radioactive isotopes that emit radiation as they decay. The most significant radionuclides in drinking water are radium, uranium, and radon, all of which occur in nature. Ingestion of uranium and radium in drinking water can cause cancer of the bone and kidney. Radon can be ingested and inhaled. The main health risk due to inhalation is lung cancer.

Microbiological contaminants such as bacteria, viruses, and protozoa may be found in water. Although some organisms are harmless, others may cause disease. Microbiological contamination continues to be a national concern because contaminated drinking water systems can rapidly spread disease.

Disinfection by-products are produced during water treatment by chemical reactions of disinfectants with naturally occurring or synthetic materials. These by-products may pose health risks, and these risks are related to long-term exposure to low levels of contaminants.

Critical Aquatic Habitats   *Wetlands* are the most productive of all ecosystems, but the United States is slowly losing them. There are many positive effects of wetlands: converts sunlight into plant material or biomass that serve as food for aquatic animals that form the base of the food chain, habitats for fish and wildlife, and spawning grounds; maintains and improves water quality in adjacent water bodies; removes nutrients to prevent eutrophication; filters harmful chemicals; traps suspended sediments; controls floods; prevents shoreline erosion with vegetation; and contributes $20 to 40 billion annually to the economy.

Coastal waters are home to many ecologically and commercially valuable species of fish, birds, and other wildlife. Coastal waters are susceptible to contamination because they act as sinks for the large quantities of pollution discharged from industry. The effects include toxic contamination, eutrophication, pathogen contamination, habitat loss and alteration, and changes in living resources. Coastal fisheries, wildlife, and bird populations have been declining, with fewer species being represented.

The Great Lakes are all being affected by toxics that are contaminating fish and the water. Lake Ontario and Lake Erie are also being affected by eutrophication.

Oceans are being polluted with sediments dredged from industrialized urban harbors that are often highly contaminated with heavy metals and toxic synthetic organic chemicals. The contaminants can be taken up by marine organisms. In addition, persistent disposal of plastics from land and sea have become serious problems. The most severe effect of the debris floating in the ocean is injury and death of fish, marine animals, and birds. Debris on beaches can affect the public safety, the beauty of the beach, and the economy.

Surface Waters   *Municipal wastewater and industrial discharges* produce nutrients in sewage that foster excessive growth of algae and other aquatic plants. Plants then die and decay, depleting the dissolved oxygen needed by fish. Wastewater that is poorly treated may contain chemicals harmful to human and aquatic life.

Non-point-source pollution consists of sediment, nutrients, pesticides, and herbicides. Sediment causes decreased light transmission through water resulting in decreased plant reproduction, interference with feeding and mating patterns, decreased viability of aquatic life, decreased recreational and commercial values, and increased drinking water costs. Nutrients promote the premature aging of lakes and estuaries. Pesticides and herbicides hinder photosynthesis in aquatic plants, affect aquatic reproduction, increase organism susceptibility to environmental stress, accumulate in fish tissues, and present a human health hazard through fish and water consumption.

Humans   Humans are not affected similarly by the presence of water pollution as they may be by the presence of polluted air. Humans are affected by water pollution through consuming contaminated water or animals (fish). Due to contaminated drinking water, lakes, and oceans, humans are inflicted with diseases, impaired blood formation, brain damage, increased blood pressure, premature birth, low birth weight, nervous system disorders, and cancer (bone, kidney, and lung).

Plants   Plants are affected by wastewater, sewage, sediments, pesticides, and herbicides found mainly in surface water. Effects on plants in these areas are:

1. Decreased plant reproduction
2. Hinderance of photosynthesis in aquatic plants
3. Excessive growth of algae and other aquatic plants
4. Ultimate death of plants

Animals Animals, especially those that live in or near the water, are directly affected by water pollution. Chemical and solid waste disposal in water can affect animals in many ways, varying from waste/pollutant accumulation effects to death. Animals such as fish, marine mammals, and birds can be injured or killed due to floating debris in the ocean. Contaminants can be taken up by marine organisms and accumulate there. The accumulations increase as the larger fish consume contaminated smaller fish. This cycle interferes with animals feeding and mating patterns, affects aquatic reproduction, and decreases the viability of aquatic life.

International Effects In Japan, contamination of seawater with organic mercury became concentrated in fish and produced a severe human neurologic disorder called Minamata disease. The epidemic occurred in the mid-1950s. Almost 10 years passed before it was realized that there was an accompanying epidemic of congenital cerebral palsy due to a transplacental effect; e.g., pregnant women who ate contaminated fish gave birth to infants who were severely impaired neurologically.[25]

Land Pollution

Land has been used as dumping grounds for wastes. Improper handling, storage, and disposal of chemicals can cause serious problems. Several types of wastes that are placed in the land are:

1. Industrial hazardous wastes
2. Municipal wastes
3. Mining wastes
4. Radioactive wastes
5. Leakage from underground storage tanks

Humans Potential health effects in humans range from headaches, nausea, and rashes to acid burns, serious impairment of kidney and liver functions, cancer, and genetic damage. Underground storage tank leaks may contaminate local drinking water systems or may lead to explosions and fires causing harm and injury to the people in the vicinity.

Plants Trees are usually not planted around landfills, and if they were they would have difficulty growing due to the contaminated soil in the vicinity of the landfill. Vegetative plants also have difficulty growing around landfills. This is due to the fact that the hazardous wastes from industry are usually dumped in the landfills. Flowers also do not normally grow near landfills for similar reasons.

Animals Animals are essentially affected in the same ways as humans. They may experience the effects of drinking contaminated water and suffer from acid burns, kidney, liver, and genetic damage, and cancer.

1.9 TEXT CONTENTS

This project was a unique undertaking. Rather than prepare a textbook on nanotechnology, the authors considered writing a book that was concerned with the environmental implications of nanotechnology. Because of the dynamic nature of this emerging field, it was also decided to write an overview of this subject rather than to provide a comprehensive treatise. One of the key features of this book is that it could serve both academia (students) and industry. As such, it can be used as a text in either a general engineering or environmental engineering/science course or (perhaps primarily) as a training tool for industry.

The book contains 10 chapters. This first chapter provides an introduction to background issues surrounding nanotechnology and the environmental implications associated with the new technology. Chapter 2 covers nanotechnology. Chapters 3, 4, and 5 address air, water, and land (solid waste) issues, respectively, particularly as they relate to control and monitoring. The general subject of multimedia approaches is reviewed in Chapter 6. Chapters 7 and 8 address risk concerns—health in the former and hazard in the later chapter. A brief introduction to ethics can be found in Chapter 9, and the text concludes with a short chapter (10) on future trends.

1.10 SUMMARY

1. Nanotechnology is concerned with the world of invisible miniscule particles that are dominated by forces of physics and chemistry that cannot be applied at the microscopic or human scale level.

2. Two very recent nanotechnology applications include the development of a nanoscale palladium that can efficiently and economically clean groundwater contaminated with trichloroethylene and the synthesis (assembled molecule by molecule) of a novel metal rubber material that has the qualities of both metal and rubber.

3. The most pertinent of intellectual property rights for nanotechnology inventions are patents. It is incumbent upon those engaged in any area of technological development to acquire a basic understanding of patent law.

4. Nanotechnology patents have increased by over 600 percent in the 1997 to 2002 periods. Most of these applications are filed by large private corporations.

5. There are both positive and negative implications associated with nanotechnology. The primary negative impacts are the potentially unknown and unforeseen environmental effects associated with this technology.

6. Environmental regulations are an organized system of statutes, regulations, and guidelines that minimize, prevent, and punish the consequences of damage to the environment. The recent popularity of environmental issues has brought about changes in legislation and subsequent advances in technology.

7. All of the criteria pollutants (ozone, carbon monoxide, airborne particulates, sulfur dioxide, lead, nitrogen oxide) except lead and nitrogen oxide are currently

a major concern in a number of areas in the country. Indoor air pollutants of special concern include radon, environmental tobacco smoke, asbestos, formaldehyde and other VOCs, and pesticides. The U.S. EPA focuses its water pollution control efforts on three themes: maintaining drinking water quality, preventing further degradation and destruction of critical aquatic habitats (wetlands, nearshore coastal waters, oceans, and lakes), and reducing pollution of free-flowing surface waters and protecting their uses. Land pollutants arise because of industrial hazardous wastes, municipal wastes, mining wastes, radioactive wastes, and leaking underground storage tank pollutants.

8. Humans and animals are affected by air pollution. Some sources of pollution affecting them are sulfur dioxide, carbon monoxide, nitrogen oxides, ozone, carcinogens, fluorides, aeroallergens, radon, cigarette smoke, asbestos, and noise. Those affecting only animals are insecticides. Plants are also affected by air pollutants such as sulfur dioxide, ethylene, acid deposition, smog, ozone, and fluoride. Pollution in waterways impairs or destroys aquatic life, threatens human health, and fouls the water such that recreational and aesthetic potential are lost. Land has been used as dumping grounds for wastes. Improper handling, storage, and disposal of chemicals can cause serious problems.

9. This book contains 11 chapters. Hopefully, it offers material not only to individuals with limited technical background, but also to those with extensive industrial experience.

REFERENCES

1. L. Theodore, personal notes, 2004.
2. Excerpted with special permission from *Chemical Engineering*, Copyright 2002, by Chemical Week Associates, New York City, S. Shelley, "Nanotechnology: The Sky's the Limit," *Chemical Engineering* (December, 2002).
3. R. G. Kunz, personal notes, 2004.
4. Source unknown.
5. M. K. Theodore and L. Theodore, *Major Environmental Issues Facing the 21st Century*, Theodore Tutorials, East Williston, NY, 1996.
6. G. Burke, B. Singh, and L. Theodore, *Handbook of Environmental Management and Technology*, 2nd ed., Wiley, Hoboken, NJ, 2000.
7. Anonymous, *Nanotechnology Opportunity* Report, 2nd ed., location unknown, 2003.
8. U.S. EPA, "Cancer Risk from Outdoor Exposure, Air Toxics," US EPA, Washington, DC, 1990.
9. J. Mycock, J. McKenna, and L. Theodore, *Handbook of Air Pollution Control Technology*, Lewis/CRC, Boca Raton, FL, 1995.
10. R. Dupont, L. Theodore, and R. Ganesan, *Pollution Prevention: The Waste Management Approach for the 21st Century*, Lewis/CRC, Boca Raton, FL, 2000.
11. U.S. EPA, Report to Congress, Office of Solid Waste, U.S. EPA Washington, DC, 1530-SW-86-033, 1986.

12. National Center for Environmental Research, "Nanotechnology and the Environment: Applications and Implication, STAR Progress Review Workshop," Office of Research and Development, National Center for Environmental Research, Washington, DC, 2003.

13. Drawn, in part, from the *Federal Register*, date unknown.

14. L. L. Bergeson, "Nanotechnology and TSCA," *Chemical Processing* (Nov. 2003).

15. L. L. Bergeson, "Nanotechnology Trend Draws Attention of Federal Regulators," *Manufacturing Today* (March/April 2004).

16. L. L. Bergeson, and B. Auerbach, "The Environmental Regulatory Implications of Nanotechnology," *BNA Daily Environment Reporters*, pp. B-1 to B-7 (April 14, 2004).

17. L. L. Bergeson, "Expect a Busy Year at EPA," *Chemical Processing* **17** (Feb. 2004).

18. L. L. Bergeson, "Genetically Engineered Organisms Face Changing Regulations," *Chemical Processing* (Mar. 2004).

19. U.S. EPA Update, "Environmental Progress and Challenge," U.S. EPA, Washington, DC, August, 1988.

20. A. Stern, *Air Pollution: The Effects of Air Pollution Vol. II.* Academic Press, New York, 1977.

21. H. Parker, *Air Pollution.* National Research Council, Washington, DC, 1977.

22. "Meeting the Environmental Challenge" EPA's Review of Progress and New Directions in Environmental Protection, U.S. EPA, Washington, DC, December 1991.

23. E. Shaheen, *Environmental Pollution: Awareness and Control*, Washington, DC, 1974.

24. E. Blaustein, R. Blaustein, and J. Greenleaf, *Your Environment and You: Understanding the Pollution Problem.* 1974.

25. B. Zoeteman, *Aquatic Pollutants and Biological Effects; With Emphasis on Neoplasia*, New York, New York Academy of Sciences, 1977.

CHAPTER 2

NANOTECHNOLOGY: TURNING BASIC SCIENCE INTO REALITY

SUZANNE A. SHELLEY, Managing Editor, *Chemical Engineering* magazine

2.1 INTRODUCTION

Having burst onto the scene in the mid-1980s or so, the scientific discipline called nanotechnology is still in its relative infancy, but in some circles, nanotechnology has already become something of a household word. The past two decades has been witness to a veritable explosion of research and development activity worldwide, with considerable effort put forth by universities large and small, companies ranging from small entrepreneurial startups to corporate leaders in all segments of the chemical process industries, and various government and military researchers.

Today, awareness of the science and engineering community's ongoing breakthroughs on nanometer-scaled materials and systems is no longer confined to the academic and industrial research community. In fact, over the past few years, many of the more promising and highly publicized breakthroughs have helped to vault nanotechnology into the public consciousness. Today, along with the business and investment communities, many consumers and the popular press are tuning in when they hear the word *nanotech.*[1]

As described earlier, the prefix nano- (10^{-9}) refers to a billionth of something, and in the case of nanotechnology, the basic unit of measurement is the nanometer (nm), which is one billionth of one meter. Nanotechnology refers to the ability to synthesize, manipulate, and characterize matter at particle diameters of 100 nm or less.

To give a sense of perspective to just how small this is, consider that 1 micrometer (μm) equals 1000 nm, and the diameter of an average human hair is about 10,000 nm.[2] Similarly, 10 hydrogen atoms lined up in a row would fit within a single nanometer. Human DNA (deoxyribonucleic acid) molecules are 2.5 nm wide. A typical bacterium, say *Escherichia coli* (*E. coli*), is a thousand times bigger, measuring between 1000 and 2000 nm, while certain viruses, like the ones that cause the common cold, measure around 20 nm.[3]

Nanotechnology: Environmental Implications and Solutions, L. Theodore and R. G. Kunz
ISBN 0-471-69976-4   Copyright © 2005 John Wiley & Sons, Inc.

To illustrate the relative scale of things and lend context to the discussion, Table 2.1[4-7] offers some examples of familiar items that range in size from 1 km (10^3 m) to 1 nm (10^{-9} m) and even 1 Å (10^{-10} m, or one-tenth of a nanometer).

Over the last 20 years, the desire to produce and use matter whose particle size dimensions are so small that they were previously unimagined has been accelerated by the discovery and verification of a broad array of novel, size-dependent properties and phenomena that occur or are vastly improved in the nanometer range.[8] In this nanoscaled range, it is not only the chemical composition but also the size, shape, and surface characteristics of ultrafine particles that determine the properties of various materials.[7]

For instance, when produced at infinitesimally small particle sizes, materials as varied as metals, metal oxides, polymers and ceramics (all discussed in greater detail in Section 2.3), and carbon derivatives such as carbon nanotubes and fullerenes, or "buckyballs," (also discussed in detail in Section 2.4) have extraordinary ratios of surface area to particle size. (The importance of the extraordinary surface area of nanoscaled particles is discussed in Section 2.2.) And, at these minute sizes, such particles are small enough that they show quantum effects, and their optical, electrical, and magnetic properties, as well as hardness, toughness, and melting point, can differ markedly from the properties exhibited by macroscopic particles of the same materials.[7]

Continuous advances made in the field of atomic-scale microscopy over the past 20 years have been a key factor in the ongoing evolution of nanotechnology research and development. Such new microscopy techniques (see Section 2.6) allow researchers to conduct precise investigation, measurement, visualization, and even manipulation of materials and behaviors on an atomic scale, and help them to reliably resolve ever-smaller details of the nanoscaled particles and structures that they are investigating.[7]

Future Appears Bright

For nanotechnology's most ardent supporters, the scope of this emerging field seems to be limited only by the imaginations of those who would dream at these unprecedented dimensions. However, considerable technological and financial obstacles still need to be reconciled before nanotechnology's full promise can be realized.

Ranking high among the challenges is the ongoing need to develop and perfect reliable techniques to produce (and mass produce) nanoscaled particles that have not just the desirable particle sizes and particle size distributions but also a minimal number of structural defects and acceptable purity levels, since these latter attributes can drastically alter the anticipated behavior of the nanoscaled particles. Experience to date shows that the scaleup issues associated with moving today's promising nanotechnology-related developments from laboratory- and pilot-scale demonstrations to full-scale commercialization can be considerable. The prevailing technologies for producing nanoscaled particles and carbon nanotubes are discussed in Section 2.3.

TABLE 2.1 Comparisons of Size Dimensions and Common Items

Power of 10	Unit	Symbol	Examples
3	Kilometer	km	A little over 0.6 miles
2	Hectometer	hm	Length of football field or distance from home plate to outfield fence (both ball park figures)
1	Dekameter	dam	Approximate width of a tennis court for doubles play
0	Meter	m	1 m ≅ 3 ft; height of adult male close to 2 m; the dreaded 3-ft putt
−1	Decimeter	dm	Approximate width of standard 2-by-4 lumber
−2	Centimeter	cm	2.54 cm = 1 inch; diameter of U.S. 1-cent coin ≅2 cm
−3	Millimeter	mm (one-thousandth of a meter)	Typical mechanical pencil lead 0.5−1 mm; large biological cell is about 0.2 mm; human eye resolves to about 0.2 mm
−4	Hundreds of microns	100 μm (or 100 um)	Diameter of human hair on the order of 100 μm
−5	Tens of microns	10 μm	Diameter of hair of wool-bearing animal ≅20 μm
−6	Micron or micrometer	μm (one-thousandth of a millimeter; one-millionth of a meter)	Nucleus of a human cell; red blood cells; bacterium
−7	Tenths of a micron, hundreds of nanometers	0.1 μm, 100 nm	Typical upper limit of nanoparticle dimensions
−8	Hundredths of a micron, tens of nanometers	0.01 μm, 10 nm	Typical nanoparticle dimensions; a virus (50 nm)
−9	Nanometer, tens of Ångstroms	nm (one-billionth of a meter; one-thousandth of a micron; 10 Å)	10 hydrogen atoms in line; the distance across a medium-sized molecule; the lower limit of nanoparticle area
−10	Ångstrom (one-tenth of a nanometer)	1 Å, 0.1 nm	Distance between atoms in a molecule

Source: Adapted and expanded by R. G. Kunz from information contained in Refs. 4–6 and by S. A. Shelley from Ref. 7.

In addition to the inherent challenges associated with designing and scaling up various methodologies to produce nanoscaled materials that have the right particle attributes, additional challenges arise in terms of handling and using these minute particles as functional additives in other matrix materials. For instance, the extremely high surface area of these minute particles creates problems that must be reconciled related to excessive attractive or repulsive surface charges, unwanted nanoparticle agglomeration, problems with dispersion and blending, and so on.

Meanwhile, nanoscaled materials generally command very high prices compared to conventional, macroscopic particles that have essentially the same chemical composition. Such high costs result from energy that is required to reduce the particle size and from the high research and development and prototyping costs that are incurred during the discovery phase of any novel materials and the related manufacturing processes. However, many industry observers acknowledge that the current prices for many nanoscaled materials represent experimental quantities produced at pilot facilities, and predict that the prices are likely to come down, once steady-state, commercial-scale manufacturing conditions are perfected and pursued.[7]

Ultimately, premium costs associated with nanoscaled particles, devices, and systems will have to be proven and justified in performance. For instance, if a low-cost, micron-scale powder will suffice in a particular application, the end user is not likely to pay a premium price for its nanoscaled counterpart, whose cost may be orders of magnitude higher.[9] Companies need a strong motivation for why they should replace an existing approach with a newer, nanotechnology-based approach.[1] Such capacity and cost issues will be key factors that will continue to influence market development and commercial adoption of nanotechnology-based materials and processes into the future.

Extraordinary Features and Novel Properties of Nanoscaled Materials

When familiar materials such as metals, metal oxides, ceramics and polymers, and novel forms of carbon, such as carbon nanotubes and fullerenes, or buckyballs, are manipulated into infinitesimally small particle sizes (and, in the case of carbon nanotubes and buckyballs, unique structural geometries, as well), the resulting particles have an orders-of-magnitude increase in available surface area. It is this remarkable surface area of particles in the nanometer range that confers upon them some unique material properties, especially when compared to macroscopic particles of the same material. For instance, nanoscaled particles tend to have greatly improved thermal and electrical conductivity, surface chemistry (which affects particle dispersibility and reactivity), photonic behavior (which is functionality that changes in the presence of light of varying wavelengths), and catalytic conversion rates, compared to larger particles of the same materials.[2]

Today, nanometer-scaled particles, carbon nanotubes, and fullerenes are already being incorporated into many polymeric matrix materials to produce composite materials that demonstrate improved electrical conductivity and catalytic activity, increased hardness and scratch resistance, and self-cleaning capabilities and antimicrobial properties. Meanwhile, other types of nanoparticles are also being widely used today in the slurries that are used for precision polishing of semiconductor chips during manufacturing.

And nanomaterials are being exploited to improve the performance of gas sensors and other analytical devices, to improve the way drugs reach their intended targets in the human body, to improve the aesthetic appeal and efficiency of consumer products, to produce environmentally friendly fuel cells and highly effective catalysts, and to produce smaller and better photonic devices for use in both hard-wired fiber-optic, and wireless, data and voice transmission applications.[2] Many of the most promising market applications for nanoscaled materials are discussed later in this chapter.

Nanotechnology's More Futuristic Aspirations

Beyond just their efforts to produce and use nanometer-sized particles of various materials, some nanotechnology-related scientists and engineers are pursuing far more ambitious—and some would say fantastic or futuristic—applications of this powerful new technological paradigm. For instance, the research community is working toward being able to design and manipulate nanoscaled objects, devices, and systems by the manipulation of individual atoms and molecules.[3] Such forward-looking researchers hope that by using atom-by-atom construction techniques, they will someday be able to create not just substances with remarkable functionality but also tiny, bacterium-sized devices and machines (thus far dubbed *nanobots*), that could be programmed, say, to repair clogged arteries, kill cancer cells, and even fix cellular damage caused by aging. Advanced development is also underway to use nanotechnology to develop advanced sensors to improve the detection of chemical, biological, radiological, and nuclear hazards.[10,11]

Most concede, however, that real, commercial-scale success in this arena is still years, if not decades, away. Nonetheless, despite the fact that the quickly evolving field of nanotechnology integrates the well-established disciplines of chemistry, physics, biology, materials science, and all branches of engineering, it also represents a brave new world—a futuristic, imaginative journey that some might say is worthy of a Jules Verne novel.

At a time when most industrialized nations are investing heavily in nanotechnology research and development, and a plethora of potential applications—both practical and fantastic—are being considered, cautious observers note that the potential environmental, health, and safety risks associated with nanotechnology are not being adequately studied and that the complex ethical, legal, and societal implications of this powerful new technological paradigm are not being explored or debated on a large enough scale.[12] These issues were introduced in Chapter 1, and are discussed in greater detail in Section 2.6 of this chapter.

While some expectations from nanotechnology may be highly hyped and overestimated in the short-term, many feel that the long-term implications for health care, productivity, and the environment, among others, are underestimated when one considers the depth and breadth of technological breakthroughs recorded to date, and the pace at which further research and developments are being undertaken.[8]

2.2   BASIC CHEMISTRY AND SIZE-RELATED PROPERTIES

The Atom

Here on Earth, human beings continue to struggle to understand—and continue to try to control—many aspects of their environment. Experience is often gained by the costly process of trial and error. To be glib, those who ate the wrong nuts or berries were quickly weeded out of the gene pool.

In the beginning, knowledge of chemistry was inferred from circumstantial evidence of how the Earth's materials behave. The ancient Greeks postulated that all matter is composed of four basic elements: earth, air, fire, and water, which when combined in the correct proportions make up all of the "stuff" on Earth (see Ref. 6, pp. 12–13).

Although the concept of an element has changed, today one has reason to believe that there are 109 such building blocks from which all of the matter in the world is assembled. Each element is composed of only one type of so-called *atom*, which cannot be further divided and still retain the unique chemical properties of that element.

In 1807, an English schoolteacher, John Dalton, proposed an atomic theory based on the scientific evidence at the time (see Ref. 6, pp. 12–13).

- Elements are made up of atoms (thought at the time not to be further divisible; see below).
- All atoms of a given element are identical (this turned out not to be totally true because of the later discovery of isotopes).
- Individual atoms combine with one another to form molecules of chemical compounds, each of which always contains the same types of atoms in the same ratios.
- The atoms participating in a chemical reaction are simply rearranged from one combination to another but are otherwise unchanged.

Dalton's postulates still hold true in essence despite the further discovery of subatomic particles, such as protons, neutrons, electrons, and others, and nuclear fission reactions, which were first reported in 1939 (see Ref. 6, pp. 12–13).

The identity of an element and its atomic number is determined by the number of positively charged protons in the nucleus of the atom. Those protons are balanced by an equal number of orbiting electrons, which have equal and opposite charge but negligible mass compared to the protons. A number of additional particles termed *neutrons* may also appear in the nucleus; the neutron is similar in mass to the proton but carries no charge.

The atomic weight of an element is determined by the total number of protons and neutrons. Atoms that have the same number of protons—thereby constituting the same element—but a different number of neutrons are known as *isotopes*.

Periodic Table

Around 1868 to 1870, Dmitri Mendeleev in Russia and Lothar Meyer in Germany independently succeeded in rearranging the then-known elements into a two-dimensional matrix called the Periodic Table.[13] In a modern version of the Periodic Table (Table 2.2), the elements are arranged by atomic number (number of protons). Elements in the vertical columns, known as *groups*, exhibit similar but gradually changing properties as one proceeds from top to bottom.

For example, the column at the extreme left of the Periodic Table consists of hydrogen and the alkali metals—a highly reactive set of elements. Except for hydrogen, all of these elements are soft metals that are good conductors of heat and electricity.

In contrast, the elements occupying the extreme right-hand column of the Periodic Table are called the *inert*, *rare*, or *noble* gases. As such, these elements are relatively unreactive and rarely react or combine with other elements.

To their left lie the halogens, a highly reactive set of nonmetals that are poor conductors of heat and electricity. Each of the groups occupying other vertical columns in between these extremes shows a progressive change in metallic character and exhibits its own unique sets of properties.

Meanwhile, each horizontal row in the Periodic Table is called a *period*. The first period contains only hydrogen and helium. Within a given group (vertical column), the size of the atom increases from the top period to the bottom, and with this changing atom size come changes in other properties that are related to atomic size. For example, the halogens proceed from gas to liquid to solid as one moves down the table.

In general, size increases from right to left in a given period (see Ref. 6, p. 24). Ionization energy, electron affinity, and electronegativity increase from the lower left-hand corner of the table to the upper right; atomic and ionic radii increase from upper right to lower left.[14] Atomic or ionic radii differ considerably according to the environment of neighboring atoms and ions and the type of chemical bond involved (see Ref. 6, p. 24).

When the atomic theory of matter was first postulated, the existence of the atom had to be taken on faith because it was impossible to see an individual atom. However, today it is no longer just a theory since individual atoms and molecules can now be analyzed and even manipulated, thanks to the latest advances in atomic force microscopy and scanning tunneling microscopy (both of these are discussed in greater detail in Section 2.6). This ability to analyze and manipulate matter at its most fundamental particle size has provided the scientific community a meaningful opportunity to pursue the long-dreamed-of developments related to nanotechnology.

For instance, in an early demonstration of their newly acquired ability to manipulate matter on an atomic level, one group of scientists carefully moved 35 xenon atoms into place on a nickel crystal at cryogenic temperatures to spell out "IBM" (see Ref. 4 and Ref. 6, p. 45), while another group wrestled 112 carbon monoxide molecules into place to spell out "NANO USA."[15]

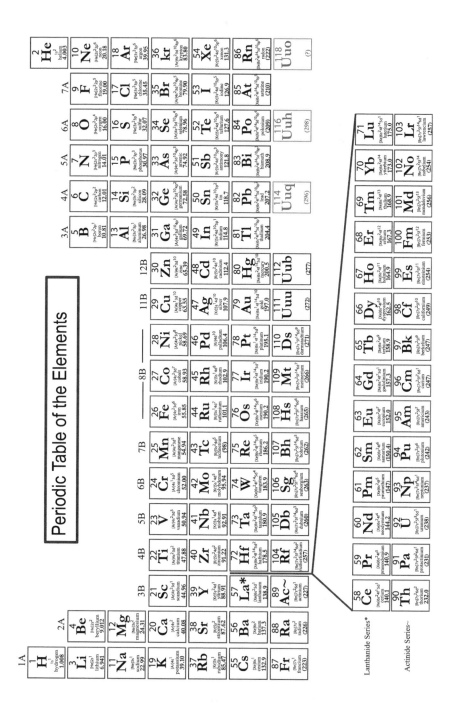

Periodic Table of the Elements

Relationship Between Particle Size and Surface Area

There is an inverse relationship between particle size and surface area, and one of the hallmarks of nanotechnology is the desire to produce and use nanometer-sized particles of various materials in order to take advantage of the remarkable characteristics and performance attributes that many materials exhibit at these infinitesimally small particle sizes.

To illustrate the relationship between shrinking particle size and increasing surface area, envision a child's alphabet block that starts out being just the right size to fit into the chubby hand of a curious toddler. Now run an imaginary knife through the block, along its horizontal, vertical, and lateral axes, to divide the original playing piece into eight smaller blocks of equal size.

While the original block had just enough surface area to hold 6 colorful pictures, three quick swipes of the imaginary knife produces 8 smaller blocks, which now have additional—previously unavailable—surface area for picture display. With each of the 8 smaller blocks now having 6 sides of its own, the newly size-reduced blocks can now display 48 little pictures of circus animals, letters, or numbers—much to the delight of the appreciative child.

Continue to divide each of these smaller blocks with three quick swipes of the imaginary knife, and you can see the exponential relationship between particle

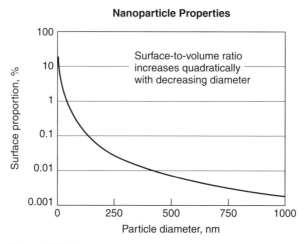

Source: Ref. 2.

Figure 2.1 The extreme surface-to-volume ratio of nanoscaled particles is a key attribute that contributes to the range of superior performance characteristics demonstrated by them. As shown here, while a particle that is 1 μm in diameter (i.e., 1000 nm) only has about 1.5-thousandths of a percent of all its atoms located on its surface, a particle with a diameter of just 10 nm has about 15 percent of its atoms on the surface. This has implications in terms of the material's attributes, such as reactivity, and optical, electrical, magnetic and other properties. (*Source*: BASF Future Business, Ref. 2)

size and surface area. This inverse relationship between particle size and surface area is a key underpinning of the field of nanotechnology.

2.3 NANOTECHNOLOGY: PRIME MATERIALS AND MANUFACTURING METHODS

Researchers in pursuit of nanoscaled materials that bring functional advantages to their end-use applications have left no stone unturned. Today, the range of elements and compounds that have been successfully produced and deployed as nanometer-sized particles includes:

- Metals such as iron, copper, gold, aluminum, nickel, and silver
- Oxides of metals such as iron, titanium, zirconium, aluminum, and zinc
- Silica sols, and fumed and colloidal silicas
- Clays such as talc, mica, smectite, asbestos, vermiculite, and montmorillonite
- Carbon compounds, such as fullerenes, nanotubes, and carbon fibers (this unique class of nanoscaled materials is discussed in Section 2.4)

Each of these types of materials, along with the manufacturing methods used to render them into nanoscaled particles, is discussed in this section. Much of the detailed technical information presented here was excerpted or adapted with permission from a comprehensive 188-page report produced by SRI Consulting.[7]

Metals

While the use of metals in catalyst systems is not new, the ability to render a large number of atomic species and alloys into particle sizes below 100 nm, reliably and with good control over the size distribution, dispersion, and surface characteristics, is a new and evolving field.

When metal particles are produced in this range, they exhibit properties not found in the standard particle sizes. This includes quantum effects, the ability to sinter at temperatures significantly below their standard melting points, increased catalytic activity due to higher surface area per mass, and more rapid chemical reaction rates. And, when nanoscaled particles of metals are consolidated into larger structures, they often exhibit increased strength, hardness, and tensile strength, when compared to structures made from conventional micron-sized powders.

However, reducing some metals to nanoscaled powders presents problems, not the least of which is that when reduced to sufficiently small particle sizes, many metals become more reactive and subject to oxidation—often explosively. Copper and silver powders are among the few metallic nanoparticles that are not pyrophoric and thus can be handled in air. Many others must be stabilized with a passivation layer or handled in an inert, blanketed environment. And in some applications, effort is required to minimize unwanted particle agglomeration.

A range of methods are available or under development to manufacture nanoparticle metals; while the details of many remain closely guarded, proprietary secrets, the technologies generally fall into two classes—chemical methods or physical methods.

The chemical methods used to produce metal nanoparticles include decomposition of metal carbonyls (thermally and with sound) and reduction of metal ions and the synthesis of nanopowders from aqueous solution precursors. For instance, metal chloride solutions can be reduced with sodium trialkylborohydride to form 2- to 5-nm metal powders, such as nanonickel and nanoiron.[7]

Nanopowders can also be produced by ultrasonically spraying an aqueous solution of mixed salts into a tank containing a reductive solution. This method permits solution dispersal of nanoparticles in a nonagglomerated form and continuous removal of reaction by-products. Some metal nanoparticles have been formed by inverse micelles (where the solvent is nonpolar and the metal ions cluster in the middle of the micelle; this is a variation of ion reduction).

Physical methods include grinding, various vaporization regimes, and electroexplosion. In chemical vapor reactions, precursor compounds, which are typically halides, and hydrogen gas are introduced into a heated reactor. The reaction occurs at temperatures between 500 to 2000°C. In the electroexplosion process, a metal wire is subjected to an intense current pulse, producing a metal aerosol.

Iron Needle-shaped iron particles (with diameters on the order of 50 to 100 nm and lengths of 200 nm or so) are already widely used in magnetic recording for analog and digital data. Finer size particles (on the order of 30 to 40 nm in length) are also available. Preliminary work suggests that such particles can provide magnetic recording media with 5 to 10 times greater recording capacity per unit space.[7] Meanwhile, when iron nanoparticles (100 to 200 nm in diameter) are coated with palladium, they show promise in the decontamination of groundwater, by reducing chlorinated hydrocarbons to nontoxic hydrocarbons and chloride.

Iron–platinum alloy nanoparticles are available in particle size as low as 3 nm. Such nanoparticles are expected to be used in the magnetic storage media of the future. Because such particles have stronger magnetization than the currently used tiny crystals of cobalt–chromium alloys, a 10-fold increase in information density is predicted.[7]

Aluminum Nanoaluminum powder, with a particle size range from 10 to 100 nm, can be made with a plasma reactor. In a few thousandths of a second, a rod of solid material with a massive pulse of electrical energy is pulverized, heating it to 50,000°C, followed by rapid cooling of the gas of vaporized material. This rapid cooling, or quenching, is how the size of the resulting nanoparticles is controlled.

Applications of nanoscaled aluminum powder include various electronics circuits, optical applications such as scratch-resistant coatings for plastic lenses, biomedical applications such as antimicrobial agents, as well as new tissue-biopsy tools. Energy-related applications include fuel cells, improved batteries, and solar energy applications.

Nickel   Traditional nickel powders—which are valued for their high conductivity and high melting point characteristics—have an average diameter of 2 to 7 μm and are consumed in powder metallurgy, nickel–cadmium batteries, and welding rods. Today, finer-scale nickel powders (with a particle size range of 100 to 500 nm) are being produced commercially by various processes, including chemical vapor deposition, wet-chemical processeses, and gas-phase reduction. Increasingly, multiplayer ceramic capacitors have become the major market for nickel nanoparticles as a lower-cost alternative to conventional pastes made from fine powders of palladium and silver.

Silver   Silver is well known for its excellent conductivity and its antimicrobial effects, and silver powder is already approved for use as a biocide by the U.S. Food and Drug Administration. However, nanoscaled silver particles have a much larger surface area, offering an opportunity to gain higher efficacy while using less material. To date, nanoscaled silver powders (with particle diameters of 10 to 90 nm) have been used as an ingredient in a biocide, in transparent conductive inks and pastes, and in various consumer and industrial products that need enhanced antimicrobial properties.

Gold   Production of gold nanoparticles is easier in comparison to other metal nanoparticles, primarily because of that element's inherent chemical stability. Colloidal gold has been used in medical applications for some time, and additional medical end uses for gold nanoparticles are under development (this is discussed in greater detail in Section 2.5).

Gold nanoparticles are also in commercial use in a wide array of catalytic applications (e.g., for low-temperature oxidation processes, including carbon monoxide oxidation in a hydrogen stream, selective production of propylene, and the oxidation of nitrogen-containing chemicals) and optical and electrical applications (as components in various probes, sensors, and optical devices).

Copper   Nanoscaled copper powders are not available yet in commercial-scale quantities, but they are being pursued to take advantage of copper's superior ability to conduct electricity and heat.

Mixed Oxides

Nanometer-sized particles of various oxides—including the iron oxides (Fe_2O_3 and Fe_3O_4), silicon dioxide (silica; SiO_2), titanium dioxide (titania; TiO_2), aluminum oxide (alumina; Al_2O_3), zirconium dioxide (zirconia; ZrO_2), and zinc oxide (ZnO), among others—are already available in commercial quantities. These materials are already being used in a wide range of existing applications and envisioned for use in many others.

Iron Oxides In general, particles of ferric oxide (Fe_2O_3) are generally used in pigment applications, while the magnetic properties of magnetite (Fe_2O_4) lend themselves to electromagnetic uses. Ferric oxide nanoparticles may be translucent to visible light but opaque to ultraviolet (UV) radiation, and ultrasmall particle size allows for the creation of rugged yet ultrathin transparent coatings with enhanced UV-blocking capabilities.

When suspended in fluids, nanoparticles of magnetic magnetite create so-called ferrofluids, which can be made to respond to electromagnetic energy in many useful ways. Ferrofluids have been used in many industrial and medical applications for decades. Magnetite nanoparticles are also being used to improve various electromagnetic media for data storage, such as magnetic tapes and computer hard drives, and to produce advanced magnets and supercapacitors, and various medical diagnostic devices.

Aluminum Oxide Nanosized powders of Al_2O_3 (also known as alumina) have a lower melting point, increased light absorption, improved dispersion in both aqueous and inorganic solvents, and ultrahigh surface area, compared to conventionally sized alumina crystals.[7] They are used in chemical mechanical planarization (CMP) slurries to polish semiconductor chips during manufacturing (nanoparticle use in CMP slurries is discussed in Section 2.5) and as components in both advanced ceramics (such as those used to make catalysts, refractory materials, ceramic filtration membranes, and substrates for microelectronic components) and advanced composites (nanocomposites are discussed in Section 2.5).

Today, a major end use for nanoscaled alumina is in the coating of lightbulbs and fluorescent tubes. The addition of aluminum oxide results in a more uniform emission of light and better flowability of the fluorescent material mixture during the electrostatic coating.[7] Such nanoparticles are also used today as a component in clear coatings that boast increased hardness and improved scratch and abrasion resistance, as a flame-retardant agent, as a performance filler in tires, as a rheology control agent, as a surface friction agent in vinyl flooring, as a detackifying agent in paints, and as a coating of high-quality inkjet papers.

Zirconium Dioxide Nanoscaled particles of ZrO_2 (also called zirconia) are valued for their ability to impart improved fracture toughness and resistance to fracture and chipping. Such nanoparticles are already being used as a component in structural ceramics (in wear parts, extrusion dies, pump components, and cutting edges), in electronic ceramics (such as those used for oxygen sensors, dielectrics, and piezoelectric components), and in thermal spray coatings used to protect components in jet engines and combustion turbines from extreme heat. Such thermal-spray coatings based on nanosized particles of zirconia have sintering temperatures that are as much as $400°C$ lower than comparable, micrometer-sized coatings, with no sacrifice in toughness or flexibility.[7]

Zirconia nanoparticle use is also being explored for use in fuel cell power generation, gas-stream purification applications, in optical connectors (ferrules),

in high-end orthopedic and dental prostheses, and as carrier particles in pharmaceutical applications.

Titanium Dioxide TiO_2 is already the largest-volume inorganic pigment produced in the world, and micrometer-sized TiO_2 powders are widely used in
surface coating, paper and plastic applications, and as a filler and whitening agent.

In applications where white pigmentation is not the aim, smaller, nanometer-
scale particles of TiO_2 are finding a host of commercial uses. For instance, at a
diameter of 50 nm or less, TiO_2 particles still maintain a strong UV light-blocking
capability, but such miniscule particles transmit—rather than scatter—visible light,
so they do not impart any white pigmentation or opacity to the matrix or substrate
into which they are incorporated. This had made such titania nanoparticles a
sought-after UV-blocking additive for sunscreens and cosmetics and even for
varnishes for the preservation of wood, textile fibers, and packaging films.

And, because nanosized particles of TiO_2 demonstrate catalytic, photocatalytic,
and electrical properties, their use is also being explored in novel applications,
such as the development of self-sanitizing tiles for restaurants and hospitals, and catalytic coatings on glass that catalyze the decomposition of organic buildup (essentially giving glass, such as car windshields, a "self-cleaning" capability).

The use of nanosized TiO_2 powders is also being explored in photoelectrochem-
ical solar cells and various types of improved thermal coatings for corrosion protection and as a component in various polymer composites to yield a product with a
tunable refractive index and improved mechanical properties for photonic and electronic applications. Since light scattering is significantly reduced in such advanced
nanocomposites, they are considered an attractive building block for optical network
component applications;[7] again, more information on nanocomposites can be found
in Section 2.5.

Zinc Oxide   ZnO is valued for its UV opacity and fungicidal action. As is the case
with the other metal oxides, nanoparticles of zinc oxide are differentiated from
larger particles by their increased surface area and transparency to visible light,
making them essentially invisible when added to other matrices, such as cosmetics,
sunscreens, and antifungal foot powders. And, whereas TiO_2 blocks only UV-B radiation from the sun, zinc oxide particles absorb both UV-A and UV-B radiation, so
they offer broader protection and improved aesthetic appeal when formulated into
sunscreens and cosmetics, and into clear-coat film polymers to protect automobiles,
furniture, and fabrics from sun damage.

An evolving use for ZnO is in the field of photoelectronics, where nanowires of
ZnO are being developed for use in UV nanolasers.[7] Meanwhile ZnO nanoparticles
are also being used in ceramics and rubber processing, where they are said to
improve elastomeric toughness and abrasion resistance.

Silica Products   The main forms of nanosized silica are precipitated silica, silica
gels, colloidal silica, silica sols, and fumed or pyrogenic silica. Today, precipitated

silicas and silica gels considered commodity chemicals, so they are not discussed further here, but ample discussion can be found in Fink et al.[7]

Colloidal silica or silica sol products are stable suspensions of independent, nonagglomerated, and nonporous spherical SiO_2 particles. They can be obtained by a liquid-phase process that involves passing a solution of sodium silicate through an ion exchange column to partially remove the sodium ions. Under alkaline conditions, silica particles start to grow, forming a stable and homogeneous silica suspension. The pH is adjusted to control particle size, and another counterion, such as an ammonium ion, may be introduced to stabilize the suspension. The resulting particle size is very narrow and the process is said to have excellent batch-to-batch consistency.[7] Most commercial sols consist of discrete spheres with a diameter range between 5 and 200 nm.

The stability of colloidal silicas is improved via the addition of bases that generate negative charges on the particle surface. When the particles are forced to repel each other, a stable sol is formed. Silica sols are considered to be either anionic (when they are stabilized with cations, such as Na^+, K^+, NH_4^+) or cationic (when the stabilizer is an aluminum derivative). Meanwhile, by lowering the pH value, the stability decreases and the particles react to form a gel. The gelling rate between pH 4 and 7 is very high and acidification must take place very quickly.

Chemical mechanical planarization (CMP), a precision polishing technique used during the production of semiconductor chips, is the largest market application that is already in use for nanosized silica, in particular fumed silica and silica sols (a detailed discussion of CMP can be found in Section 2.5).

In addition to the widely used oxides mentioned above, nanoparticle versions of other compounds, such as antimony (III) oxide, chromium (III) oxide, iron (III) oxide, germanium (IV) oxide, vanadium (V) oxide, and tungsten (VI) oxide, are also being developed and their possible uses are being explored.[7]

Production Routes for Various Oxides

In general, there are six widely used methods for producing nanoscaled particles— on the order of 1 to 100 nm in diameter—of various materials (Ref. 6, pp. 56–86, and Ref. 7, p. 16):

- Plasma-arc and flame-hydrolysis methods (including flame ionization)
- Chemical vapor deposition (CVD)
- Electrodeposition techniques
- Sol–gel synthesis
- Mechanical crushing via ball milling
- Use of naturally occurring nanomaterials

High-Temperature Processes, Such as Plasma-Based and Flame-Hydrolysis Methods These production routes involve the use of a high-temperature plasma or flame ionization reactor. As an electrical potential difference is imposed across two electrodes in a gas, the gas, electrodes, or other materials ionize and vaporize if necessary and then condense as nanoparticles, either as separate structures or as surface deposits. An inert gas or vacuum is used when volatilizing the electrodes.

During flame ionization, a material is sprayed into a flame to produce ions.[6] Using flame hydrolysis, highly dispersed oxides can be produced via high-temperature hydrolysis of the corresponding chlorides. Flame hydrolysis produces extremely fine, mostly spherical particles with diameters in the range of 7 to 40 nm and high specific surface areas (in the range of 50 to 400 m^2/g).[7]

In general high-temperature flame processes for making nanoparticles are divided into two classifications—gas-to-particle or droplet-to-particle methods—depending on how the final particles are made.[16]

In gas-to-particle processes, individual molecules of the product material are made by chemically reacting precursor gases or rapidly cooling a superheated vapor. Depending on the thermodynamics of the process, the molecules then assemble themselves into nanoparticles by colliding with one another or by repeatedly condensing and evaporating into molecular clusters. Gas-to-particle processes involve the use of flame, hot-wall, evaporation-condensation, plasma, laser, and sputtering-type reactors.

In droplet-to-particle processes, liquid atomization is used to suspend droplets of a solution or slurry in a gas at atmospheric pressure. Solvent is evaporated from the droplets, leaving behind solute crystals, which are then heated to change their morphology. Spray drying, pyrolysis, electrospray, and freeze-drying equipment are typically used in the droplet-to-particle production process.

Chemical Vapor Deposition   During CVD, a starting material is vaporized and then condensed on a surface, usually under vacuum conditions. The deposit may be the original material or a new and different species formed by chemical reaction.

Electrodeposition   Using this approach, individual species are deposited from solution, with an aim to lay down a nanoscaled surface film in a precisely controlled manner.

Sol–Gel Processing   This is a wet-chemical method that allows high-purity, high-homogeneity nanoscale materials to be synthesized at lower temperatures compared to competing high-temperature methods. A significant advantage that sol–gel science affords over more conventional materials-processing routes is the mild conditions that the approach employs.

Two main routes and chemical classes of precursors have been used for sol–gel processing:[7]

- The inorganic route ("colloidal route"), which uses metal salts in aqueous solution (chloride, oxychloride, nitrate) as raw materials. These are generally less costly and easier to handle than the precursors discussed below for the metal-organic route, but their reactions are more difficult to control and the surfactant that is required by that process might interfere later in downstream manufacturing and end use.

- The metal-organic route ("alkoxide route") in organic solvents. This route typically employs metal alkoxides M(OR)Z as the starting materials, where M is Si, Ti, Zr, Al, Sn, or Ce; OR is an alkoxy group, and Z is the valence or the oxidation state of the metal. Metal alkoxides are preferred because of their commercial availability and the high lability of the M−OR bond. They are available for nearly all elements, and cost-effective methods of synthesis from cheap feedstocks have been developed for some. The selection of appropriate OR groups (bulky, functional, fluorinated) allows developers to fine-tune their properties. Other precursors are metal diketonates and metal carboxylates. A larger range of mixed-metal nanoparticles can be produced under mild conditions, often at room temperature, by mixing metal alkoxides (or oxoalkoxies) and other oxide precursors.

In general, the sol−gel process consists of the following steps:

- Sol formation
- Gelling
- Shape forming
- Drying
- Densification

First, after mixing the reactants, the organic or inorganic precursors undergo two chemical reactions: hydrolysis, and condensation or polymerization, typically with an acid or base as a catalyst, to form small solid particles or clusters in a liquid (either an organic or aqueous solvent).

The resulting solid particles or clusters are so small (1 to 1000 nm) that gravitational forces are negligible and interactions are dominated by van der Waals, coulombic, and steric forces. These sols—colloidal suspensions of oxide particles—are stabilized by an electric double layer, or steric repulsion, or a combination of both. Over time, the colloidal particles link together by further condensation and a dimensional network occurs. As gelling proceeds, the viscosity of the solution increases dramatically.

The sol−gel can then be formed into three different shapes: thin film, fiber, and bulk. Thin (100 nm or so) uniform and crack-free films can readily be formed on various materials by lowering, dipping, spinning, or spray coating techniques.

Sol−gel chemistry is promising, yet it is still in its infancy and a better understanding of the basic inorganic polymerization chemistry has to be reached.[7] Some drawbacks include the high cost for the majority of alkoxide precursors, relatively long processing times, and high sensitivity to atmospheric conditions. And the batch nature of present sol−gel processing leaves cost and scaleup issues associated with the development of viable continuous production routes.

Mechanical Crushing via High-Energy Ball Milling Progressive particle-size reduction or pulverization using a conventional ball mill is one of the primary methods for preparing nanoscaled particles of various metal oxides. High-energy ball milling is in use today, but its use is considered by some to be limited because of the potential for contamination problems. Today, the availability of tungsten carbide components and the use of an inert atmosphere and high-vacuum processes has helped operators to reduce impurities to acceptable levels for many industrial applications.[7] Other common drawbacks, however, include the highly polydisperse size distribution and partially amorphous state of nanoscaled powders prepared using the pulverization method.

Naturally Occurring Materials Certain naturally occurring materials, such as zeolites, can be used as found or synthesized and modified by conventional chemistry. A zeolite is a caged molecular structure containing large voids that can admit molecules of a certain size and deny access to other, larger molecules (Ref. 6, p. 99). They find application as catalysts and adsorbents, among other uses.

In addition to the proven techniques discussed above, some promising technologies are emerging to produce nanoscaled particles of various materials.[7] These include:

- Flame or jet-flame reactors that introduce an additional flame behind the reaction zone, in order to transform the aggregates into spherical particles more effectively.
- Plasma processes that are designed to promote more rapid cooling, in order to produce fewer agglomerates.
- Sonochemical processing routes, in which an acoustic cavitation process generates a transient localized hot zone with an extremely high temperature gradient and pressure. Sudden changes in temperature and pressure assist in the destruction of the precursor material (e.g., organometallic solution) and promote the formation of nanoparticles.
- Hydrodynamic cavitation processes, in which nanoparticles are generated through the creation and release of gas bubbles inside a sol–gel solution. Erupting hydrodynamic bubbles are responsible for nucleation, growth, and quenching of the nanoparticles. Particle sizes can be controlled by adjusting the pressure and the solution retention time in the cavitation chamber.
- Microemulsion techniques, which show promise for the synthesis of metallic semiconductor silica, barium sulfate, and magnetic and superconductor nanoparticles.

The ongoing challenge for the research community is to continue to devise, perfect, and scale up viable production methodologies that can cost effectively and reliably produce the desired nanoparticles with the desired particle size, particle size distribution, purity, and uniformity in terms of both composition and structure.

2.4   CARBON NANOTUBES AND BUCKYBALLS

One particular type of nanometer-scaled structure that is generating considerable interest is the carbon nanotube. Today, carbon nanotubes are among the most hotly pursued type of nanoparticle, and in some applications they are leading the charge in terms of nanoscaled particles that are making their way into commercial-scale use.

Carbon nanotubes are seamless cylinders composed of carbon atoms in a regular hexagonal arrangement, closed on both ends by hemispherical endcaps. As shown in Figure 2.2, they can be produced as single-wall nanotubes (SWNTs) or multiwall nanotubes (MWNTs). While carbon nanotubes can be difficult and costly to produce, and challenges remain when it comes to distributing and incorporating them homogeneously within the matrices of other materials, they have been an object of desire among researchers working in the nanosphere since their discovery in 1991, due to their excellent intrinsic physical properties and promising potential applications in a wide array of fields.[17]

The following list summarizes some of the unprecedented structural, mechanical, and electronic properties that are exhibited by carbon nanotubes:[7,17–19]

- SWNTs can have diameters ranging from 0.7 to 2 nm, while MWNTs can have diameters ranging from 10 to 300 nm. Individual nanotube lengths can be up to 20 cm.

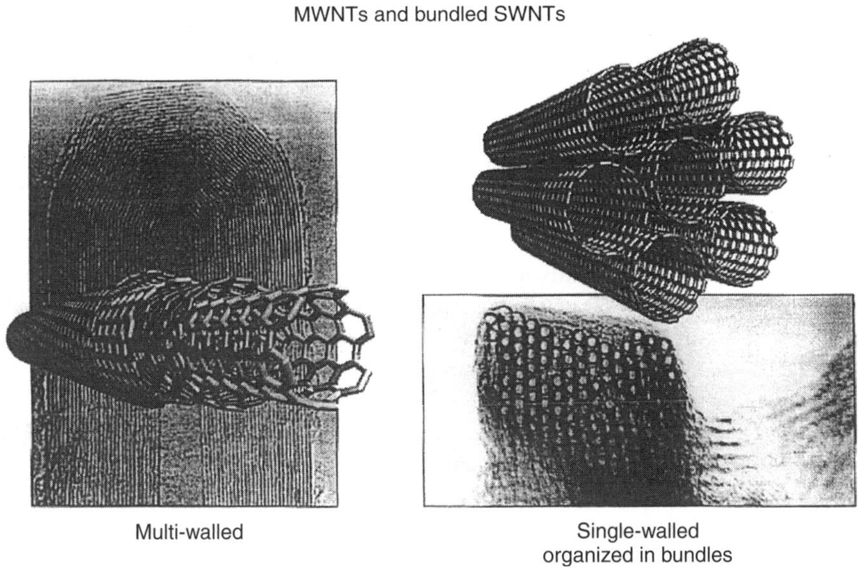

MWNTs and bundled SWNTs

Multi-walled   Single-walled
                                              organized in bundles

Figure 2.2   Carbon nanotubes can be produced as single-wall nanotubes (SWNTs) or multi-wall nanotubes (MWNTs), and each type displays distinctive material characteristics. (*Source*: A. Hirsch, Friedrich-Alexander University, Erlangen-Nuremberg, Germany, Ref. 18)

- Carbon nanotubes have a surface area of up to 1500 m^2/g and a density of 1.33 to 1.40 g/cm^3.
- Depending on their structure, nanotubes can function as either conductors or semiconductors of electricity and heat; thermal conductivity has been demonstrated to 2000 W/mK (with some published work suggesting values as high as 6000 W/mK).
- They exhibit extremely high thermal and chemical stability.
- They are extremely elastic, with a modulus of elasticity on the order of 1000 gigapascals (GPa).
- They have a demonstrated tensile strength greater than 65 GPa, with a predicted value as high as 200 GPa; a widely cited comparison is that carbon nanotubes have a tensile strength 100 times that of steel, but have only one-sixth the weight of steel.
- Some SWNTs can withstand 10 to 30 percent elongation before breakage.
- As nanocapillaries, they can accommodate guest molecules and form dispersions with surfactants in water.
- They have the potential for further chemical functionalization via various types of surface treatments.

As a result of this distinctive suite of material properties, carbon nanotubes are being incorporated into various matrices to produce ultralightweight materials that have exceptional strength and other functional advantages over conventional materials and to create advanced thin-film membranes, fibers, foams, and coatings.

For example, nanotube-reinforced plastics, including fluoropolymers such as ethylene tetrafluoroethylene (ETFE) and polyvinylidene fluoride (PVDF), are already being used in various automotive, electronics, and materials-handling applications, particularly those that require precise control of static electricity, improved chemical resistance, a higher barrier to chemical permeation, inherent lubricity, and better resistance to sloughing compared to the conventional, non-nanotube-reinforced materials.

The conductivity of nanotubes makes them an ideal additive for imparting electrical conductivity to inherently nonconductive materials. This is already being done using conventional carbon black, but by comparison, the addition of carbon nanotubes for this purpose requires extremely small volumes to achieve comparable or even better changes in electrical properties.

When blending polymers with carbon nanotubes (or any other nanoparticles, for that matter), it is important to use as small a loading as possible, both from a cost and processing standpoint and to ensure the mechanical integrity and other physical properties (such as color, brittleness, or surface smoothness) of the finished part. The greater the aspect ratio (length-to-diameter ratio) of the conductive nanotubes, the lower the loading level required to provide a desired level of conductivity.

One company's carbon nanotubes have a very high aspect ratio (1 : 1000 or greater). When compounded into an inherently nonconductive fluoropolymer matrix, the end result is the formation of an electrically conductive network at loading levels that are significantly lower than those required when using conventional fillers, such as graphite fibers or carbon black. For instance, the addition of only 1 to 3 wt % nanotubes (10 to 12-nm diameter) achieves target electrostatic discharge (ESD) values, compared to 7 to 9 wt % for carbon fibers or 12 to 15 wt % for carbon black.[17]

In automotive applications, the ability to produce electrically conductive parts from inherently nonconductive materials allows these parts to be painted using electrostatic spray painting—an environmentally friendly process that is increasingly popular among automakers who are eager to phase out solvent-borne painting processes.[20] Also, nanotube-bearing nylon is being used to make fuel lines for cars, where it is critical to dissipate static charges that could spark an explosion.[9]

Meanwhile, researchers at the U.S. National Aeronautical and Space Administration (NASA) are developing polyimide nanocomposites that incorporate SWNTs to improve radiation and tear resistance, and thermal and electrical conductivity—all of which are important for aircraft and spacecraft construction.

In addition to their use as key components in advanced composite materials, carbon nanotubes are also being explored as key components in tomorrow's advanced sensors, electronic and optical devices, catalysts, batteries, and fuel cells.[7,19]

In fuel cell applications, for example, the use of carbon nanotubes has been shown to improve the performance of industrial fuel cells while greatly reducing the amount of platinum catalyst required. In some recently reported work, a conventional fuel cell anode (made of platinum-impregnated carbon black with binder) was replaced with a membrane made solely of SWNTs.[17] Run under identical conditions, the SWNT fuel cell provided more than twice the electrical current output at 0.6 V (a typical voltage) compared to the standard version. Where the standard fuel cell gives 350 mA current at 0.6 V, the SWNT cell gives more than 800 mA current. Ongoing work seeks to bring this level of improvement to the cathode side, as well.

Reconciling the Remaining Challenges

As is the case with the other types of nanoscaled materials discussed in Section 2.2, some fundamental technical and financial hurdles associated with the production and use of carbon nanotubes remain, and these must be reconciled before the full commercial-scale potential of these tiny carbon structures can really be known.

Until now, the production methods for making carbon nanotubes (discussed below in this section) have been limited to minute quantities, so this has made carbon nanotubes prohibitively expensive for most applications. Transforming these laboratory wonders into viable commercial products will require substantial reductions in their production costs.

However, while the high cost associated with carbon nanotubes is often cited as a barrier for commercialization, this high cost may be something of a red herring. As one observer notes, today's premium tag for carbon nanotubes is largely a function of limited production, unproven process technologies and manufacturing routes, and remaining technical issues—factors that plague the introduction of nearly all new materials.[20] Once more cost-effective production routes are developed, proven, and scaled up, the cost of the materials should ultimately come down, opening up the doors for a wide array of potential applications.

Meanwhile, the prevailing nanotube-manufacturing processes often yield products that preferentially aggregate into bundles of different diameters, nanotubes that vary greatly in length or diameters, nanotubes that have a range of helicities, and nanotubes that have defects in both the sidewalls and tube ends. Ongoing work is underway to improve the various production techniques and to develop routes that yield carbon nanotubes with improved and more predictable uniformity and structural integrity.

Researchers are also working to develop techniques to effectively manipulate and functionalize carbon nanotubes and other nanoscale particles, to overcome issues that can hinder their ease of use of processability, particularly when it comes to integrating carbon nanotubes into other matrix materials. For instance, intrinsic van der Waals attraction among the tubes, working in concert with their high surface area and aspect ratio (length-to-diameter ratio), often causes nanotubes to agglomerate, which decreases the effective dispersion and solubility in the host matrix.

Functionalization of nanotube sidewalls by covalent bond formation will only be successful if a highly reactive reagent is used; initial work using elemental fluorine and such reactive organic groups such as nitrenes, nucleophilic carbenes, and radicals, shows promise. While further discussion is beyond the scope of this work, ample discussion can be found in Hirsch.[18]

Meanwhile, today's nanotechnology researchers are also seeking better ways to purify nanotubes themselves, to separate one desired type of carbon nanotube from a mixed batch. For instance, exclusively semiconducting versions are needed to produce electronic components, but in many batches, the semiconducting nanotubes cannot yet be reliably separated from their metallic counterparts.[18] When it comes to removing unwanted contaminants such as amorphous carbon and metal catalyst particles from manufactured nanotubes, these can be removed by treatment with oxidizing agents, microfiltration, or chromatographic procedures. However, under harsh oxidative conditions, the carbon nanotubes themselves may also be damaged.

Manufacturing Routes for Carbon Nanotubes

Today, a variety of production methods exist for making carbon nanotubes, although in many cases, the process specifics are closely guarded by companies hoping to retain a competitive advantage by developing and patenting more effective or less costly production routes. In general, most of the prevailing methods are variations of the following two basic process types:[7,17,19,21]

1. Carbon is subjected to high heat (such as through the use of a 3000-K electric arc), and the resulting carbon nanotubes and other carbon nanostructures are isolated from the soot mixture that is formed. Arc discharge processes can make either SWNTs or MWNTs, using a plasma-based process whose feedstock carbon typically comes from burning either a solid carbon electrode

(to produce MWNTs) or a composite of carbon and a transition-metal catalyst (to produce SWNTs). Another high-temperature route involves the use of pulsed-laser vaporization (PLV), which makes only SWNTs. This process uses a high-power, pulsed Nd:YAG laser (Nd:YAG is an acronym for Neodymium-doped Yttrium Aluminum Garnet) aimed at powdered graphite that is loaded with metal catalyst (a later variation of this process uses a continuous, high-energy CO_2 laser, focused on a composite, carbon–metal feedstock source).

2. Alternately, carbon-containing compounds (such as graphite rods) are decomposed using heat or catalysis to yield carbon nanotubes and other structures. This latter method, which typically uses CVD, is increasingly being used since it offers better control over the type of product manufactured. CVD-based processes can be used to make either SWNTs or MWNTs, and the technique produces them by heating a precursor gas and flowing it over a metallic or oxide surface with a prepared catalyst (typically nickel, iron, molybdenum, or cobalt).

Fullerenes or Buckyballs

Still another nanoscaled carbon derivative has been making headlines in recent years. Until the late 1980s, graphite and diamond were the only known and macroscopically available modifications of carbon.[18] This changed when chemistry professor Richard Smalley and his colleagues at Houston's Rice University found that by vaporizing carbon and allowing it to condense in an inert gas, it formed highly stable crystals, composed of 60 atoms of carbon apiece (C-60). Table 2.3 presents a comparison of some of the key attributes of carbon nanotubes and fullerenes, or buckyballs, (and presents other information on a class of materials, carbon nanofibers, that is not discussed further here).

Since the novel, cagelike molecule was suggestive of the soccer-ball structure used by the architect R. Buckminster Fuller in his futuristic geodesic buildings, the researchers named their discovery *buckminsterfullerene*, which was quickly shortened to *fullerence* or *buckyball*. The buckyball remains nanotechnology's most famous discovery. It not only earned Smalley and his colleagues the 1996 Nobel Prize in Chemistry but it also cemented nanotechnology's reputation as a cutting-edge research field.[22]

Today, buckyball use is being considered for a vast array of potential applications. For example, research is already underway to demonstrate the advantages that buckyballs may confer when used as functional additives in cosmetics, as proton-conducting membranes, as components of lithium battery anodes and industrial grinding media, as alternative raw materials for nanocrystalline diamond film, as a solar cell component, as a component in photoresist to yield finer chip resolution and smaller circuits, as a shielding material in protective eyewear, and as components in high-speed photonic switching devices and signal beam amplifiers.[7]

One of the most promising applications for buckyballs is in the area of pharmaceuticals. Since the C-60 molecule is about the same size as many proteins and enzymes, and theoretically has 60 sites onto which a functional group may be

TABLE 2.3 Nanoscaled Carbon Products[a]

Product Group	Dimensions	Description
Fullerenes or buckyballs	1–3 nm by 1–3 nm by 1–3 nm	Elemental carbon in spherical C-60 structure, other similar spherical/ovoid structures with higher (or lower) number of carbons, chemically functionalized fullerenes (e.g., oxides) and fullerenes with entrained species (e.g., endohedral metal complexes). Fullerenes have also been produced with multiple shells.
Carbon nanotubes	1–20 nm by 1–20 nm by up to 5–10 microns or more[b]	There are two major groups, single-wall nanotubes (SWNTs) and multi-wall nanotubes (MWNTs). A SWNT is essentially a C-60 fullerene extended indefinitely into a hollow cylinder with a graphite-like tube. MWNTs have one or more additional cylinders of carbon concentric to the central tube. These products can be open or closed, can have modified ends (e.g., nanohorns), or can be chemically functionalized.
Carbon nanofibers	20–200 nm by 20–200 nm by indefinite length (depends on the process)	Similar to nanotubes but significantly thicker and usually produced by somewhat different technologies than carbon nanotubes. Their surfaces may also be modified and their physical properties are also substantially different from carbon nanotubes, especially when sufficiently large in diameter to negate quantum effects.
Other carbon nanomaterials		Includes other carbon products with at least one dimension in the nanometer range (e.g., diamond film).

Source: Ref. 7.

[a]A quick comparison is offered to showcase the differences among the fullerenes and buckyballs and carbon nanotubes, which are the primary types of nanoscaled carbon derivatives that are being pursued today.

[b]Individual researchers have produced carbon nanotubes up to 20 cm in length.[17] (See also Section 2.4.)

attached, buckyball use for the delivery of pharmaceutical moieties or use in medical testing is being avidly pursued. Diseases that appear to be susceptible to these treatment strategies include HIV/AIDS, Parkinson's disease, amyotrophic lateral sclerosis (ALS, or Lou Gehrig's disease), osteoporosis, and cancer. One treatment mode could be the inclusion of radioactive isotopes within the fullerene and attaching tumor-finding moieties to the outside. Meanwhile, the antioxidant properties of buckyballs also make them attractive for medical applications.[7]

As the functional advantages of ultrasmall nanotubes and buckyballs continue to be deciphered, and cost-effective processes are perfected to manufacture and manipulate them, there seems to be no limit to what these novel carbon derivatives can do.

2.5 CURRENT AND FUTURE MARKET APPLICATIONS

As discussed earlier, when materials such as metals, metal oxides, ceramics, and polymers are manipulated into increasingly smaller and smaller particle sizes, the resulting increase in available surface area leads to a direct improvement in a variety of material properties, including thermal and electrical conductivity, surface chemistry (which affects particle dispersibility and reactivity), photonic behavior (which is functionality that changes in the presence of light of varying wavelengths), and catalytic conversion rates, among others. Today, nanoscaled particles are already in use in a diverse array of viable commercial applications, while many other promising laboratory- and pilot-scale developments are being groomed for commercial-scale production.

Ultimately, the high cost associated with nanoscaled particles and carbon nanotubes will have to be justified in terms of enhanced performance. Nanotechnology researchers are under increasing pressure to devise viable manufacturing methods that both bring down production costs and exploit to the fullest extent possible the broadest array of functional advantages associated with a given type of nanoscaled material.

Today, nanoparticle use is the most robust in the manufacture of semiconductor devices and in the production of both advanced composite materials for a wide variety of end uses and improved consumer products, such as cosmetics and sunscreen. In addition, nanoparticle-related developments are being actively pursued to improve fuel cells, batteries and solar devices, advanced data-storage devices such as computer chips and hard drives, magnetic audio- and videotapes, and sensors and other analytical devices. Meanwhile, nanotechnology-related developments are also being hotly pursued in various medical applications, such as the development of more effective drug-delivery mechanisms and improved medical diagnostic devices, to name just a few. All of these are discussed in detail below.

Semiconductor Manufacturing

Perhaps considered the grandfather of all nanoparticle-related applications is chemical mechanical planarization (CMP), a precision polishing process that is used during the production of integrated electronic circuits on semiconductor chips. Today's semiconductor chips involve increasingly fine and exacting geometries, and increased numbers of metal layers are used to create the integrated circuits. According to one report, circuits designed at 25 nm can incorporate as many as 10 million transistors and 50 million connections.[7]

During CMP, nanoscaled particles of abrasive materials—typically oxides of aluminum and zirconium, and colloidal or fumed silica, but increasingly cerium, as well—with diameters of 20 to 300 nm are formulated into a polishing slurry. Slurries for polishing silicon chip surfaces are typically made from a fumed or colloidal silica suspension stabilized with dilute potassium or ammonium hydroxide. Slurries for treating metal surfaces typically use aluminum oxide suspensions plus an oxidizer such as ferric nitrate, potassium iodate, or hydrogen peroxide.[7]

Such polishing slurries combine both mechanical and chemical action to smooth the topography of the semiconductor device, reduce height variations across the dielectric region, and scavenge metal deposits and excess oxide that is used during fabrication. In doing so, the slurry ensures that the metal and dielectric layers on the silicon wafers are smooth and essentially defect free.

As the drive toward miniaturization continues to spur the development of smaller and smaller chip geometries, the exacting demands placed on CMP slurries will continue to grow.[1] For instance, while line widths of 130 nm are common in today's semiconductor devices, 90-nm line widths will be the standard for next-generation devices.[17] Such intricate chip assemblies have no tolerance for even the smallest defects.

Today, CMP is considered the "gold standard" for polishing and cleaning complex semiconductor chip geometries, and this has created a viable commercial-scale market for related nanoparticles. In the coming years, growth in demand for various oxide nanoparticles for CMP will be driven primarily by three factors: an overall increase in the number of semiconductor wafers produced, a rise in the percentage of wafers that are processed using CMP, and an increase in the number of layers per wafer that must be planarized during manufacture. Because the technology is "absolutely critical during semiconductor manufacturing," nanoparticle use for CMP applications is projected by one market analyst to grow by 20%/year through 2005.[1,17]

Advanced Composites

Nanocomposites are formed when nanometer-sized particles of useful additives are blended with a polymer matrix resin. The idea of blending functional additives with plastics is not new, but the ability to produce and use nanoscaled versions of the traditional additive materials brings many previously unknown benefits to this class of materials. Today, nanometer-sized clay platelets and carbon nanotubes are the most widely used additives that are incorporated into commercially available nanocomposites, along with various oxides. For instance, today, nanometer-sized oxides of alumina (aluminum oxide) are being used to improve scratch and abrasion resistance, nanoparticles of zinc oxide and cerium oxide improve ultraviolet light attenuation, nanoscaled oxides of cerium improve oxygen storage, antimony/tin oxides (ATO) improve near-infrared light attenuation, and indium/tin oxides and ATO improve conductivity.

One of the earliest nanocomposites was pioneered by Toyota Central R&D labs in 1987.[7] By combining nylon-6 with just 5 wt % nanoscaled clay platelets, the nanocomposite had the following property improvements compared to the neat resin:

- 40 percent higher tensile strength
- 68 percent higher tensile modulus
- 60 percent higher flexural strength
- 126 percent higher flexural modulus
- 65 to 125°C higher heat distortion temperature

Today, many different types of nanocomposites are commercially available. An exhaustive range of polymers has been used in various nanocomposite applications, including but not limited to nylon (6, 6-6, 11, 12, and aromatic), polystyrene polypropylene, polyethylene, polyimides, ethylene vinyl acetate copolymers, poly(styrene-*b*-butadiene) copolymers, poly(dimethyl-siloxane) elastomers, ethylene propylene diene monomer, acrylonitrile butadiene rubber, polyurethane polypyrrole, polyethylene oxide, polyvinyl chloride, polyaniline, epoxies, and phenolic resins.[7]

One of the big advantages of using ultrasmall versions of the key additive is that much lower loadings can be used, compared to conventional additives. For instance, a nanocomposite typically contains just 3 to 5 wt % nanoclay or other nanoparticles, while conventional filled or reinforced composites typically require the addition of 20 to 40 wt % fillers, such as talc, mica, calcium carbonate, asbestos, graphite, and various oxides.[7] While the use of conventional, micron-sized fillers can also increase the mechanical properties of a polymer composite, traditionally high loading rates lead to trade-offs in terms of decreased polymer clarity (micrometer-sized inclusions lead to a loss of transparency because they scatter light), increased brittleness, and higher density. This is not the case when relatively minute proportions of nanoscaled particles are used instead.

In general, using much smaller loadings of nanoscaled additives, today's nanocomposites demonstrate the following improved properties, compared with both the neat resin and conventional composites that are made using micron-sized additive particles:[7]

- Better mechanical properties, such as higher tensile and flexural strength and moduli, improved toughness, improved scratch resistance, and surface appearance even at low additive loading levels
- Better barrier properties in terms of decreased liquid and gas permeability and improved resistance to organic solvents
- Improved fire retardancy

- Better optical clarity
- Lower coefficient of thermal expansion and higher heat distortion temperature
- Improved electroconductivity and catalytic activity
- Self-cleaning capabilities and antimicrobial properties
- The ability to make lower-weight plastic parts because of the lower density of the nanoscaled fillers

By offering an excellent combination of improved mechanical properties, heat resistance, optical properties, and barrier properties, nanocomposites are being used in both flexibile and rigid films and sheets for various food-packaging applications. For instance, nanoscaled clay platelets (measuring 1 nm by 100 nm) are impermeable to O_2 and CO_2, so that when they are dispersed in a polymeric matrix, these nanoplatelets impart improved gas-barrier and oxygen-scavenging properties to the composite.

According to process developers, during the polymerization process, the nanoplatelets form an obstacle course in the solidified nylon around which gas molecules have difficulty maneuvering.[23] From a practical standpoint, this increased resistance to gas diffusion allows the food and pharmaceutical industry to design novel packaging schemes that, for example, keep out unwanted O_2 to reduce product spoilage, while keeping CO_2 inside the packaging where it belongs, to prevent carbonated beer and soft drinks from going flat.

At the same time, in automotive and aviation applications, nanocomposites are valued not just for their improved physical properties but for their ability to produce parts with reduced weight (which leads to improved fuel efficiency) and increased recyclability. And, as detailed in the discussion on carbon nanotubes (Section 2.4), the addition of nanometer-sized carbon nanotubes to inherently non-conductive polymeric materials allows the resulting nanocomposite parts to be painted using electrostatic spray painting. This is finding favor among automakers who are eager to phase out solvent-borne painting processes.[20] Similarly, nanotube-bearing nylon is being used to make fuel lines for cars, where it is critical to dissipate static charges that could spark an explosion.[9] Thanks to their improved fire retardancy, nanocomposites are also being widely used today in the electrical and electronics industries, in the manufacture of wire and cable and TV housings.

As is the case when working with other types of nanoparticles, the ability to get nanoparticles, carbon nanotubes, and nanoplatelets of clays to disperse effectively in the polymer matrix is no easy task. Ongoing research and development efforts are geared toward perfecting various surface treatments to mitigate excessive surface charges in order to minimize agglomeration and improve dispersion, and to get the nanoparticles to bond to the polymer rather than to one another.

Advanced Ceramics

In general, ceramics are inorganic, nonmetallic materials that are consolidated at high temperatures, usually starting as powder particles and ending as solid, usable forms. A typical ceramic contains complex crystal structures based on the various oxides (described in Section 2.3) and may involve both covalent and ionic bonding.[7]

Structural ceramics are defined as stress-bearing components or ceramic coatings that show special resistance to corrosion, wear, and high temperature. Today, advanced ceramics based on aluminum oxide, or alumina (Al_2O_3), and zirconium dioxide, or zirconia (ZrO_2), are used for many industrial applications, such as cutting tools, bearings and seals, filters and membranes, refractory materials, catalysts and catalyst supports, and automotive engine components, and in bone implants used in dental and orthopedic prostheses. In addition, electronic uses of advanced ceramics include substrates and packaging for integrated circuits, capacitors, transformers, inductors, piezoelectric devices, and chemical and physical sensors.

Finished ceramic components are typically resistant to high temperatures and corrosion, but they are also brittle and hard to work with. Traditionally, high-performance ceramics have been made from powders whose constitutent particles have a diameter of just under 1 micron (μm), or 1000 nm. Increasingly, however, improvements have been demonstrated, by producing ceramics from powers consisting of much smaller particles, say, 100 nm in diameter or less.[7]

The size and form of the particles have a big influence on the properties of the ceramic. In high-density components, for example, they determine the component's strength, while in porous ceramics, they can be used deliberately to control the size and form of the pores. In microelectronics applications nanoceramics can be used to create structures whose dimensions are far smaller than the limit allowed by conventional techniques.

Nanocrystalline ceramics also allows a higher level of homogeneity to be attained with more well-tailored grain sizes in the final sintered product, and this leads to enhanced properties, such as increased fracture toughness, transparency for use in lamps and windows, superplasticity for near-set shaped parts, very high hardness and wear resistance, enhanced gas sensitivities, diffusion bonding for ceramic seals, improved tribological properties such as higher resistance to friction and wear, and increased high-temperature resistance.[7]

Meanwhile, since there is a strong relationship between sintering temperature and particle size, the ability to reduce the particle size of the initial material to about 20 nm, allows the sintering temperature for zirconia, for instance, to be lowed from 1400 to 1110°C.[7]

As with other nanoparticles, the reduction in the size of the powder particles drastically increases their total surface area. For example, a 30-g sample of a zirconium powder with a 700-nm diameter has a surface area of only 200 m^2 while the same amount of zirconium powder with 15-nm particles has 1500 m^2 of surface area.[7]

One additional feature of nanoscale materials is the high fraction of atoms that reside at particle surfaces or grain boundaries. At a grain size of less than 20 nm, as much as 30 percent of the atoms can be present at the grain boundaries. This is important, since these grain-to-grain interfaces play a crucial role in determining the mechanical, optical, and electrical properties of the final material.

Typically, ceramic nanopowders are synthesized using either precipitation from organometallic or aqueous salt solution (including sol–gel processing) or condensation from the vapor phase, using spray pyrolysis and gas condensation.

However, the use of nanoscaled ceramic powders creates three sets of challenges that are not encountered when working with oxide powders at the micron or sub-micron scale:[7]

- Conventional ceramic powder is produced by progressively finer grinding of the material. However, nanocrystalline powder is typically chemically precipitated from solutions or condensed from gaseous states. Both of these alternative production methods are costly and more elaborate compared to traditional production routes. As a result, issues related to viable, cost-effective production capacities and batch-to-batch consistency remain with these nanoparticle-based production routes for making ceramics.

- Shaping nanoceramic powders into homogeneous solids remains another big challenge. Nanopowders tend to clump together unevenly because they have large surface areas with excessive surface charge.

- During the sintering process, grain size may be difficult to control. The higher reactivity of the fine powder can lead to increased crystal growth resulting in coarse and in non-homogeneous grain structures.

Catalytic and Photocatalytic Applications

In another area, researchers are using nanotechnology to harness the photocatalytic capabilities of certain nanoscaled substances and use these capabilities for novel commercial applications. When exposed to UV light, photocatalytic substances, such as the anatase form of titanium dioxide (TiO_2; but not the rutile form), strongly absorb UV radiation. In the presence of water, oxygen, and UV light, such substances generate free radicals that can be used to decompose unwanted chemical substances and reduce the adhesive forces that bind dirt (both organic and inorganic matter) and algae to various surfaces.

This photocatalytic effect can be exploited for various commercial applications, such as water and air purification, or to impart self-cleaning, antimicrobial, and anti-algae properties to surfaces. Since all such reactions take place very close to the surface (of the TiO_2 particles or coating), nanoparticles with extremely high surface area allow these reactions to proceed at orders-of-magnitude faster rates than ordinary grades of the oxide.[2] Such photocatalytic reactions can be used to treat, sanitize, and deodorize air and water and various surfaces and fabrics.

In addition, the unique physical interaction of water droplets with photocatalytic surfaces has led to the development and marketing of TiO_2 films that impart self-cleaning or antifogging properties to windows, mirrors, and other substrates. During exposure to light, the contact angle of a water droplet on a photocatalytic surface is gradually reduced to zero; in other words, the water cannot maintain the form of a droplet on such a surface, so it spreads out in a sheetlike form. This results in a superhydrophilic or extremely water-loving surface.[1] Today, the use of photocatalytic substances are being explored in the development of antifogging bathroom mirrors and self-cleaning tiles for use in hospitals and restaurants as well as air cleaners, water purifiers, and germicidal lamps, among others.

Conventional photocatalytic TiO_2 films can be produced using chemical vapor deposition (CVD) techniques, although more recently, nanotechnology researchers have devised ways to produce them from dispersions of nanoparticles. Anatase-phase TiO_2 is currently the most popular semiconducting oxide for photocatalysis, although zinc oxide, tin oxide, and other oxides also exhibit photocatalytic behavior, so their use in such applications is also being pursued.

Meanwhile, in search of improved catalysts to destroy air pollutants, one group of researchers is focusing its attention on gold nanoparticles layered on TiO_2. The group has shown that the addition of gold (a notoriously unreactive element) to TiO_2 (a widely used industrial catalyst) changes the electronic properties of both materials. The result is a catalyst that is said to be 5 to 10 times more reactive than ordinary TiO_2 in terms of dissociating sulfur dioxide (SO_2).[2]

Gas Sensors and Other Analytical Devices

The research community is hard at work to exploit the extraordinary surface area and increased reactivity and catalytic properties of many nonoscaled materials, and to use these attributes to develop highly sensitive sensors and other analytical devices (such as the so-called *lab-on-a-chip* applications). Such devices are envisioned to improve industrial monitoring devices, check food quality, improve medical diagnostics such as disease detection, and improve the detection of chemical, biological, radiological, and nuclear hazards.[24]

Resistive, metal oxide gas sensors (using, for example, nanoscaled oxides of zinc, tin, titanium, and iron) rely on a change in electrical conductivity at the surface of the sensor as it comes in contact with the target gas. Thus, reaction rate is dictated by the amount of surface area on the probe. Ongoing advances in nanoparticle engineering have given manufacturers the opportunity to greatly increase surface area on the sensor probe in order to improve the gas detection sensitivity and selectivity. For example, one company has patented a new fabrication process for preparing metal oxide sensors that may overcome some for the disadvantages of conventional thick-film fabrication techniques (including considerable labor, limited reproducibility, and high costs).[2]

Comparing prototypes of sensors made from conventional, coarse-grained oxides ($>1\,\mu$m) with those produced using nanopowders (<100 nm), the nanostructured sensors demonstrate enhanced sensitivity and respond more quickly—which gives them a distinct advantage for certain monitoring applications. Using the new approach, different layers are deposited to form multilayer, chip-style sensors with a series of internal electrodes. For example, as reported in the literature,[2] one company has already developed prototype devices for detecting low-level (10 to 1000 ppm) and high-level ($>1\%$) concentrations of hydrogen and is developing improved sensors for NH_3, NO_X, volatile organic compounds (VOCs), and CO_2 as well.

Consumer Products

Makers of sunscreens, cosmetics, and other personal-care products have discovered that the use of nanometer-scaled versions of common additives can improve the effectiveness and aesthetic appeal of many products, compared to conventional formulations. For instance, the recent trend in sunscreen development has been to devise products that offer broader-spectrum protection against not just ultraviolet-B (UV-B) radiation (which is to blame for sunburns), but shorter-wavelength ultraviolet-A (UV-A) radiation (the chief culprit in skin cancer and other skin damage) as well. With the advent of affordable methods to produce and use nanoscaled particles of the common UV blockers titanium oxide (TiO_2) and zinc oxide (ZnO), sunscreen manufacturers can now use these broad-spectrum UV-blocking agents to produce transparent lotions that are aesthetically superior to the opaque white oxide creams that are the hallmark of surfers and lifeguards.[2] Because the wavelength of visible light is 400 to 800 nm, nanoscaled particles are able to transmit—rather then scatter—light. This allows transparent (rather than opaque) sunscreen, cosmetic, and other personal-care product formulations to be made.

Drug Delivery Mechanisms and Medical Therapeutics

The intersection of nanotechnology with the life sciences has the potential to revolutionalize medicine and health care. Today, a new era in biotechnology and biomedical engineering is emerging with the use of nanoscaled structures for disease diagnosis, gene sequencing, and improved drug delivery.[24] Meanwhile, a host of biocompatible nanomaterials are being pursued to allow for the development of more robust artificial tissues and organs, and novel materials, such as alumina ceramic toughened with nanosized zirconia, are being developed to extend the life of ceramic hip and knee replacement materials.[23] Nanotechnology-based materials and devices are also being used to develop high-performance medical instruments and diagnostic devices.

One of the most promising nanotechnology-based developments in the medical arena is the use of nanoscaled materials and devices to improve the way often toxic drugs both reach their intended target within the human body and are sequestered preferentially in and around a tumor, while avoiding capture by the body's immune system or accumulation in healthy organs and tissues.[2] For example, the systemic, intravenous administration of chemotherapy drugs causes devastating damage to healthy cells and severe side effects for the patient. One industry observer notes that, to minimize the collateral damage associated with chemotherapy, lower-than-optimal drug doses are often used, because administering many cancer drugs at the doses needed to induce the desired anticancer effect "could kill the patient" (Ref. 2, p. 25).

Some developmental efforts may sound as though they are right out of the pages of a science fiction novel. For instance, one research team is developing superparamagnetic iron oxide, Fe_2O_3, as a potential tool for treating tumors. About 20 nm in diameter and dispersed within an amorphous silicon dioxide matrix, the tiny iron oxide particles are magnetic when exposed to a magnetic field, yet lose their magnetism when the field is removed. And, in an alternating magnetic field, the particles heat up. Researchers are developing a process by which the particles are injected into the bloodstream of a patient and guided to the tumor site using a magnet, and then, by applying a very localized alternating magnetic field, the particles are heated to 50 to 60°C to destroy the tumor without harming the surrounding healthy tissue.[23]

Scientists elsewhere have been able to demonstrate that semiconductor nanocrystals (quantum dots) coated with homing peptides can target specific types of cancer cells in live mice. As a next step, the group is working to synthesize quantum dots that are functionalized with both homing peptides and cancer treatment drugs in order to target and destroy the cancerous tissue.[2]

Meanwhile, while many drug-targeting techniques have focused on the use of two chemically distinct colloid particles—nanoscale liposomes and biodegradable polymers—that carry the drug payload inside a hollow shell, one company has turned its attention to a fundamentally different carrier—colloidal gold nanoparticles. Gold particles are inert, and they are said to bind protein biologics avidly and preferentially on their surface through available thiol groups.[2]

Thanks to a "leaky hose effect" that results when blood vessel networks develop hastily to feed a fast growing tumor, nanometer-sized gold particles (typically 35 nm), carrying cancer drugs on their surface, are able to sneak through the porous blood vessel network, bringing the drugs right to the cancerous site.

In work to date using the cancer drug TNF (tumor necrosis factor, an anticancer protein biologic that has shown promise in killing a variety of solid tumor types) bound to colloidal gold nanoparticles, one company has found that the drug was preferentially sequestered in tumors in mice and dogs at higher rates, compared to testing using unbound TNF. However, initial work also showed that the gold–TNF also aggregated preferentially in the livers and spleens (the chief organs of the immune system) of the tested animals. To make the gold "more invisible" to these organs, the developers added a linear form of polyethylene glycol (PEG) to the colloidal gold nanoparticles—just like adding brush bristles to the surface. This improvement resulted in "ten times more TNF getting to the tumor than with the non-PEG version, and with no active sequestration in the liver and spleen," say the process developers (Ref. 2, p. 25).

Of course, in addition to overcoming potential manufacturing, price-related, and performance hurdles, researchers working in the medical arena must get their nanotechnology-based routes through the arduous approval process of the U.S. Food and Drug Administration and comparable regulatory-approval processes elsewhere. These regulatory hurdles will ultimately dictate which of today's encouraging nanotechnology-related medical breakthroughs will ever be able to fulfill their commercial promise.

Microelectronics Applications

Progressive miniaturization of materials continues to be the hallmark of the electronics and microelectronics industry. Ongoing size reduction and increasing packaging density of microelectronic components has led to an increasing demand for smaller particle sizes of various materials that are used to manufacture these components.

As a result, researchers are developing ways to use nanotechnology to improve the fabrication and functionality of photonic devices that are used in today's broadband optical networks (for both traditional fiber-optic and novel wireless applications). Growing demand for low-cost, high-speed voice and data transmission components, ultrahigh-density memory media, ultrahigh-frequency electronic devices, integrated devices with ultralow energy consumption, and the overall trend toward miniaturization of photonic components are all factors that are driving research activity in this area.

In microelectronics applications, the use of nanoscaled powders of precious metals provides significant advantages during the production of smaller and smaller electronic circuits with decreasing layer thickness, and such an approach can also reduce the overall consumption of precious metals.

2.6 ANALYTICAL METHODS

The scientific community's ability to reliably observe and manipulate infinitesimally small matter has been a fundamental underpinning in the progress of nanotechnology-related research and development and commercial-scale development efforts. While nanotechnology concepts may have been evolving in the minds and imaginations of forward-thinking scientists and engineers for a long time, they did not emerge meaningfully until the 1980s, in large part because the scientific community had yet to develop adequate instruments, techniques, and methods to visualize and study materials and phenomena in the nanometer range. This changed in the mid-1980s, with the invention of two powerful new microscopy techniques—atomic force microscopy (AFM; which is also referred to as *scanning force microscopy*, or SFM) and scanning tunneling microscopy (STM), both of which permit accurate, atomic-scale measurements.[22]

Both of these relatively new techniques offer a radical departure from conventional types of microscopy, which work by reflecting either light (in the case of optical microscopes) or an electron beam (in the case of electron microscopes) off a surface and onto a lens. While conventional microscopy techniques are sufficient for detailed characterization and visualization of large macromolecular assemblies, no reflective microscope, not even the most powerful one, can provide precise imaging of individual atoms or nanometer-scaled structures.

Atomic Force Microscopy (AFM)

By comparison, the surface topology maps created using AFM can portray in explicit detail the surface features of nanoscaled materials and biological materials such as protein, DNA, and the membrane surfaces of cells.[25] AFM works by scanning an ultrafine ceramic or semiconductor probe over the surface of the material being analyzed. The tip is positioned at the end of a cantilever spring, shaped much like a diving board on a swimming pool. As the ultrasharp, ultrafine probe tip moves over the sample surface in a raster scan—whereby it is systematically and slowly advanced across the sample surface, from left to right, and top to bottom—it measures the forces (at the atomic level) between the probe tip and the sample.

As the probe tip is repelled by or attracted to the surface being studied, the elastic cantilever beam deflects ever so slightly. The magnitude of this infinitesimal deflection is measured as changes in the light intensity from a laser beam that bounces off the cantilever and is recorded by a photodiode sensor (see Figure 2.3). The data from the photodiode sensor are translated into digital form and processed by specialized software. A plot of the laser deflection versus probe tip position on the sample surface provides a highly resolved picture of the peaks and valleys of the sample surface, which constitutes, essentially, an atom-by-atom topographical visualization of the material surface.[25–28]

There are three ways in which the probe tip and the sample can interact—contact mode, tapping mode, and noncontact mode. In contact mode (the most common method of operation of AFM), the probe tip touches the sample continuously as

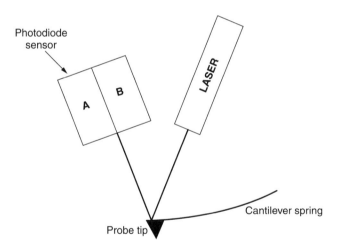

Source: University of Bristol (Bristol, UK)
Ref. [28]

Figure 2.3 Using atomic force microscopy (AFM), an ultrafine, ultrasharp probe tip is positioned on a cantilever spring, above the sample being examined. As the tip moves over the sample surface in a raster scan, atomic level forces between the probe tip and the sample cause the cantilever to deflect ever so slightly. The magnitude of this infinitesimal deflection is measured as changes in the light intensity from a laser beam that bounces off the cantilever and is recorded by a photodiode sensor. The device uses these data to produce an atom-by-atom topographical visualization of the material surface. (*Source*: University of Bristol, Bristol, UK, Ref. 28)

the scanning proceeds. However, one drawback of remaining in contact with the sample is that large lateral forces are exerted on the sample as the tip is "dragged" over the specimen.[28] As a result, this method is used primarily with inert, hard specimens and may not be appropriate for softer biological material, where the continuous contact of the sharp probe tip can potentially "tear" the specimen or indirectly affect its topology (via friction, adhesion, electrostatic forces, and so on).[25]

To circumvent this problem, the newer, intermittent-contact "tapping mode" AFM method was developed. With this approach, the cantilever is oscillated at its resonant frequency to produce a tapping motion. Once positioned over the surface, the probe tip taps the sample surface periodically, for what amounts to a very small fraction of its oscillation period, and—much like the cane of a blind person—is able to decipher and produce a visualization of the surface topography of the sample.

Finally, a third option is noncontact AFM. This provides a mechanism to analyze samples that cannot be immersed in liquid, are especially prone to physical damage by any probe contact, or whose hardness is such that the probe might be easily damaged by contact with it.[25] Using this approach, the probe rides slightly above the surface of the sample, and attractive van der Waals forces that act between the sample and the probe are monitored and translated into a topologic surface of the map.

Noncontact AFM is less sensitive than either of the two contact modes of operation. It is often difficult to preclude any contact between the probe tip and the sample surface at these close ranges, so what often happens is that moisture contamination on the surface of the sample causes a small capillary bridge to form between the probe tip and the surface, causing the tip periodically to "jump to contact," and function in tapping mode after all.[28] Most atomic force microscopes in operation today are capable of operating in each of the modes mentioned above.

One of the most important factors influencing the resolution that may be achieved with AFM is the sharpness of the scanning tip. Today's commercially fabricated tips typically have a radius of curvature of about 5 nm.

Atomic force microscopy can be used for the precise study of nanoscaled materials, and of such surface-related phenomena as abrasion, adhesion, cleaning, corrosion, etching, friction, lubrication, plating, and polishing in the electronics, telecommunications, biological, chemical, automotive, aerospace, and energy industries.

Scanning Tunneling Microscopy (STM)

Scanning tunneling microscopy relies on the so-called tunneling current, which starts to flow when a sharp probe tip approaches a conducting surface at a distance of approximately 1 nm. In an STM assembly, the probe tip is mounted on a piezoelectric tube, which uses a voltage applied at its electrodes to precisely maintain the probe position in the x, y, and z directions (Fig. 2.4). The electronics of the STM

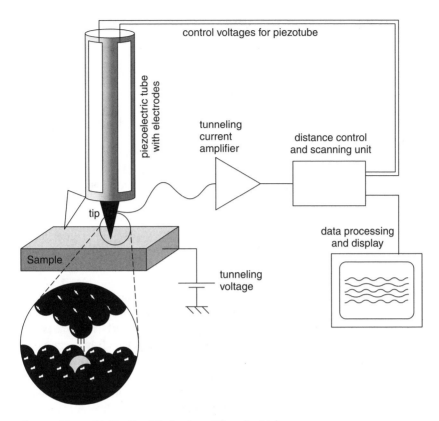

Source: Vienna University of Technology (Wien, Austria)
Ref. [29]

Figure 2.4 Using scanning tunneling microscopy (STM), the probe tip is mounted on a piezoelectric tube, which uses a voltage applied at its electrodes to precisely maintain the probe position in the x, y, and z axes. Analysis of the conductance between the probe tip and the sample is used to produce an atomic-scale image of the surface topography of the sample. (*Source*: Vienna University of Technology, Wien, Austria, Ref. 29)

system control the tip position in such a way that the tunneling current, and hence, the tip–surface distance is kept constant while a small area of the sample surface is scanned, and this movement is recorded and can be displayed as an image of the surface topography. Under ideal circumstances, individual atoms of a surface can be resolved and displayed.[29,30]

With STM, sample–probe conductance is used to image the physical landscape of the sample indirectly. The mapping is accomplished by measuring the conductance between the probe tip and the sample, and by monitoring the probe-to-surface distance that is continuously modulated to maintain a constant tunneling current between the probe and the sample surface.[25]

2.7 HEALTH AND SAFETY ISSUES: ETHICAL, LEGAL, AND SOCIETAL IMPLICATIONS

Today, despite two decades or so of rapidly advancing research, development, and commercialization of nanotechnology-related discoveries, and the considerable attention that such developments has garnered in the scientific and popular press, the general public has shown a mix of both optimism and skepticism toward nanotechnology. Eager supporters predict that—once the proverbial kinks are worked out—the use of nanoscaled materials, devices, and systems will have the potential to revolutionize many aspects of life as we now know it. For such supporters, every new breakthrough inspires confidence and enthusiasm about the potential of this powerful new technological paradigm.

For others, however, the rapid progress of nanotechnology-related developments in recent years brings uncertainty. For instance, early studies on the transport and uptake of nanoscaled materials into living systems suggest that there may be harmful effects on living organisms. This has prompted many to call for further study to identify all of the potential environmental and health risks that might be associated with nanosized materials.[24]

Meanwhile, the uncharted territory that this nascent technology represents also incites tremendous suspicion and opposition among worried critics. In some quarters, the growing public resistance has had increasingly vocal critics, who have the potential to shape the future of nanotechnology research and development. Nanotechnology is, according to one set of observers, "a new frontier with few sheriffs" (Ref. 24, p. 10).

The public fear is fueled in part by an extrapolation of some of the most imaginative claims of today's ardent nanotechnology enthusiasts. For instance, nanoscientists are seen as dabbling with dangerous forces they cannot control. And the case is not helped by the hype that surrounds nanotechnology today, whose most overzealous boosters act like a cult of futurists who foresee [nanotechnology] as a pathway to a technological utopia.[31] When claims about the potential of a technological innovation are so grand, the backlash can be exaggerated, too.

In particular, opponents of nanotechnology voice concern that the pace of research into the production and use of nanoscaled materials and systems is greatly outstripping that of research into the potential short-term and long-term environmental, health, and safety impacts associated with the manipulation of matter at previously unattainable small scales. And, they argue that there has been woefully little meaningful study and public debate related to the societal, legal, and ethical implications associated with this powerful new technology.[3] Critics point out that, compared to the explosion of academic, government, and industry publications covering ongoing nanotechnology-related research and development efforts, there has been very little concomitant increase in the number of publications related to the health and safety, or the ethical, legal, and societal implications of nanotechnology.[31]

Environmental, Health, and Safety Concerns

As for the potential environmental, health, and safety risks associated with nanotechnology, critics note that since the plethora of nanosized materials (such as metals and metal oxides, polymers, ceramics, and carbon derivatives) are not biodegradable, rigorous ongoing investigation is required to determine what their behavior will be in various ecosystems, in terms of absorption or desorption, biotic uptake, and accumulation in plants and animals. Similarly, questions remain about the potential toxicity of nanoscaled materials to humans, in terms of all potential modes of exposure to such ultrafine particles. Short-term and long-term modes of exposure, such as skin absorption, ingestion, and inhalation, among others, must be systematically studied to determine any potential for organ or tissue damage, inflammation, a triggering of autoimmune diseases, and other health-related consequences.[24] Says one researcher, "if such ultrafine particles are to be used in products coming in direct contact with the body, we'd better know what effect these particles have on the cells being exposed" (Ref. 2, p. 27).

In one 2004 study, it was reported that exposure to fullerenes, or buckyballs, (C-60 molecules) can cause extensive brain damage and alter the behavior of genes in the liver cells of juvenile largemouth bass.[32] Meanwhile, in an unrelated 2003 study, researchers found that carbon nanotubes can damage the lungs if inhaled. The animal studies show that the nanotubes are so small that the cells that normally resist other air contaminants are not equipped to handle them.[33]

These reported findings are among several studies that raise questions about the potential health and environmental effects of nanoscaled materials, and while the initial toxicological data are preliminary, they underscore the need to learn more about how buckyballs and other nanoscaled materials are absorbed, how they might damage living organisms, and what levels of exposure create unacceptable hazards. (Note that additional discussion on the environmental, health, and safety aspects of nanoparticles can be found in Chapter 7, Health Risk Assessement.)

Ethical, Legal, and Societal Implications

In addition to the lingering environmental, health, and safety questions associated with nanotechnology, a wide-ranging array of ethical, legal, and societal questions also arise when on considers the wide-ranging nature of the many nanotechnology applications currently under development. Researchers from the University of Toronto Joint Centre for Bioethics suggest that considering the following questions can help to shape the public debate:[12]

Equity Who will benefit from advances in nanotechnology? Will it be just another way for wealthy nations to get richer or should we be taking steps now to ensure that developing countries participate too?

Privacy and security How will personal privacy be protected in an age of invisible microphones, cameras, and tracking devices? Will these technologies increase security or usher in a new era of nanoterrorism? Who will regulate military nanotechnology research?

Environment Where do the new nanomaterials go when they enter the environment and what are their effects?

Human or machine   With many people feeling skeptical about any modification of living systems, how will they view the prospect of implanting artificial materials or machines into humans?

Meanwhile, the societal and ethical implications grow even more thorny when one considers some of the more fantastic or futuristic claims that are put forth by today's vociferous nanotechnology enthusiasts. Says one set of observers: "The revenge of unintended consequences" is a common aspect of any new technology development, whose inventors are not always able to predict how a change in one system might impact or influence changes in others.[24]

These ardent critics fear the unimagined—some might say outlandish—consequences that could result from accidental or unintended outcomes associated with nanotechnology. And, they point out that any intentional abuse of this powerful technology—should terrorists, criminals, dictators, and irresponsible users ever be able to co-opt or misappropriate nanotechnology's unprecedented technological capabilities and use them for evil purposes—could have menacing or diabolical consequences.[34]

For instance, nanotechnology opponents point out that the invisible nature of nanoscaled materials, devices, and systems could lead to a significant invasion of privacy, while sophisticated nanoscaled devices could be used as artificial disease agents, with terrifying consequences. Notes one observer, "what could be more scary than a foe we cannot even see?"[10]

Others fear that the futuristic capabilities envisioned for tomorrow's nanoscaled devices (so-called *nanobots*)—such as microlocomotion, autonomous operation, and self-replication—"could spin out of control with dire consequences for society" (Ref. 24, p. 2).

Nanoscaled manufacturing also raises the possibility of producing horrifically effective weapons. For instance, the ability to usher in an era of stronger, more powerful, extremely compact chemical and biological weapons and delivery systems, which are increasingly deadly and easier to conceal, could inflict serious damage to society. Other envision terrifying devices, such as remote assassination weapons, that would be difficult to detect or avoid.[36]

Fueled by these fears and spurred by doomsday prophecies, nanotechnology's most outspoken critics have called for action ranging from sweeping legislative restrictions to a complete moratorium on all nanotechnology research and development efforts until standard protocols governing the handling of nanoparticles can be implemented.[3,37] Specific steps must be taken to address the growing backlash, before it becomes entrenched. One set of observers notes that, unless the scientific community takes the lead in investigating some of the troubling aspects of nanotechnology, there is a real possibility that the public and various governmental entities will latch onto the fictitious dangers and try to reign in the discipline.[12]

Similarly, others fear that severe restrictions placed on the technology could create prices that are artificially inflated, causing mainstream consumers to explore any means available to circumvent the restrictions. And, if the technology is restricted to the point where people are literally dying for lack of it—which seems possible, considering the number of potential life-saving spinoffs that today's nanotechnology enthusiasts are seeking to develop—then even governments, humanitarians, and some of the technology controllers will have strong motivation to break any imposed restrictions.[38]

The scientists and engineers working on the tiny particles say that to stop the research would be unethical and even immoral because the potential gains from nanotechnology, especially in the fields of medicine and energy management, are so great. They argue that a moratorium on nanotechnology-related research and developmental efforts would rob future generations of the great potential benefits of this powerful technology.[31] Others note that a moratorium is also impractical since such a ban would only serve to push the research underground—as has happened to some extent since the public mobilized its efforts to restrict research and development efforts related to cloning, stem-cell research, and genetically modified foods. This would create a pent-up demand for nanotechnology, which could lead to espionage and the theft of intellectual property, covert parallel research and development efforts, and, eventually, the creation of a black market that is no longer within the reach of regulators.[36,37]

What Can Be Done to Win the Public Trust?

To some, it is not surprising that a technology that promises to make massive changes to people's lives would be viewed with suspicion and perhaps outright fear.[12,39] Perhaps the "willingness of the general public to believe science fiction over fact" represents the scientific community's failure to adequately inform the public about both the anticipated benefits and the potentially unexpected consequences of nanotechnology and to engage in a meaningful discussion of the ethics and impacts of this work (Ref. 39, p. 1).

Nanotechnology supporters argue that the irrational hostility of some should not be allowed to hinder scientific progress as it relates to nanotechnology. Among the ways for society to address the schism between nanotechnology's supporters and opponents, various industry observers offer these recommendations (this list should be considered representative of the prevailing ideas but is certainly not all-inclusive):

- Stimulate ongoing balanced public discussion and debate among the scientific and engineering community, various legislators, stakeholder groups, and other opinion makers to showcase the potential benefits and the potential risks associated with this new technology, and to close the gap between the science and ethics of nanotechnology.[12,39] The public should be given factual elements to help them differentiate science from science fiction.[24]
- Carry out educational campaigns to promote better scientific literacy in the population at large; this will be particularly useful to assuage fears that arise from "illogical extrapolation of the ongoing research," says one observer (Ref 39, p. 1). Mechanisms for enhancing science literacy should be explored at all levels of education, from primary school to secondary school and university and through continuous training of the workforce.[24] (Theodore claims that there is a need to prepare materials that can be disseminated to educational institutions.)

- Stress the need for open reporting of research, with regulatory and peer review in scientific journals. However, some feel that this may be at odds with competitive pressures to protect proprietary discoveries, particularly those that have large potential commercial upside associated with them.[37]
- Develop better models, analytical tools, and methods to study complex, nano-scaled systems so that harmful consequences can be anticipated and avoided.[24]
- Carry out systematic studies to investigate the environmental, health, and safety aspects associated with exposure to nanoscaled materials and systems.
- Have various government take a strong lead in handling risk issues associated with nanotechnology.
- Conduct surveys to properly assess what the real issues are among the concerned populace. Gut feelings and subjective impressions should not be relied upon. It is no good to make assumptions about what the public fears or misunderstands.

As the pace of nanotechnology-related research, development, and commercial-scale progress continues to move forward, the need to analyze the problems, debate the issues, and design and implement workable solutions is urgent. It is increasingly important that scientists and engineers take the lead in thinking about the ethical dimensions of nanotechnology-related efforts, rather than simply reacting to dissenting voices or claiming that these aspects are outside the domain of their expertise and responsibilities. A separate discussion of ethical considerations keying on case studies is provided in Chapter 9.

2.8   FUNDING FUTURE DEVELOPMENTAL EFFORTS

The unbridled promise of nanotechnology-based solutions has motivated academic, industrial, and government researchers throughout the world to investigate nanoscaled materials, devices, and systems with the hope of commercial-scale production and implementation. Today, the private-sector companies that have become involved run the gamut from established, global leaders throughout the chemical process industries, to countless small, entrepreneurial startup companies, many of which have been spun off from targeted research and development efforts at universities.

The governments of many industrialized nations are also keenly interested in nanotechnology. This stems in part from their desire to maintain technological superiority in an important evolving field, and from the military recognition that some applications of nanotechnology could have significant implications for national security.[7]

As one old sage—my esteemed father, a seasoned mechanical engineering professor, text, book author, and consultant—is fond of saying: "Science is what is; engineering is what you'd like it to be." Despite the promise of many of the early nanotechnology-related breakthroughs, the ability to develop cost-effective, commercially and technically viable applications for these laboratory wonders will ultimately be predicated on the research community's ability to bridge the gap—some might say chasm—between the science involved and engineering required, particularly during scaleup.

In the United States, a primary unifying force in government spending in this arena has been the National Nanotechnology Initiative (NNI), a long-term research and development program that coordinates the nanotechnology-related activities of 16 departments of the U.S. government and other independent agencies. During 2000, the Clinton administration established the NNI to focus the nation's efforts and funding initiatives related to the precise manipulation of matter at its most fundamental level.

Since the NNI was established, about 40 other countries have announced priority nanotechnology programs. This signaled an important phase transition in the history of this new burgeoning area of scientific research: What was once perceived as blue sky research of limited interest—or in the view of several groups, as science fiction or even pseudoscience—is now being treated as a key technology of the 21st century.[8]

Since then, the NNI has evolved into law. After initially passing the U.S. House with a vote of 405 to 19 (H.R. 766), and then the U.S. Senate with unanimous support (S. 189) in November 2003, the 21st Century Nanotechnology R&D Act was signed into law by President George W. Bush on December 3, 2003. Such bipartisan support is noteworthy because it signals that nanotechnology progress is seen as having "a higher purpose," beyond party affiliation.[9]

The new law institutionalizes federal nanotechnology R&D and authorizes nearly $3.7 billion in spending for fiscal years 2005–2008. The activities will revolve around—among other things—nanoscale applications for materials, information technology, and microelectromechanical systems (MEMS), as well as life science, pharmaceutical, and biotechnology applications.

The main goals of various funding initiatives will be:[8]

- To extend the frontiers of nanoscale science and engineering through ongoing support for research and development.
- To establish a balanced and flexible infrastructure, including the development of a skilled workforce.
- To address the societal implications of nanotechnology, including actions and anticipatory measures that should be undertaken in the society to help realize the advantages of the new technology in a responsible way.
- To establish a grand coalition of academia, industry, and government to realize the full potential of the new technology. The goal is to foster collaboration among all the key stakeholders, including the scientists and engineers who are developing fundamental nanoscaled materials and systems (universities and national labs), those making commercial nanotechnology products (related to various industries, medicine, and the environment), and various nanotechnology funding sources (federal agencies, state, and local organizations).

The new law focuses on, among other things, nine specific R&D areas that are directly related to applications of nanotechnology. In each case, one or more U.S. government agencies (noted in parentheses) is taking the lead:[8]

- Nanostructured materials by design (National Science Foundation)

- Manufacturing at the nanoscale (National Institutes of Health and National Science Foundation)
- Chemical-biological-radiological-explosive detection and protection (Department of Defense)
- Nanoscale instrumentation (National Institutes of Health and National Science Foundation)
- Nanoelectronics, nanophotonics, and nanomagnetics (Department of Defense)
- Health care, therapeutics, and diagnostics (National Institutes of Health)
- Efficient energy conversion and storage (Department of Energy)
- Microcraft and robotics (National Aeronautical and Space Administration)
- Nanoscale process for environmental improvement (Environmental Protection Agency and National Science Foundation)

These have been identified as having the potential to realize the greatest economic, governmental, and societal impact over the next decade.

2.9 SUMMARY

1. Today, awareness of the science and engineering community's ongoing breakthroughs on nanometer-scaled materials and systems is no longer confined to the academic and industrial research community. In fact, over the past few years, many of the more promising and highly publicized breakthroughs have helped to vault nanotechnology into the public consciousness. Today, along with the business and investment communities, many consumers and the popular press are tuning in when they hear the word *nanotech.*

2. There is an inverse relationship between particle size and surface area, and one of the hallmarks of nanotechnology is the desire to produce and use nanometer-sized particles of various materials in order to take advantage of the remarkable characteristics and performance attributes that many materials exhibit at these infinitesimally small particle sizes.

3. Researchers in pursuit of nanoscaled materials that bring functional advantages to their end-use applications have left no stone unturned. Today, the range of elements and compounds that have been successfully produced and deployed as nanometer-sized particles includes:

- Metals such as iron, copper, gold, aluminum, nickel, and silver
- Oxides of metals such as iron, titanium, zirconium, aluminum, and zinc
- Silica sols, and fumed and colloidal silicas
- Clays such as talc, mica, smectite, asbestos, vermiculite, and montmorillonite
- Carbon compounds, such as fullerenes, nanotubes, and carbon fibers

4. One particular type of nanometer-scaled structure that is generating considerable interest is the carbon nanotube, and today, carbon nanotubes are among the most hotly pursued type of nanoparticle, and in some applications, they are

leading the charge in terms of nanoscaled particles that are making their way into commercial-scale use.

5. Ultimately, the high cost associated with nanoscaled particles and carbon nanotubes will have to be justified in terms of enhanced performance. Nanotechnology researchers are under increasing pressure to devise viable manufacturing methods that both bring down production costs and exploit to the fullest extent possible the broadest array of functional advantages associated with a given type of nanoscaled material.

6. Nanotechnology did not emerge meaningfully until the 1980s, in large part because the scientific community had yet to develop adequate instruments, techniques, and methods to visualize and study materials and phenomena in the nanometer range. This changed in the mid-1980s, with the invention of two powerful new microscopy techniques—atomic force microscopy (AFM; which is also referred to as *scanning force microscopy*, or SFM) and scanning tunneling microscopy (STM), both of which permit accurate, atomic-scale measurements.

7. Opponents of nanotechnology voice concern that the pace of research into the production and use of nanoscaled materials and systems is greatly outstripping that of research into the potential short-term and long-term environmental, health, and safety impacts associated with the manipulation of matter at previously unattainable small scales. And, they argue that there has been woefully little meaningful study and public debate related to the societal, legal, and ethical implications associated with this powerful new technology.

8. Despite the promise of many of the early nanotechnology-related breakthroughs, the ability to develop cost-effective, commercially and technically viable applications for these laboratory wonders will ultimately be predicated on the research community's ability to bridge the gap—some might say chasm—between the science involved and engineering required, particularly during scaleup.

REFERENCES

1. M. N. Rittner, World Market Overview for Nanoparticulate Materials, Presented at the Nanoparticles 2002 Conference, October 29, 2002, New York, N.Y., Business Communications Co. (Norwalk, Conn.).

2. S. A. Shelley, with G. Ondrey, "Nanotechnology—The Sky's the Limit," *Chemical Engineering*, December 2002, pp. 23–27, continues on p. 72.

3. R. Bailey, "The Smaller the Better," *Reason*, December 1, 2003.

4. National Science and Technology Council (NSTC), *Nanotechnology: Shaping the World Atom by Atom*, Brochure for the Public, NSTC, Washington, DC, 1999.

5. C. P. Poole, Jr., and F. J. Owens, *Introduction to Nanotechnology*, Wiley-Interscience, Hoboken, NJ, 2003, pp. 72–74.

6. M. Wilson, K. Kannangara, G. Smith, M. Simmons, and B. Raguse, *Nanotechnology Basic Science and Emerging Technologies*, Chapman & Hall/CRC, Boca Raton, FL, 2002, reprinted 2004.

7. U. Fink, R. E. Davenport, S. L. Bell, and Y. Ishikawa, Nanoscale Chemicals and Materials—An Overview on Technology, Products and Applications, Specialty Chemicals Update Program, SRI Consulting, Menlo Park, CA, December, 2002.

8. M. C. Roco, "Nanoscale Science and Engineering: Unifying and Transforming Tools," *AIChE Journal*, **50**(5), 890–897 (2004).

9. A. Wood and A. Scott, "Nanomaterials," *Chemical Week*, October 16, 2002, pp. 17–21.

10. G. H. Reynolds, "The Science of the Small," *Legal Affairs*, July 1, 2003.

11. K. E. Drexler, "Machine-Phase Nanotechnology," *Scientific American*, September 16, 2001.

12. A. Mnyusiwalla, A. S. Daar, and P. A. Singer, "Mind the Gap: Science and Ethics in Nanotechnology," *Nanotechnology*, **14**, R9–R14 (Feb. 13, 2003).

13. M. J. Sienko and R. A. Plane, *Chemistry*, 4th ed., McGraw-Hill, New York, 1971, pp. 46–49.

14. B. S. Mitchell, *An Introduction to Materials Engineering and Science for Chemical and Materials Engineers*, Wiley-Interscience, Hoboken, NJ, 2004, p. 7.

15. W. Schulz, "Crafting a National Nanotechnology Effort," *Chemical & Engineering News (C & E News)*, **78**(42), Special Report—Government (Oct. 16, 2000).

16. D. Hairston, "Nano-primed," *Chemical Engineering*, April 1999, pp. 39–41.

17. S. A. Shelley, "Carbon Nanotube: A Small-Scale Wonder," *Chemical Engineering*, January 2003, pp. 27–29.

18. A. Hirsch, Nanotubes: Small Tubes with Great Potential, Remarks made at a BASF Aktiengessellschaft event entitled "Journalists and Scientists in Dialogue: Nanotechnology in Chemistry—Experience meets Vision," October 28–29, 2002, Mannheim, Germany.

19. S. Brauer, The Carbon Nanotube Industry, Presented at The Nanoparticles 2002 Conference, October 29, 2002, New York, Business Communications Co. (Norwalk, CT).

20. P. Fairley, "The Start of Something Big," *Chemical Week*, December 12, 2001, pp. 23–26.

21. G. Ondrey, T. Kamiya, and D. Hairston, "Buckyballs Stretch Out in Nanotubes," *Chemical Engineering*, January 2002, pp. 41–43.

22. A. Ghosh, Nanotechnology Will Initiate a Quantum Leap in Manufacturing Efficiency, ARC Insights #2002-49M, October 23, 2002, ARC Advisory Group, Dedham, MA.

23. C. Crabb and G. Parkinson, "The Nanosphere: A Brave New World," *Chemical Engineering*, February 2002, pp. 27–31.

24. M. Roco and R. Tomelini, eds., Nanotechnology: Revolutionary Opportunities and Societal Implications, Summary of Proceedings of the 3rd Joint European Commission—National Science Foundation Workshop on Nanotechnology, Lecce, Italy, January 21–February 1, 2002.

25. C. Wright-Smith and C. M. Smith, "Atomic Force Microscopy: Researchers Map the Topography of Biological Macromolecules," *Scientist*, **15**(2), 23 (2001).

26. Essay: What Is an Atomic Force Microscope? from the web site of the University of Toledo, Toledo, Ohio; *www.che.utoledo.edu/nadarajah/webpages/whatsafm.html*.

27. Essay: The Atomic Force Microscope, from the web site of the Condensed Matter Theory Group, Imperial College, London, UK; *www.sst.ph.ic.ac.uk/photonics/intro/AFM.html*.

28. Essay: Atomic Force Microscopy, from the web site of the H.H. Wills Physics Laboratory, University of Bristol, Bristol, UK; *http://spm.phy.bris.ac.uk/techniques/AFM/*.

29. Essay: The Scanning Tunneling Microscope: What It Is and How It Works; from the web site of Institut für Allgemeine Physik, Vienna University of Technology, Wien, Austria; *www.iap.tuwien.ac.at/www/surface/STM_Gallery/stm_schematic.html.*

30. Essay: Scanning Tunneling Microscopy, from the web site of the University of Hamburg, Hamburg, Germany; *www.physnet.uni-hamburg.de/home/vms/pascal/stm.htm.*

31. G. Pascal Zachary, "Ethics for a Very Small World," *Nanotechnology*, **14**(3), (2003).

32. B. J. Feder, "Study Raises Concerns about Carbon Particles," *New York Times*, March 29, 2004, p. C5.

33. K. Sissell, "Studies Question Health Effects of Nanotechnology," *Chemical Week*, April 7, 2004, p. 12.

34. Essay: The Coming Technological Revolution, from the web site of the Center for Responsible Nanotechnology; *www.crnano.org/magic/htm.*

35. A. Scott and A. Wood, "Big Firms Bulk up in Nanoscale Materials," *Chemical Week*, October 8, 2003.

36. Essay: Dangers of Molecular Manufacturing, from the web site of the Center for Responsible Nanotechnology; *www.crnano.org/dangers/htm.*

37. S. Graham, "Nanotech: It's Not Easy Being Green," *Scientific American*, July 28, 2003.

38. Essay: No Simple Solutions for Nanotechnology Risks, from the web site of the Center for Responsible Nanotechnology; *www.crnano.org/solutions.htm.*

39. C. Sealy, "Science Fact or Fiction?" *Materials Today*, May 2003, p. 1.

40. T. Agres, "Opportunity Awaits Small Thinkers," *Drug Discovery & Development*, February 1, 2004.

CHAPTER 3

AIR ISSUES

3.1 INTRODUCTION

Referring to Chapter 1, there are two classes (or forms) of pollution emitted into the atmosphere: (1) gases and (2) particulates. Liquid aerosols or fine droplets/mists are considered and treated in class 2. Details on the description, design, prediction of performance, and operation/maintenance of traditional control equipment for these pollutants, routinely employed in practice for both control and recovery purposes, are available in the literature.[1-3] An overview on these control devices is presented below in the next section.

Of primary concern in evaluating unknown "air" situations are atmospheric health and safety hazards. Concentrations (or potential concentrations) of vapors, gases, and particulates, low oxygen content, explosive potential, the possibility of radiation exposure, and any potential time variation of these variables, all present immediate atmospheric concerns. Initial resolutions will unquestionably be based on experience, judgment, and professional knowledge.

Extending standard procedures and design of equipment to control emissions from nanotechnology processes is subject to question at this time. The chemical and physical properties of these nanoemissions are not fully known, thus rendering judgments regarding equipment essentially mute. However, information is critical to the advancement and development of new air pollution control technologies. This is especially true of ultrafine particles, that is, at the nanosize, since their mass, volume, and size are extremely small.

Nanotechnology: Environmental Implications and Solutions, L. Theodore and R. G. Kunz
ISBN 0-471-69976-4   Copyright © 2005 John Wiley & Sons, Inc.

Historically, there has been relatively little to a few changes in control equipment in the past century. Unless some new control methodology is developed in the near future, it is safe to assume at this time that similar type control equipment will probably be employed for most, if not all, nanoemissions. In effect, it is anticipated that "new" gaseous and particulate (no matter how small) emissions arising from nanoapplications will probably be "managed" by employing control technology as it exists today. The reduction and control of these discharges will be difficult, but not impossible. Thus, it can be concluded that the air pollution equipment presently in place and available will see probably extensive use in the control of emissions from nanotechnology processes.

The reader should note that, regarding the chemical industries, there has been relatively little change in the past 100 years on how chemicals are processed, separated, purified, and packaged. In fact, the teaching of unit operations[4] to chemical engineers has remained stagnant during this period. This is in stark comparison to advances that have occurred, e.g., in medicine, travel, and computers. Hopefully, some new and innovative changes will soon occur in the environmental management field, particularly as they apply to nanotechnology.

This chapter includes discussions on air pollution control equipment (both particulate and gases), dispersion modeling, stack design, indoor air quality, and monitoring methods. Most of this material has been drawn from the literature.[5,6]

3.2 AIR POLLUTION CONTROL EQUIPMENT

Controlling the emission of pollutants from industrial and domestic sources is important in protecting the quality of air. Air pollutants can exist in the form of particulate matter or as gases. Air cleaning devices have been reducing pollutant emissions from various sources for many years. Originally, air cleaning equipment was used only if the contaminant were highly toxic or had some recovery value. Now, with recent legislation, control technologies have been upgraded and more sources are regulated in order to meet the National Ambient Air Quality Standards (NAAQS). In addition, state and local air pollution agencies have adopted regulations that are in some cases more stringent than federal emission standards.

Equipment used to control particulate emissions are gravity settlers (often referred to as *settling chambers*), mechanical collectors (cyclones), electrostatic precipitators (ESPs), scrubbers (venturi scrubbers), and fabric filters (baghouses). Techniques used to control gaseous emissions are absorption, adsorption, combustion, and condensation. The applicability of a given technique depends on the physical and chemical properties of the pollutant and the exhaust stream. More than one technique may be capable of controlling emissions from a given source. For example, vapors generated from loading gasoline into tank trucks at large bulk terminals are controlled by using any of the above four gaseous control techniques. Most often, however, one control technique is used more frequently than others for a given source–pollutant combination. For example, absorption is commonly used to remove sulfur dioxide (SO_2) from boiler flue gas.

It should be noted that the environmental concerns with nanoparticle emissions repeatedly reported in the literature are probably not justified. In applying the classic work of Cunningham[7] and Einstein[8] to the recovery and/or control of particulates, it has been demonstrated that air pollution control devices for particulates operate at near 100 percent efficiency for particles in the nano- to submicrometer-size range.[1] Both these effects are described below.[2]

At very low values of the Reynolds number, when particles approach sizes comparable to the mean free path of the fluid molecules, the medium can no longer be regarded as continuous, since particles can move between the molecules at a faster rate than predicted by the aerodynamic theories that led to the standard drag coefficients. To allow for this "slip," Cunningham introduced a multiplying correction factor that alters Stokes' law:

$$C = \left(1 + \frac{2A\lambda}{d_p}\right) \tag{3.1}$$

where λ is the mean free path of the fluid molecules; A is the $1.257 + 0.40 \exp(-1.10 d_p/2\lambda)$; and C is the Cunningham correction factor. The modified Stokes' law equation, which is usually referred to as the Stokes–Cunningham equation, is then

$$F_D = 3\pi\mu d_p v / C g_c \tag{3.2}$$

As shown in Table 3.1, the Stokes–Cunningham correction is less than 1 percent for particles larger than 16 μm moving freely in air at ambient conditions. The correction factor however, should definitely be included in the drag-force term when dealing with submicron and nanomicron-sized particles.

As a result of bombardment by the molecules of a fluid medium, suspended particles are subjected to a random motion known as Brownian movement. This effect becomes significant only when the particles are very small and their mass approaches that of the fluid molecules. Einstein[8] showed that the root mean square displacement of a particle, $\bar{l}$, in an interval of time t is given as

$$\bar{l} = (2RTCt/3N_A\pi\mu d_p)^{0.3} \tag{3.3}$$

TABLE 3.1   Stokes–Cunningham Correction Factors for Particles in Ambient Air (70°F, 14.7 psia, $\lambda = 6.53 \times 10^{-6}$ cm)

d_p (μm)	$1 + (2A\lambda/d_p)$
0.01	22.350
0.10	2.870
1.00	1.160
10.00	1.016
16.00	1.010
20.00	1.008

where R is the gas constant, T is the absolute temperature, C is the Stokes–Cunningham correction factor, and N_A is the number of molecules per unit volume. Brownian movement, in general, becomes significant for particles less than about 0.05 μm or 50 nm.

A typical size–efficiency curve for the three high-efficiency control devices, i.e., baghouses, electrostatic precipitators, and venturi scrubbers, is presented in Figure 3.1.[7] This graph clearly demonstrates that submicrometer particles are easier to capture that their submicrometer-plus larger counterparts. Although much of this has been extracted from theoretical physics, there is limited experimental evidence confirming this hypothesis.

The material presented in this section regarding air pollution control equipment is, at best, an overview of each control device. Equipment diagrams and figures, operation and maintenance procedures, and so forth have not been included in this discussion. More details, including calculational predictive and design procedures, are available in the literature.[1-3] Applications of control equipment to specific processes can also be found in the literature.[1-3]

Air Pollution Control Equipment for Particulates

The five major types of particulate air pollution control equipment are:

1. Gravity settlers
2. Cyclones
3. Electrostatic precipitators
4. Venturi scrubbers
5. Baghouses

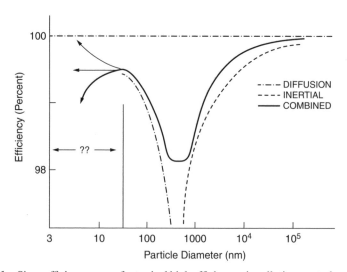

Figure 3.1   Size–efficiency curve for typical high efficiency air pollution control equipment.

Each of these devices is briefly described below. Note that the latter three devices are the equipment of choice when dealing with extremely small (e.g., submicron) particles, and requiring high collection efficiencies.

Gravity Settlers Gravity Settlers, or gravity settling chambers, had been earlier utilized industrially for the removal of solid and liquid waste materials from gaseous streams. Advantages, accounting for their use are simple construction, low initial cost and maintenance, low pressure losses, and simple disposal of waste materials. Gravity settlers are usually constructed in the form of a long, horizontal parallelepipeds with suitable inlet and outlet ports. In its simplest form the settler is an enlargement (large box) in the duct carrying the particle-laden gases: The contaminated gas stream enters at one end, while the cleaned gas exits from the other end. The particles settle toward the collection surface at the bottom of the unit with a velocity at or near their settling velocity. One advantage of this device is that the external force leading to separation is provided free by nature. Its use in industry is generally limited to the removal of large particles, e.g., those larger than 40 μm.

Cyclones Centrifugal separators, commonly referred to as cyclones, are widely used in industry for the removal of solid and liquid particles (or particulates) from gas streams. Typical applications are found in mining and metallurgical operations, the cement and plastics industries, pulp and paper mill operations, chemical and pharmaceutical processes, petroleum production (cat-cracking cyclones) and combustion operations (fly ash collection).

Particulates suspended in a moving gas stream possess inertia and momentum and are acted upon by gravity. Should the gas stream be forced to change direction, these properties can be utilized to promote centrifugal forces to act on the particles. In the conventional unit the entire mass of the gas stream with the entrained particles enter the unit tangentially and is forced into a constrained vortex in the cyclindrical portion of the cyclone. Upon entering the unit, a particle develops an angular velocity. Because of its greater inertia, it tends to move across the gas streamlines in a tangential rather than rotary direction; thus, it attains a net outward radial velocity. By virtue of their rotation with the carrier gas around the axis of the tube (main vortex) and its high density with respect to the gas, the entrained particles are forced toward the wall of the unit. Eventually the particles may reach the outer wall, where they are carried by gravity and assisted by the downward movement of the outer vortex and/or secondary eddies toward the dust collector at the bottom of the unit. The flow vortex is reversed in the lower (conical) portion of the unit, leaving most of the entrained particles behind. The cleaned gas then passes up through the center of the unit (inner vortex) and out of the collector.

Multiple-cyclone collectors (multicones) are high-efficiency devices that consist of a number of small-diameter cyclones operating in parallel with a common gas inlet and outlet. The flow pattern differs from a conventional cyclone in that instead of bringing the gas in at the side to initiate the swirling action, the gas is brought in at the top of the collecting tube and the swirling action is then imparted by a stationary vane positioned in the path of the incoming gas. The diameters of the collecting tubes usually range from 6 to 24 inches. Properly designed units can be constructed and operated with a collection efficiency as high as 90 percent for particulates in the 5- to 10-μm (5000 to 10,000 nm) range. The most serious problems encountered with these systems involve plugging and flow equalization.

Electrostatic Precipitators Electrostatic precipitators (ESPs) are satisfactory devices for removing small particles from moving gas streams at high collection efficiencies. They have been used almost universally in power plants for removing fly ash from the gases prior to discharge.

Two major types of high-voltage ESP configuration currently used are tubular and plate. Tubular precipitators consist of cylindrical collection tubes with discharge electrodes located along the axis of the cylinder. However, the vast majority of ESPs installed are the plate type. Particles are collected on flat parallel collection surfaces spaced 8 to 12 inches apart, with a series of discharge electrodes located along the centerline of the adjacent plates. The gas to be cleaned passes horizontally between the plates (horizontal flow type) or vertically up through the plates (vertical flow type). Collected particles are usually removed by rapping.

Depending on the operating conditions and the required collection efficiency, the gas velocity in an industrial ESP is usually between 2.5 and 8.0 ft/s. A uniform gas distribution is of prime importance for precipitators, and it should be achieved with a minimum expenditure of pressure drop. This is not always easy since gas velocities in the duct ahead of the precipitator may be 30 to 50 ft/s in order to prevent dust buildup. It should be clear that the best operating condition for a precipitator will occur when the velocity distribution is uniform. When significant maldistribution occurs, the higher velocity in one collecting plate area will decrease efficiency more than a lower velocity at another plate area will increase the efficiency of that area.

The maximum voltage at which a given field can be maintained depends on the properties of the gas and the dust being collected. These parameters may vary from one point to another within the precipitator, as well as with time. In order to keep each section working at high efficiency, a high degree of sectionalization is recommended. This means that the many separate power supplies and controls will produce better performance on a precipitator of a given size than if there were only one or two independently controlled sections. This is particularly true if high efficiencies are required.

Venturi Scrubbers Wet scrubbers have found widespread use in cleaning contaminated gas streams because of their ability to effectively remove both particulate and gaseous pollutants. Specifically, wet scrubbing involves a technique of bringing a contaminated gas stream into intimate contact with a liquid. Wet scrubbers include all the various types of gas absorption equipment (to be discussed later). The term "scrubber" is restricted to those systems that utilize a liquid, usually water, to achieve or assist in the removal of particulate matter from a gas stream.

The use of wet scrubbers to remove gaseous pollutants from contaminated streams is considered later in this section.

Another important design consideration for the venturi scrubber (as well as absorbers) is concerned with suppressing the steam plume. Water-scrubber systems removing pollutants from high-temperature processes (i.e., combustion) can generate a supersaturated water vapor that becomes a visible white plume as its leaves the stack. Although not strictly an air pollution problem, such a plume may be objectionable for aesthetic reasons. Regardless, there are several methods to avoid or eliminate the steam plume. The most obvious way is to specify control equipment that does not use water in contact with the high-temperature gas stream, (i.e., ESP, cyclones, or fabric filters). Should this not be possible or practical, a number of suppression methods are available:

1. Mixing with heated and relatively dry air
2. Condensation of moisture by direct contact with water, then mixing with heated ambient air
3. Condensation of moisture by direct contact with water, then reheating the scrubber exhaust gas

Baghouses The basic filtration process may be conducted in many different types of fabric filters in which the physical arrangement of hardware and the method of removing collected material from the filter media will vary. The essential differences may be related, in general, to:

1. Type of fabric
2. Cleaning mechanism
3. Equipment
4. Mode of operation

Gases to be cleaned can be either pushed or pulled through the baghouse. In the pressure system (push through), the gases may enter through the cleanout hopper in the bottom or through the top of the bags. In the suction type (pull through), the dirty gases are forced through the inside of the bag and exit through the outside.

Baghouse collectors are available for either intermittent or continuous operation. Intermittent operation is employed where the operational schedule of the particulate-generating source permits halting the gas cleaning function at periodic intervals (regularly defined by time or by pressure differential) for removal of collected material from the filter media (cleaning). Collectors of this type are utilized primarily for the control of small-volume operations such as grinding and polishing, and for aerosols of a very coarse nature. For most air pollution control installations and major particulate control problems, however, it is desirable to use collectors that allow for continuous operation. This is accomplished by arranging several filter areas in a parallel flow system and cleaning one area at a time according to some preset mode of operation.

Baghouses may also be characterized and identified according to the method used to remove collected material from the bags. Particle removal can be accomplished in a variety of ways, including shaking the bags, blowing a jet of air on the bags, or rapidly expanding the bags by a pulse of compressed air. In general, the various types of bag cleaning methods can be divided into those involving fabric flexing and those involving a reverse flow of clean air. In pressure-jet or pulse-jet cleaning, a momentary burst of compressed air is introduced through a tube or nozzle attached at the top of the bag. A bubble of air flows down the bag, causing the bag walls to collapse behind it.

A wide variety of woven and felted fabrics are used in fabric filters. Clean felted fabrics are more efficient dust collectors than are woven fabrics, but woven materials are capable of giving equal filtration efficiency after a dust layer accumulates on the surface. When a new woven fabric is placed in service, visible penetration of dust within the fabric may occur, however. This normally takes from a few hours to a few days for industrial applications, depending on the dust loadings and the nature of the particles.

Baghouses are constructed as single units or compartmental units. The single unit is generally used on small processes that are not in continuous operation, such as grinding and paint-spraying processes. Compartmental units consist of more than one baghouse compartment and are used in continuously operating processes with large exhaust volumes such as electric melt steel furnaces and industrial boilers. In both cases, the bags are housed in a shell made of rigid metal material.

It is anticipated that baghouses will become the equipment of choice for treating nanoparticles.

Air Pollution Control Equipment for Gaseous Pollutants

As described earlier, the four generic types of gaseous control equipment include:

1. Absorption
2. Adsorption
3. Combustion
4. Condensation

Absorption   Absorption is a mass transfer operation in which a gas contacts a liquid. A contaminant (pollutant exhaust stream) contacts a liquid, and the contaminant diffuses (is transported) from the gas phase into the liquid phase. The absorption rate is enhanced by (1) high diffusion rates, (2) high solubility of the contaminant, (3) large liquid–gas contact area, and (4) good mixing between liquid and gas phases (turbulence).

The liquid most often used for absorption is water because it is inexpensive, is readily available, and can dissolve a number of contaminants. Reagents can be added to the absorbing water to increase the removal efficiency of the system. Certain reagents merely increase the solubility of the contaminant in the water.

Other reagents chemically react with the contaminant after it is absorbed. In reactive scrubbing the absorption rate is much higher, so in some cases a smaller, economical system can be used. However, the reactions can form precipitates that could cause plugging problems in the absorber or in associated equipment.

If a gaseous contaminant is very soluble, almost any of the wet scrubbers will adequately remove this contaminant. However, if the contaminant is of low solubility, the packed tower or the plate tower is more effective. Both of these devices provide long contact time between phases and have relatively low pressure drops. The packed tower, the most common gas absorption device, consists of an empty shell filled with packing. The liquid flows down over the packing, exposing a large film area to the gas flowing up the packing. Plate towers consist of horizontal plates placed inside the tower. Gas passes up through the orifices in these plates while the liquid flows down across the plate, thereby providing the desired contact.

Adsorption   Adsorption is a mass transfer process that involves removing a gaseous contaminant by adhering to the surface of a solid. Adsorption can be classified as physical or chemical. In physical adsorption, a gas molecule adheres to the surface of the solid due to an imbalance of natural forces (electron distribution). In chemisorption, once the gas molecule adheres to the surface, it reacts chemically with it. The major distinction is that physical adsorption is readily reversible whereas chemisorption is not.

All solids physically adsorb gases to some extent. Certain solids, called adsorbents, have a high attraction for specific gases; they also have a large surface area that provides a high capacity for gas capture. By far the most important adsorbent for air pollution control is activated carbon. Because of its unique surface properties, activated carbon will preferentially adsorb hydrocarbon vapors and odorous organic compounds from an airstream. Most other adsorbents (molecular sieves, silica gel, and activated aluminas) will preferentially adsorb water vapor, which may render them useless to remove other contaminants.

For activated carbon, the amount of hydrocarbon vapors that can be adsorbed depends on the physical and chemical characteristics of the vapors, their concentration in the gas stream, system temperature, system pressure, humidity of the gas stream, and the molecular weight of the vapor. Physical adsorption is a reversible process; the adsorbed vapors can be released (desorbed) by increasing the temperature, decreasing the pressure, or using a combination of both. Vapors are normally desorbed by heating the adsorber with steam.

Adsorption can be a very useful removal technique since it is capable of removing very small quantities (a few parts per million) of vapor from an airstream. The vapors are not destroyed; instead, they are stored on the adsorbent surface until they can be removed by desorption. The desorbed vapor stream is normally highly concentrated. It can be condensed and recycled, or burned as an ultimate disposal technique.

The most common adsorption system is the fixed-bed adsorber. This system consists of two or more adsorber beds operating on a timed adsorbing/desorbing cycle. One or more beds are adsorbing vapors, while the other bed(s) is/are being regenerated. If particulate matter or liquid droplets are present in the vapor-laden airstream, this stream is sent to pretreatment to remove them. If the temperature of the inlet vapor stream is high (much above 120°F), cooling may also be required. Since all adsorption processes are exothermic, cooling coils in the carbon bed itself may also be needed to prevent excessive heat buildup. Carbon bed depth is usually limited to a maximum of 4 ft, and the vapor velocity through the adsorber is held below 100 ft/min to prevent an excessive pressure drop.

Combustion Combustion is defined as a rapid, high-temperature gas-phase oxidation. Simply, the contaminant (a carbon–hydrogen substance) is burned with air and converted to carbon dioxide and water vapor. The operation of any combustion source is governed by the three T's of combustion: temperature, turbulence, and time. For complete combustion to occur, each contaminant molecule must come in contact (turbulence) with oxygen at a sufficient temperature, while being maintained at this temperature for an adequate time. These three variables are dependent on each other. For example, if a higher temperature is used, less mixing of the contaminant and combustion air or shorter residence time may be required. If adequate, turbulence cannot be provided, a higher temperature or longer residence time may be employed for complete combustion.

Combustion devices can be categorized as flares, thermal incinerators, or catalytic incinerators. Flares are direct combustion devices used to dispose of small quantities or emergency releases of combustible gases. Flares are normally elevated (from 100 to 400 ft) to protect the surroundings from the heat and flames. Flares are often designed with steam injection at the flare tip. The steam provides sufficient turbulence to ensure complete combustion; this prevents smoking. Flares are also very noisy, which can cause problems for adjacent neighborhoods.

Thermal incinerators are also called afterburners, direct flame incinerators, or thermal oxidizers. These are devices in which the contaminant airstream passes around or through a burner and into a refractory-lined residence chamber where oxidation occurs. To ensure complete combustion of the contaminant, thermal incinerators are designed to operate at a temperature of 700 to 800°C (1300 to 1500°F) and a residence time of 0.3 to 0.5 s. Ideally, as much fuel value as possible is supplied by the waste contaminant stream; this reduces the amount of auxiliary fuel needed to maintain the proper temperature.

In catalytic incineration the contaminant-laden stream is heated and passed through a catalyst bed that promotes the oxidation reaction at a lower temperature. Catalytic incinerators normally operate at 370 to 480°C (700 to 900°F). This reduced temperature represents a continuous fuel savings. However, this may be offset by the cost of the catalyst. The catalyst, which is usually platinum, is coated on a cheaper metal or ceramic support base. The support can be arranged to expose a high surface area, which provides sufficient active sites on which the reaction(s) occur. Catalysts are subject to both physical and chemical deterioration. Halogens and sulfur-containing compounds act as catalyst suppressants and decrease the catalyst usefulness. Certain heavy metal such as mercury, arsenic, phosphorous, lead, and zinc are particularly poisonous.

Condensation Condensation is a process in which the volatile gases are removed from the contaminant stream and changed into a liquid. Condensation is usually achieved by reducing the temperature of a vapor mixture until the partial pressure of the condensable component equals its vapor pressure. Condensation requires low temperatures to liquify most pure contaminant vapors. Condensation is affected by the composition of the contaminant gas stream. The presence of additional gases that do not condense at the same conditions—such as air—hinders condensation.

Condensers are normally used in combination with primary control devices. Condensers can be located upstream of (before) an incinerator, adsorber, or absorber. These condensers reduce the volume of vapors that the more expensive equipment must handle. Therefore, the size and the cost of the primary control device can be reduced. Similarly, condensers can be used to remove water vapors from a process stream with a high moisture content upstream of a control system. A prime example is the use of condensers in rendering plants to remove moisture from the cooker exhaust gas. When used alone, refrigeration is required to achieve the low temperatures required for condensation. Refrigeration units are used for successful control of gasoline vapors at large gasoline dispensing terminals.

Condensers are classified as being either contact condensers or surface condensers. Contact condensers cool the vapor by spraying liquid directly on the vapor stream. These devices resemble a simple spray scrubber. Surface condensers are normally shell-and-tube heat exchangers. Coolant flows through the tubes, while vapor is passed over and condenses on the outside of the tubes. In general, contact condensers are more flexible, simpler, and less expensive than surface condensers. However, surface condensers require much less water and produce nearly 20 times less wastewater that must be treated than do contact condensers. Surface condensers also have an advantage in that they can directly recover valuable contaminant vapors.

Hybrid Systems Hybrid systems are defined as those types of control devices that involve combinations of control mechanisms—for example, fabric filtration combined with electrostatic precipitation. Unfortunately, the term hybrid system has come to mean different things to different people. The two most prevalent definitions employed today for hybrid systems are:

1. Two or more different air pollution control devices connected in series, for example, a baghouse followed by an absorber.
2. An air pollution control system that utilizes two or more collection mechanisms simultaneously to enhance pollution capture, for example, an ionizing wet scrubber, which will be discussed shortly.

The two major hybrid systems found in practice today include ionizing wet scrubbers and dry scrubbers. These are briefly described next.

Ionizing Wet Scrubbers

The ionizing wet scrubber (IWS) is a relatively new development in the technology of the removal of particulate matter from a gas stream. These devices have been incorporated in commercial incineration facilities.[9,10] In the IWS, high-voltage ionization in the charging section places a static electric charge on the particles in the gas stream, which then passes through a crossflow packed-bed scrubber. The packing is normally polypropylene in the form of circular-wound spirals and gearlike wheel configurations, providing a large surface area. Particles with sizes of 3 μm or larger are trapped by inertial impaction within the bed. Smaller charged particles pass close to the surface of either the packing material or a scrubbing water droplet. An opposite charge on that surface is induced by the charged particle, which is then attracted to an ion attached to the surface. All collected particles are eventually washed out of the scrubber. The scrubbing water also can function to absorb gaseous pollutants.

According to Celicote (the IWS vendor), the collection efficiency of the two-stage IWS is greater than that of a baghouse or a conventional ESP for particles in the 0.2 to 0.6 μm (200 to 600 nm) range. For 0.8 μm and above, the ESP is as effective as the IWS. Scrubbing water can include caustic soda or soda ash when needed for efficient absorption of acid gases. Corrosion resistance of the IWS is achieved by fabricating its shell and most internal parts with fiberglass-reinforced plastic (FRP) and thermoplastic materials. Pressure drop through a single-stage IWS is approximately 5 in H_2O (primarily through the wet scrubber section). All internal areas of the ionizer section are periodically deluge-flushed with recycled liquid from the scrubber system.

Dry Scrubbers

The success of fabric filters in removing fine particles from flue gas streams has encouraged the use of combined dry scrubbing/fabric filter systems for the dual purpose of removing both particulates and acid gases simultaneously. Dry scrubbers offer potential advantages over their wet counterparts, especially in the areas of energy savings and capital costs. Furthermore, the dry-scrubbing process design is relatively simple, and the product is a dry waste rather than a wet sludge.

There are two major types of so-called dry scrubber systems: spray drying and dry injection. The first process is often referred to as a wet–dry system. When compared to the conventional wet scrubber, it uses significantly less liquid. The second process has been referred to as a dry–dry system because no liquid scrubbing is involved. The spray-drying system is predominately used in utility and industrial applications.

The method of operation of the spray dryer is relatively simple, requiring only two major items: a spray dryer similar to those used in the chemical food-processing and mineral-preparation industries, and a baghouse or ESP to collect the fly ash and entrained solids. In the spray dryer, the sorbent solution, or slurry, is atomized into the incoming flue gas stream to increase the liquid–gas interface and to promote the mass transfer of the SO_2 from the gas to the slurry droplets where it is absorbed. Simultaneously, the thermal energy of the gas evaporates the water in the droplets to produce a dry powdered mixture of sulfite–sulfate and some unreacted alkali.

Because the flue gas is not saturated and contains no liquid carryover, potentially troublesome mist eliminators are not required. After leaving the spray dryer, the solids-bearing gas passes through a fabric filter (or ESP), where the dry product is collected and where a percentage of unreacted alkali reacts with the SO_2 for further removal. The cleaned gas is then discharged through the fabric-filter plenum to an induced draft (ID) fan and to the stack.

Among the inherent advantages that the spray dryer enjoys over the wet scrubbers are:

1. Lower capital cost
2. Lower draft losses
3. Reduced auxiliary power
4. Reduced water consumption
5. Continuous, two-stage operation, from liquid feed to dry product

The sorbent of choice for most spray-dryer systems is a lime slurry.

Dry-injection processes generally involve pneumatic introduction of a dry, powdery alkaline material, usually a sodium-base sorbent, into the flue gas stream with subsequent fabric filter collection. The injection point in such processes can vary form the boiler-furnace area all the way to the flue gas entrance to the baghouse, depending on operating conditions and design criteria.

Factors in Control Equipment Selection

There are a number of factors to be considered prior to selecting a particular piece of air pollution control hardware.[11] In general, they can be grouped into three categories: environmental, engineering, and economic. These are detailed next.

Environmental

1. Equipment location
2. Available space
3. Ambient conditions
4. Availability of adequate utilities (i.e., power, water, etc.) and ancillary system facilities (i.e., waste treatment and disposal, etc.)
5. Maximum allowable emissions (air pollution regulations)
6. Aesthetic considerations (i.e., visible steam or water vapor plume, impact on scenic vistas, etc.)
7. Contribution of air pollution control system to wastewater and solid waste
8. Contribution of air pollution control system to plant noise levels

Engineering

1. Contaminant characteristics (i.e., physical and chemical properties, concentration, particulate shape and size distribution, etc., and in the case of particulates, chemical reactivity, corrosivity, abrasiveness, toxicity, etc.)

2. Gas stream characteristics (i.e., volume flow rate, temperature, pressure, humidity, composition, viscosity, density, reactivity, combustibility, corrosivity, toxicity, etc.)
3. Design and performance characteristics of the particular control system (i.e., size and weight, fractional efficiency curves, mass transfer and/or contaminant destruction capability, pressure drop, reliability and dependability, turndown capability, power requirements, utility requirements, temperature limitations, maintenance requirements, flexibility toward complying with more stringent air pollution regulations, etc.)

Economic

1. Capital cost (equipment, installation, engineering, etc.)
2. Operating cost (utilities, maintenance, etc.)
3. Expected equipment lifetime and salvage value

Proper selection of a particular system for a specific application can be extremely difficult and complicated. In view of the multitude of complex and often ambiguous pollution regulations, it is in the best interest of the prospective user to work closely with regulatory officials as early in the process as possible. Finally, previous experience on a similar application cannot be overemphasized.

Comparing Control Equipment Alternatives

The final choice in equipment selection is usually dictated by that equipment capable of achieving compliance with the regulatory codes at the lowest uniform annual cost (amortized capital investment plus operation and maintenance costs). The reader is referred to the literature for details on the general subject of economics.[1,2] In order to compare specific control equipment alternatives, knowledge of the particular application and site is essential. A preliminary screening, however, may be performed by reviewing the advantages and disadvantages of each type of air pollution control equipment. For example, if water or a waste-treatment stream is not available at the site, this may preclude use of a wet scrubber system and instead focus on particulate removal by dry systems, such as cyclones or baghouses and/or ESP. If auxiliary fuel is unavailable on a continuous basis, it may not be possible to combust organic pollutant vapors in an incineration system. If the particle size distribution in the gas stream is relatively fine, cyclone collectors would probably not be considered. If the pollutant vapors can be reused in the process, control efforts may be directed to adsorption systems. There are many more situations where the knowledge of the capabilities of the various control options, combined with common sense will simplify the selection procedure. General advantages and disadvantages of the most popular types of air pollution control equipment for gases and particulates are too detailed to present here but are available in the literature.[1–3,11]

3.3 ATMOSPHERIC DISPERSION MODELING

Because of the (small) size of nanoparticles, their dispersion in the atmosphere can almost certainly be treated by the procedures already in place for gases. Theodore[12] has developed equations that can be applied to particulate deposition, as opposed to gases. But since nanoparticles approach sizes comparable to the mean free path of fluid molecules, and effectively behave as gases, the traditional approaches presented below can be assumed to apply.

This section focuses on some of the practical considerations of dispersion in the atmosphere. Both continuous and instantaneous discharges are of concern to individuals involved with environmental management. However, the bulk of the material presented here is for continuous emissions from point sources—for example, a stack. This has traditionally been an area of much concern in the air pollution field because stacks have long been one of the more common industrial methods of "disposing of" waste gases. The concentrations that humans, plants, animals, and structures are exposed to at ground level can be reduced significantly by emitting the waste gases from a process at great heights. This permits the pollutants to be dispersed over a much larger area and will be referred to as *control by dilution.* Although tall stacks may be effective in lowering the ground-level concentration of pollutants, they still do not in themselves reduce the amount of pollutants released into the atmosphere. However, in certain situations, it can be the most practical and economical way of dealing with an air pollution problem.

Air quality models describe the fate of airborne gases and particles. As these pollutants travel over their pathways, physical and chemical reactions may occur. The categories of mechanisms are nonreactive, reactive (photochemical and nonphotochemical), gas-to-particle conversions, gas/particle processes, and particle/particle processes. In addition, the gases and particle may be radioactive, in which case the models must contain some provisions for accounting for radioactive decay and the production of subsequent radioactive elements. Furthermore, for an adequate assessment of the significance of the air quality impact of a source, background concentrations must be considered. Background air quality relevant to a given source includes those pollutant concentrations due to natural sources and also distant, unidentified man-made sources. For example, it is commonly assumed that the annual mean background concentration of particulate matter is 30 to 40 $\mu g/m^3$ over much of the eastern United States. Typically, air quality data are used to establish background concentrations in the vicinity of the source under consideration.

A four-step procedure is recommended in performing dispersion calculations, particularly for health effect studies:

1. Estimate the rate, duration, and location of the release into the environment.
2. Select the best available model to perform the calculations.
3. Perform the calculations and generate "downstream" concentrations, including isopleths (i.e., lines of constant concentration) resulting from the source emission(s).

4. Determine what effect, if any, the resulting discharge has on the environment, including humans (particularly, and often the only concern), animals, vegetation, and materials of construction. These calculations often include estimates of the so-called vulnerability zones—that is, regions that may be adversely affected because of the emissions.

The U.S. EPA's *Guideline on Air Quality Models*[13,14] specifically addresses atmospheric dispersion calculations. This guideline is used by the EPA, by the states, and by private industry in the review and preparation of Prevention of Significant Deterioration (PSD) permits and State Implementation Plans (SIP) revisions. The guideline serves as a means of maintaining consistency in air quality analyses. On September 8, 1986 (51 FR 32180), the EPA proposed to include four changes to this guideline: (i) addition of a specific version of the Rough Terrain Diffusion Model (RTDM) as a screening model, (ii) modification of the downwash algorithm in the Industrial Source Complex (ISC) model, (iii) addition of the Offshore and Coastal Dispersion (OCD) model to the EPA's list of preferred models, and (iv) addition of the AVACTA 11 model as an alternative model in the guideline. Other minor modifications have been introduced since then.

Perhaps the most important consideration in dispersion applications in first to determine the acceptable ground-level concentration of the waste pollutant(s). The topography of the area must also be considered. Awareness of the meteorological conditions prevalent in the area, such as the prevailing winds, humidity, and rainfall, is also essential. Finally, an accurate knowledge of the constituents of the waste gas and its physical and chemical properties is paramount.[15]

Atmospheric contamination arises primarily from the exhausts generated by industrial plants, power plants, refuse disposal plants, domestic activities, commercial heating, and transportation. These pollutants—which are in the form of particulates, smog, odors, and other—arise mostly from combustion processes and also contain varying amounts of undesirable gases such as oxides of sulfur, oxides of nitrogen, hydrocarbons, and carbon monoxide. The expanding needs of society for more energy and advanced transportation technology, coupled with the rapid growth of urban areas, has led to ever-increasing amounts and concentrations of pollutants in the atmosphere.

Just as a river or stream is able to absorb a certain amount of pollution without the production of undesirable conditions, the atmosphere can also absorb a certain amount of contamination without "bad" effects. Dilution of air contaminants in the atmosphere is also of prime importance in the prevention of undesirable levels of pollution. In addition to dilution, several self-purification mechanisms are at work in the atmosphere, such as sedimentation of particulate matter, washing action of precipitation photochemical reactions, and absorption by vegetation and soil. The self-purification of a stream is primarily the result of biological action and dilution.

Nature of Dispersion

The release of pollutants into the atmosphere is a traditional technique for disposing of them. Although gaseous emissions may be controlled by various sorption processes (or by combustion) and particulates (either solid or aerosol) by mechanical collection, filtration, electrostatic precipitators, or wet scrubbers, the effluent from the control device must still be dispersed into the atmosphere. Fortunately, one of the important properties of the atmosphere is its ability to disperse such streams of gaseous pollutants. Of course, the atmosphere's ability to disperse such streams is not perfect and varies from quite good to quite poor, depending on the local meteorological and geographical conditions. Therefore, the ability to model atmospheric dispersion and to predict pollutant concentrations from a source are important parts of air pollution science and engineering.

A continuous stream of pollutants released into a steady wind in an open atmosphere will first rise (usually), then bend over and travel with the mean wind, which will dilute the pollutants and carry them away from the source. This plume of pollutants will also spread out or disperse both in the horizontal and vertical directions from its centerline. In doing so, the concentration of the gaseous pollutant is contained within a larger volume. This natural process of high concentration spreading out to lower concentration is the process of dispersion. Atmospheric dispersion is primarily accomplished by the wind movement of pollutants, but the character of the source of pollution requires that this action of the wind be taken into account in different ways. A pictorial representation of this phenomenon can be found later in this section.

The dilution of air contaminants is also a direct result of atmospheric turbulence and molecular diffusion. However, the rate of turbulent mixing is so many thousand times greater than the rate of molecular diffusion that the latter effect can be neglected in the atmospheric diffusion analysis. Atmospheric turbulence and, hence, atmospheric diffusion vary widely with the weather conditions and topography.

Meteorological Concerns

The atmosphere has been labeled the dumping ground for air pollution. Industrial society can be thankful that the atmosphere cleanses itself (up to a point) by natural phenomena. Atmospheric dilution occurs when the wind moves because of wind circulation or atmospheric turbulence caused by local sun intensity. As described earlier, pollution is removed from the atmosphere by precipitation and by other reactions (both physical and chemical) as well as by gravitational fallout.

The atmosphere is the medium in which air pollution is carried away from its source and diffuses. Meteorological factors have a considerable influence over the frequency, length of time, and concentrations of effluents to which the general public may be exposed. The variables that affect the severity of an air pollution problem at a given time and location are wind speed and direction, insolation (amount of sunlight), lapse rate (temperature variation with height), mixing depth, and precipitation. Unceasing change is the predominant characteristic of the atmosphere; for example temperatures and winds vary widely with latitude, season, and surrounding topography.[16,17]

Atmospheric dispersion depends primarily on horizontal and vertical transport. Horizontal transport depends on the turbulent structure of the wind field. As the wind velocity increases, the degree of dispersion increases with a corresponding decrease in the ground-level concentration of the contaminant at the receptor site. This is a result of the emissions being mixed into a larger volume of air. The dilute effluents may, depending on the wind direction, be carried out into essentially unoccupied terrain away from any receptors. Under different atmospheric conditions, the wind may funnel the diluted effluent down a river valley or between mountain ranges. If an inversion (temperature increases with height) is present aloft that would prevent vertical transport, the pollutant concentration may build up continually.

One can define atmospheric turbulence as those vertical and horizontal convection currents or eddies that mix process effluents with the surrounding air. Several generalizations can be made regarding the effect of atmospheric turbulence on the effluent dispersion. Turbulence increases with increasing wind speed and causes a corresponding increase in horizontal dispersion. Mechanical turbulence is caused by changes in wind speed and wind shear at different altitudes. Either of these conditions can lead to significant changes in concentration of the effluent at different elevations.

Topography can also have a considerable influence on the horizontal transport and thus pollutant dispersion. The degree of horizontal mixing can be influenced by sea and land breezes. It can also be influenced by man-made and natural terrain features such as mountains, valleys, or even a small ridge or a row of hills. Low spots in the terrain or natural bowls can act as sites where pollutants tend to settle and accumulate because of the lack of horizontal transport in the land depressions. Other topographical features that can affect horizontal transport are city canyons and isolated buildings. City canyons occur when the buildings on both sides of a street are fairly close together and are relatively tall. Such situations can cause funneling of emissions from one location to another. Isolated buildings or the presence of a high-rise building in a relatively low area can cause redirection of dispersion patterns and route emissions into an area in which many receptors live.

Dispersion of air contaminants is strongly dependent on the local meteorology of the atmosphere into which the pollutants are emitted. The mathematical formulation for the design of pollutant dispersal is usually associated with the open-ground terrain free of obstructions. Either natural or manmade obstructions alter the atmospheric circulation and with it the dispersion of pollutants. Thus of particular concern are the effects of mountain valley terrain, hills, lakes, shorelines, and buildings.

Plume Rise

A plume of hot gases emitted vertically has both a momentum and a buoyancy. As the plume moves away from the stack, it quickly loses its vertical momentum (owing to drag by and entrainment of the surrounding air). As the vertical momentum declines, the plume bends over in the direction of the mean wind. However, quite often the effect of buoyancy is still significant, and the plume continues to rise for a long time after bending over. The buoyancy term is due to the less-than-atmospheric density of the stack gases and may be temperature or composition induced. In either case, as the plume spreads out in the air (all the time mixing with the surrounding air), it becomes diluted by the air.

Modeling the rise of the plume of gases emitted from a stack into a horizontal wind is a complex mathematical problem. Plume rise depends not only on such stack gas parameters as temperature, molecular weight, and exit velocity, but also on such atmospheric parameters as wind speed, ambient temperature, and stability conditions.[18]

The behavior of plumes emitted from any stack depends on localized air stability. Effluents from tall stacks are often injected at an effective height of several hundred feet to several thousand feet above the ground because of the added effects of buoyancy and velocity on the plume rise. Other factors affecting the plume behavior are the diurnal variations in the atmospheric stability and the long-term variations that occur with changing seasons.[19]

Effective Stack Height

Reliance on atmospheric dispersion as a means of reducing ground-level concentrations is not foolproof. Inversions can occur with a rapid increase in ground-level pollutant concentrations. One solution to such situations is the tall stack concept. The goal is quite simple: Inject the effluent above any normally expected inversion layer. This approach is used for exceptionally difficult or expensive treatment situations because tall stacks are quite expensive. To be effective, they must reach above the inversion layer so as to avoid local plume fallout. The stack itself does not have to penetrate the inversion layer if the emissions have adequate buoyancy and velocity. In such cases, the effective stack height will be considerably greater than the actual stack height.[17] The effective stack height (equivalent to the effective height of emission) is usually considered the sum of the actual stack height, the plume rise due to velocity (momentum) of the issuing gases, and the buoyancy rise, which is a function of the temperature of the gases being emitted and the atmospheric conditions.

The effective stack height depends on a number of factors. The emission factors include the gas flow rate, the temperature of the effluent at the top of the stack, and the diameter of the stack opening. The meteorological factors influencing plume rise and wind speed, air temperature, shear of the wind speed with height, and the atmospheric stability. No theory on plume rise presently takes into account all these variables, and it appears that the number of equations for calculating plume rise varies inversely with one's understanding of the process involved. Even if such a theory were available, measurements of all of the parameters would seldom be available. Most of the equations that have been formulated for computing the effective height of an emission stack are semiempirical in nature. When considering any of these plume rise equations, it is important to evaluate each in terms of assumptions made and the circumstances existing at the time the particular correlation was formulated. Depending on the circumstances, some equation may definitely be more applicable than others.

The effective height of an emission rarely corresponds to the physical height of the source or the stack. If the plume is caught in the turbulent wake of the stack or of buildings in the vicinity of the source of stack, the effluent will be mixed rapidly downward toward the ground. If the plume is emitted free of these turbulent zones, a number of emission factors and meteorological factors will influence the rise of the plume. The influence of mechanical turbulence around a building or stack can also significantly alter the effective stack height. This is especially true with high winds when the beneficial effect of the high stack gas velocity is at a minimum and the plume is emitted nearly horizontally.

Details regarding a host of plume rise models and calculation procedures are available in the literature.[20–22]

Atmospheric Dispersion Models[22]

The initial use of dispersion modeling occurred in military applications during World War I. Both sides of the conflict made extensive use of poison gases as a weapon of war. The British organized the Chemical Defense Research Establishment at Porton Downs during the war. Research at this institute dominated the field of dispersion modeling for more than 30 years through the end of World War II.

With the advent of the potential use of nuclear energy to generate electrical power, the United States Atomic Energy Commission invested heavily in understanding the nature of atmospheric transport and diffusion processes. Since about 1950 the United States has dominated research in the field. The U.S. Army and Air Force have also studied atmospheric processes to understand the potential effects of chemical and biological weapons.

The Pasquill–Gifford model has been the basis of many models developed and accepted today.[23–25] This model has served as an atmospheric dispersion formula from which the path downwind of emissions can be estimated after obtaining the effective stack height. There are many other dispersion equations (models) presently available, most of them semiempirical in nature. Calculation details regarding the use of this equation are available in the literature.[20,21] A pictorial representation of both the effective stack height and subsequent dispersion is provided in Figure 3.2.[20,21]

The problem of having several models is that various different predictions can be obtained. In order to establish some reference, a standard was sought by the government. The *Guideline on Air Quality Models*[26] was the end result. However, in industry today, the ISC models are the preferred models for permitting and therefore are used in many applications involving normal or "after the fact" releases, depending on which regulatory agency must be answered to.

The ISC model is available as part of UNAMAP (Version 6). The computer code is available on magnetic tape from the National Technical Information Service (NTIS) or via modem through their Bulletin Board Services (BBS). It can account for the following: settling and dry deposition of particulates; downwash; area,

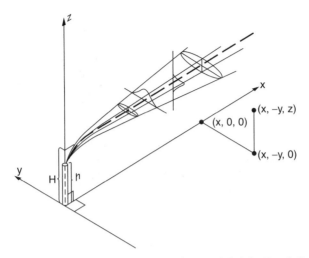

Figure 3.2 Coordinate system showing the effective stack height H and dispersion in the horizontal and vertical directions.

line, and volume sources; plume rise as a function of downwind distance; separation of point sources; and limited terrain adjustment.

In order to prepare for and prevent the worst, screening models are applied to simulate the worse-case scenario. One difference between the screening models and the refined models mentioned earlier is that certain variables are set that are estimated to be values to give the worst conditions. In order to use these screening models, the parameters of the model must be fully grasped.

In short, these models are necessary to predict somewhat the behavior of the atmospheric dispersions. These predictions may not necessarily be correct; in fact, they are rarely completely accurate and may be off by several orders of magnitude.[26] In order to chose the most effective model for the behavior of an emission, the source and the models have to be well understood.

3.4 STACK DESIGN[15,21]

As experience in designing stacks has accumulated over the years, several guidelines have evolved:

1. Stack heights should be at least 2.5 times the height of any surrounding buildings or obstacles so that significant turbulence is not introduced by these factors.
2. The stack gas exit velocity should be greater than 60 ft/s so that stack gases will escape the turbulent wake of the stack. In many cases, it is good practice to have the gas exit velocity on the order of 90 or 100 ft/s.

3. A stack located on a building should be set in a position that will assure that the exhaust escapes the wakes of nearby structures.

4. Gases from the stacks with diameters less than 5 ft and heights less than 200 ft will reach the ground early, and the ground-level concentration will be excessive. In this case, the plume becomes unpredictable.

5. The maximum ground-level concentration of stack gases subjected to atmospheric dispersion occurs about 5 to 10 effective stack heights downwind from the point of emission.

6. When stack gases are subjected to atmospheric diffusion and building turbulence is not a factor, ground-level concentrations on the order of 0.001 to 1 percent of the stack concentration are possible for a properly designed stack.

7. Ground-level concentrations can be reduced by the use of higher stacks. The ground-level concentration varies inversely as the square of the effective stack height.

8. Average concentrations of a contaminant downwind from a single stack are directly proportional to the discharge rate. An increase in discharge rate by a given factor increases ground-level concentrations at all points by the same factor.

9. In general, increasing the dilution of stack gases by the addition of excess air in the stack does not effect ground-level concentrations appreciably. Practical stack dilutions are usually insignificant in comparison to the later atmospheric dilution by plume diffusion. Addition of dilution air will increase the effective stack height, however, by increasing the stack exit velocity. This effect may be important at low wind speeds. On the other hand, if the stack temperature is decreased appreciably by the dilution, the effective stack height may be reduced. Stack dilution will have an appreciable effect on the concentration in the plume close to the stack.

These nine guidelines represent the basic design elements of a pollution control system. An engineering approach suggests that each element be evaluated independently and as part of the whole control system. However, the engineering design and evaluation must be an integrated part of the environmental control program.

3.5 INDOOR AIR QUALITY[5,6]

Concern with air quality has increased in recent years. Potential nanoemissions in the future will only add to this concern, which is why this section has been included in this chapter. For certain, nanotechnology environmental impacts will arise not only at industrial sites but also at the domestic level.

Indoor air pollution is rapidly becoming a major worldwide health issue. Although research efforts are still underway to better define the nature and extent of the health implications for the general population, recent studies have shown significant amounts of harmful pollutants in the indoor environment. Nanotechnology emissions will unquestionably add to this concern. The serious concern over pollutants in indoor air is due largely to the fact that indoor pollutants are not easily dispersed or diluted as are pollutants outdoors. Thus, indoor pollutant levels are frequently higher than outdoors, particularly where buildings are tightly constructed to save energy. In some cases, these indoor levels exceed the (EPA) standards already established for outdoors. Research by the EPA in this area, called the Total Exposure Assessment Methodology (TEAM) studies, has documented the fact that levels indoors for some pollutants may exceed outdoor levels by 200 to 500 percent.[27]

Since most people spend 90 percent of their time indoors, many may be exposed to unhealthy concentrations of pollutants. People most susceptible to the risks of pollution—the aged, the ill, and the very young—spend nearly all of their time indoors. These indoor environments include such places as homes, offices, hotels, stores, restaurants, warehouses, factories, government buildings, and even vehicles. In these environments, people are exposed to pollutants emanating from a wide array of sources.

Some common indoor present-day air contaminants are:

1. Radon
2. Formaldehyde
3. Volatile organic compounds (VOCs)
4. Combustion gases
5. Particulates
6. Biological contaminants

In addition to air contaminants, other factors need to be observed in indoor air quality (IAQ) monitoring programs to understand fully the significance of contaminant measurements. Important factors to be considered in IAQ studies include:

1. Air exchange rates
2. Building design and ventilation characteristics
3. Indoor contaminant sources and sinks
4. Air movement and mixing
5. Temperature
6. Relative humidity
7. Outdoor contaminant concentrations and meteorological conditions

Designers, builders, and homeowners must make crucial decisions about the kinds and potential levels of existing indoor air pollutants at proposed house sites. Building structure design, construction, operation, and household furnishings, all rely on specific design parameters being set to handle the reduction of these pollutants at their sources.

The health effects associated with IAQ can be either short- or long-term (see Chapters 7 and 8). Immediate effects experienced after a single exposure or repeated exposures include irritation of the eyes, nose, and throat; headaches; dizziness; and fatigue. These short-term effects are usually treatable by some means, oftentimes by eliminating the person's exposure to the source of pollution.

The likelihood of an individual's developing immediate reactions to indoor air pollutants depends on several factors, including age and preexisting medical conditions. Also, individual sensitivity to a reactant varies tremendously. Some people can become sensitized to biological pollutants after repeated exposures, and it appears that some people can become sensitized to chemical pollutants as well. Other health effects may show up either years after exposure has occurred or only after long or repeated periods of exposure. These effects range from impairment of the nervous system to cancer, emphysema and other respiratory diseases, and heart disease. These can be severely debilitating or fatal. Certain symptoms are similar to those of other viral diseases and difficult to determine if it is a result of indoor air pollution. Therefore, special attention should be paid to the time and place symptoms occur.

Further research is needed to better understand which health effects can arise after exposure to the average pollutant concentrations found in homes. These can arise from the higher concentrations that occur for short periods of time. Yet, both the amount of pollutant, called the dose, and the length of time of exposure are important in assessing health effects. (See Chapter 7.) The effects of simultaneous exposure to several pollutants are even more uncertain. Indoor air quality can be severely debilitating or even fatal. Indoor air pollutants of special concern are described below in separate sections.

It is not possible to provide estimates of typical mixtures of pollutants found in residences. This is because the levels of pollutants found in homes vary significantly depending on location, use of combustion devices, existing building materials, and use of certain household products. Also, emission of pollutants into the indoor air may be sporadic, as in the case of aerosols or organic vapors that are released during specific household activities or when woodstoves or fireplaces are in use. Another important consideration regarding indoor pollutant concentrations is the interaction among pollutants. Pollutants often tend to attach themselves to airborne particles that get caught more easily in the lungs. In addition, certain organic compounds released indoors could react with each other to form highly toxic substances.

The data provided in this section consists of approximate ranges of indoor pollutants based on studies conducted around the United States. These provide an overview of several major pollutants that have been measured in residences at levels that may cause health problems ranging from minor irritations or allergies to potentially debilitating diseases.

Radon

Radon is a unique environmental problem because it occurs naturally. Radon results from the radioactive decay sequence of radium 226, a long-lived precursor to radon.

The isotope of most concern radon-222, has a half-life (time for half to disappear) of 3.8 days. Radon itself decays and produces a series of short-lived decay products called radon progeny or daughters. Polonium-218 and polonium-214 are the most harmful because they emit charged alpha particles more dangerous than x-rays or gamma rays. They also tend to adhere to other particles (attachment) or surfaces (plate out). These larger particles are more susceptible to becoming lodged in the lungs when inhaled and cause irreparable damage to surrounding lung tissue (which may lead to lung cancer).

Radon is a colorless, odorless gas that is found everywhere at very low levels. Radon becomes a cause for concern when it is trapped in buildings and concentrations build up. In contrast, indoor air has approximately 2 to 10 times higher concentrations of radon than outdoor air.

Primary sources of radon are from soil, well water supplies, and building materials. Most indoor radon comes from the rock and soil around a building and enters structures through cracks or openings in the foundation or basement. High concentrations of radon are also found in wells, where storage, or hold-up time, is too short to allow time for radon decay. Building materials, such as phosphate slag (a component of concrete used in an estimated 74,000 U.S. homes) has been found to be high in radium content.[28] Studies have shown concrete to have the highest radon content when compared to all other building materials, with wood having the least.

It is becoming increasingly apparent that local geological factors play a dominant role in determining the distribution of indoor radon concentrations in a given area. To date, no indoor radon standard has been promulgated for all residential housing in the United States. However, various organizations have proposed ranges of guidelines and standards.

Data taken from various states suggest an average indoor radon-222 concentration of 1.5 pCi/L (picocuries per liter, a radiation concentration term), and approximately 1 million homes have concentrations exceeding 8 pCi/L.[28] One curie is equal to a quantity of a material with 37 billion radioactive decays per second. One trillionth of a curie is a pCi. Assuming residents in these home spend close to 80 percent of their time indoors, their radon exposure would come close to the level for recommended, remedial action set by the U.S. National Council on Radiation Protection and Measurements. The EPA believes that up to 8 million homes may have radon levels exceeding 4 pCi/L air, the level at which the EPA recommends corrective action. In comparison, the maximum level of radon set for miners by the U.S. Mine Safety and Health Administration is as high as 16 pCi/L.

Radon may be the leading cause of lung cancer among nonsmokers. Several radiation protection groups have approximated the number of annual lung cancer deaths attributable to indoor radon. The EPA estimates that radon may be responsible for 5000 to 20,000 lung cancer deaths among nonsmokers. Also, scientific evidence indicates that smoking, coupled with the effects of exposure to radon, increases the risk of cancer by 10 times that of nonsmokers.[27]

A variety of measures can be employed to help control indoor concentrations of radon and/or radon progeny. Mitigation methods for existing homes include placing barriers between the source material and living space itself as well as using several other techniques, such as:

1. Covering exposed soil inside a structure with cement
2. Eliminating and sealing any cracks in the floors or walls
3. Adding traps to underfloor drains
4. Filling concrete block walls

Soil ventilation prevents radon from entering the home by drawing the gas away before it can enter the home. Pipes are inserted into the stone aggregate under basement floors or onto the hollow portion of concrete walls to ventilate radon gas accumulating in these locations. Pipes can also be attached to underground drain tile systems drawing the radon gas away from the house. Fans are often attached to the system to improve ventilation. Crawl space ventilation is also generally regarded as an effective and cheap method of source reduction. This allows for exchange of outdoor air by placing a number of openings in the crawl space walls.

Home ventilation involves increasing a home's air exchange rate—the rate at which incoming outdoor air completely replaces indoor air—either naturally (by opening windows or vents) or mechanically (through the use of fans). This method works best when applied to houses with low initial exchange rates. However, when indoor air pressure is reduced, pressure-driven radon entry is induced, increasing levels in the home instead of decreasing them. The benefits of increased ventilation can be achieved without raising radon exposure by opening windows evenly on all sides of the home.

Mechanical devices can also be used to help rid indoor air of radon progeny. Air cleaning systems use high-efficiency filters or electronic devices to collect dust and other airborne particles, some with radon products attached to them. These devices decrease the concentration of airborne particles, but do not decrease the concentration of smaller unattached radon decay products, which can result in a higher radiation dose when inhaled.

Formaldehyde

Formaldehyde is a colorless, water-soluble gas that has a pungent, irritating odor noticeable at less than 1 ppm. It is an inexpensive chemical with excellent bonding characteristics that is produced in high volume throughout the world. A major use is in the fabrication of urea–formaldehyde (UF) resins used primarily as adhesives when making plywood, particleboard, and fiberboard. Formaldehyde is also a component of UF foam insulation, injected into sidewalls primarily during the 1970s. Many common household cleaning agents contain formaldehyde. Other minor sources in the residential environment include cigarette smoke and other combustion sources such as gas stoves, woodstoves, and unvented gas space heaters. Formaldehyde can also be found in paper products such as facial tissues, paper towels, and grocery bags, as well as stiffeners and wrinkle resisters.[29]

Although information regarding emission rates is limited, in general, the rate of formaldehyde release has been shown to increase with temperature, wood moisture content, humidity, and with decreased formaldehyde concentration in the air.

As indicated above, UF foam was used as a thermal insulation in the sidewalls of many buildings. It was injected directly into wall cavities through small holes that were then sealed. When improperly installed, UF foam emits significant amounts of formaldehyde. The Consumer Product Safety Commission (CPSC) measured values as high as 4 ppm and imposed a nationwide ban on UF foam, but it was later overturned.

The superior bonding properties and low cost of formaldehyde polymers make them the resins of choice for the production of building materials. Plywood is composed of several thin sheets of wood glued together with UF resin. Particleboard (compressed wood shavings mixed with UF resins at high temperatures) can emit formaldehyde continuously for a long time, from several months to several years. Medium density fiberboard was found to be the highest emitter of formaldehyde.

Indoor monitoring data on formaldehyde concentrations are variable because of the wide range of products that may be present in the home. However, elevated levels are more likely to be found in mobile homes and new homes with pressed-wood construction materials. Indoor concentrations also vary with home age since emissions decrease as products containing formaldehyde age and cure. In general, indoor formaldehyde concentrations exceed levels found outdoors.

Although individual sensitivity to formaldehyde varies, about 10 to 20 percent of the population appears to be highly sensitive to even low concentrations. Its principal effect is irritation of the eyes, nose, and throat, as well as asthmalike symptoms. Allergic dermatitis may possibly occur from skin contact. Exposure to higher concentrations may cause nausea, headache, coughing, construction of the chest, and rapid heartbeat.[27]

One of the most promising techniques for reducing indoor formaldehyde concentrations is to modify the source materials to reduce emission rates. This can be accomplished by measures performed during manufacture or after installation. A variety of production changes, that is, changes in raw materials, processing times, and temperatures, are promising methods for reducing emission rates. Applying vinyl wallpaper or nonpermeable paint to interior walls, venting exterior walls, and increased ventilation are other methods employed after installation.

Volatile Organic Compounds (VOCs)

In addition to formaldehyde, many other organic compounds may be present in the indoor environment. More than 800 different compounds can be attributed to volatile vapors alone. Common sources in the home are building materials, furnishings, pesticides, gas or wood burning devices, and consumer products (cleaners, aerosols, deodorizers). In addition, occupant activities such as smoking, cooking, or arts and crafts activities can contribute to indoor pollutant levels.

Organic contaminants in the home are usually present as complex mixtures of many compounds at low concentrations. Thus, it is very difficult to provide estimates of typical indoor concentrations or associated health risks. It is likely, however, that organic compounds may be responsible for health-related complaints registered by residents where formaldehyde and other indoor pollutants are found to be low or undetectable. The sources of three major types of organic contaminants include solvents, polymer components, and pesticides.

Volatile organic solvents commonly pollute air. Exposure occurs when occupants use spot removers, paint removers, cleaning products, paint adhesives, aerosols, fuels, lacquers and varnishes, glues, cosmetics, and numerous other household products. Halogenated hydrocarbons such as methyl chloroform and methylene chloride are widely used in a variety of home products. Aromatic hydrocarbons such as toluene have been found to be present in more than 50 percent of samples taken of indoor air.[28] Alcohols, ketones, ethers, and esters are also present in organic solvents. Some of them, especially esters, emit pleasant odors and are used in flavors and perfumes, yet are still potentially harmful.

Polymer components are found in clothes, furniture, packages, and cookware. Many are used for medical purposes—for example, in blood transfusion bags and disposable syringes. Fortunately, most polymers are relatively nontoxic. However, polymers contain unreacted monomers, plasticizers, stabilizers, fillers, colorants, and antistatic agents, some of which are toxic. These chemicals diffuse from the polymers into air. Certain monomers (acrylic acid esters, toluenediisocynate, and epichlorohydrin) used to produce plastics, polyurethane, and epoxy resins in tile floors, are all toxic.

Most American households use pesticides in the home, garden, or lawn, and many people become ill after using these chemicals. According to an EPA survey, 9 out of 10 U.S. households use pesticides and another study suggests that 80 to 90 percent of most people's exposure to pesticides has been found in the air inside homes. Pesticides used in and around the home include products to control insects, termites, rodents, and fungi. Chlordane, one of the most harmful active ingredients in pesticides, has been found in structures up to 20 years after its application. In addition to the active ingredient, pesticides are also made up of inerts that are used to carry the active agent. These inerts may not be toxic to the targeted pest, but are capable of causing health problems. Methylene chloride, discussed earlier as an organic pollutant, is used as an inert.[27]

Human beings can also be significant sources of organic emissions. Human breath contains trace amounts of acetone and ethanol at 20°C and 1 atmosphere. Measurements taken in schoolrooms while people were present averaged almost twice the amount of acetone and ethanol present in unoccupied rooms. At least part of this increase for ethanol was presumed to be due to perfume and deodorant, in addition to breath emissions.[29]

As mentioned earlier, large numbers of organic compounds have been identified in residences. Studies have shown that of the 40 most common organics, nearly all were found at much higher concentrations indoors than outdoors. Another EPA study identified 11 chemicals present in more than half of all samples taken nationwide. Although individual compounds are usually present in low concentrations, which are well below outdoor air quality standards, the average total hydrocarbon concentration can exceed both outdoor concentrations and ambient air qualify standards.[28]

Little is known of the short- and long-term health effects of many organic compounds at the low levels of exposure occurring in nonindustrial environments. Yet cumulative effects of various compounds found indoors have been associated with a number of symptoms, such as headache, drowsiness, irritation of the eyes and mucous membranes, irritation of the respiratory system, and general malaise. In general, volatile organic compounds are lipid soluble and easily absorbed through the lungs. Their ability to cross the blood–brain barrier may induce depression of the central nervous system and cardiac functions. Some known and suspected human and animal carcinogens found indoors are benzene, trichlorethane, tetrachloroethylene, vinyl chloride, and dioxane.

One of the best methods to reduce health risks from exposure to organic compounds is for residents or consumers to increase their awareness of the types of toxic chemicals present in household products. Attention to warnings and instructions for storage and use are important, especially regarding ventilation conditions. In some instances, substitution of less hazardous products is possible, as in use of a liquid or dry form of a product rather than an aerosol spray. Consumers should also be wary of the simultaneous use of various products containing organic compounds, since chemical reactions may occur if products are mixed, and adverse health effects may result from the synergism between/among components.

Combustion Gases

Combustion gases, such as carbon monoxide, nitrogen oxides, and sulfur dioxide, can be introduced into the indoor environment by a variety of sources. These sources frequently depend on occupant activities or lifestyles and include the use of gas stoves, kerosene and unvented gas space heaters, woodstoves, and fireplaces. In addition, tobacco smoke is a combustion product that contributes to the contamination of indoor air. More than 2000 gaseous compounds have been identified in cigarette smoke, and carbon monoxide and nitrogen oxides are among them.[27]

This subsection focuses on nitrogen oxides (primarily nitrogen dioxide) and carbon monoxide because they are frequently occurring products of combustion often found at higher indoor concentrations that outdoors. Other combustion products such as sulfur dioxide, hydrocarbons, formaldehyde, and carbon dioxide are produced by combustion sources to a lesser degree or only under unusual or infrequent circumstances.

Unvented kerosene and gas space heaters can provide an additional source of heat for homes in cold climates or can serve as a primary heating source when needed for homes in warm climates. There are several basic types of unvented kerosene and gas space heaters that can be classified by the type of burner and type of fuel. Unvented gas space heaters can be convective or infrared and can be fueled by natural gas or propane. Kerosene heaters can be convective, radiant, two-stage, and wickless. A recent study found that emission rates from the various types of heaters fall into three distinct groups. The two-stage kerosene heaters emitted the least CO and the least NO_2. The radiant/infrared heater group emitted the most CO under well-tuned conditions. The convective group emitted the most NO_2. Many studies have also noted that some heaters have significantly higher emission rates than heaters of other brands or than models of the same type. Older or improperly used heaters will also increase emission rates.

The kitchen stove is one of the few modern gas appliances that emits combustion products directly into the home. It is estimated that natural gas is used in over 45 percent of all U.S. homes, and studies show that most of these homes do not vent the combustion-product emissions to the outside. Combustion gas emissions vary considerably and are dependent upon factors such as the fuel consumption rate, combustion efficiency, age of burner, and burner design, as well as the usage pattern of the appliance. An improperly adjusted gas stove is likely to have a yellow-tipped flame rather than a blue-tipped flame, which can result in increased pollutant emissions (mostly NO_2 and CO).

Increasing energy costs, consumer concerns about fuel availability, and desire for self-reliance are some of the factors that have brought about an upswing in the use of solid fuels for residential heating. These devices include woodburning stoves, furnaces, and fireplaces. Although woodstoves and fireplaces are vented to the outdoors, a number of circumstances can cause combustion products to be emitted to the indoor air: improper installation (such as insufficient stack height), cracks or leaks in stovepipes, negative air pressure indoors, downdrafts, refueling, and accidents (as when a log rolls out of a fireplace). The type and amount of wood burned also influences pollutant emissions, which vary from home to home. Although elevated levels of CO and NO_2 have been reported, the major impact of woodburning appears to be on indoor respirable suspended particles.

The term nitrogen oxides (NO_X) refers to a number of compounds, all of which have the potential to affect humans. NO_2 and NO have been studied extensively as outdoor pollutants, yet cannot be ignored in the indoor environment. There is evidence that suggests these oxides may be harmful at levels of exposure that can occur indoors. Both NO and NO_2 combine with hemoglobin in the blood, forming methemoglobin, which reduces the oxygen-carrying capacity of the blood. It is about four times more effective than CO in reducing the oxygen-carrying capacity of the blood. NO_2 produces respiratory illness that range from slight burning and pain in the throat and chest to shortness of breath and violent coughing. It places stress on the cardiovascular system and causes short-term and long-term damage to the lungs. Concentrations typically found in kitchens with gas stoves do not appear to cause chronic respiratory diseases but may affect sensory perception and produce eye irritation.

Carbon monoxide (CO) is a poisonous gas that causes tissue hypoxia (oxygen starvation) by binding with blood hemoglobin and blocking its ability to transport oxygen. CO has an excess of 200 times more binding affinity for hemoglobin than oxygen does. The product, carboxyhemoglobin, is an indicator of reduction in oxygen-carrying capacity. A small amount of CO is even produced naturally in the body, producing a concentration in unexposed persons of about 0.5 percent CO-bound hemoglobin. Under chronic exposure (e.g., cigarette smoking), the body compensates somewhat by increasing the concentration of red blood cells and the total amount of hemoglobin available for oxygen transport. The central nervous system, cardiovascular system, and liver are most sensitive to CO-induced hypoxia. Hypoxia of the central nervous system causes a wide range of effects in the exposure range of 5 to 15 percent carboxyhemoglobin. These include loss of alertness and impaired perception, loss of normal dexterity, reduced learning ability, sleep disruption, drowsiness, confusion, and at very high concentrations, coma and death. Health effects related to hypoxia of the cardiovascular system include decrease in exercise time required to produce angina pectoris (chest pain), increase in incidences of myocardosis (degeneration of heart muscle), and a general increase in the probability of heart failure among susceptible individuals.[28]

Population groups at special risk of detrimental effects of CO exposure include fetuses, persons with existing health impairments (especially heart disease), persons under the influence of drugs, and those not adapted to high altitudes who are exposed to both CO and high altitudes.

Proper installation, operation, and maintenance of combustion devices can significantly reduce the health risks associated with these appliances. Manufacturers' instructions regarding the proper size space heater in relation to room size, ventilation conditions, and tuning should be observed. This includes using vented range hoods when operating gas stoves. Studies have indicated reductions in CO, CO_2, and NO_2 levels as high as 60 to 87 percent with the use of range hoods during gas stove operation. Unvented forced draft and unvented range hoods with charcoal filters can be effective for removing grease, odors, and other molecules, but cannot be considered a reliable control for CO and other small molecules. Fireplace flues and chimneys should be inspected and cleaned frequently, and opened completely when in use.[29]

Particulates

Environmental tobacco smoke, ETS (smoke that nonsmokers are exposed to from smokers), has been judged by the Surgeon General, the National Research Council, and the International Agency for Research on Cancer to pose a risk of lung cancer to nonsmokers. Nonsmokers' exposure to environmental tobacco smoke is called "passive smoking," "second-hand smoking," and "involuntary smoking." Tobacco smoke contains a number of pollutants, including inorganic gases, heavy metals, particulates, VOCs, and products of incomplete burning, such as polynuclear aromatic hydrocarbons. Smoke can also yield a number of organic compounds. Including both gases and particles, tobacco smoke is a complex mixture of over 4700 compounds.[27]

There are two components of tobacco smoke: (1) mainstream smoke, which is the smoke drawn through the tobacco during inhalation, and (2) sidestream smoke, which arises from the smoldering tobacco. Sidestream smoke accounts for 96 percent of gases and particles produced.[29]

Studies indicate that exposure to tobacco smoke may increase the risk of lung cancer by an average of 30 percent in the nonsmoking spouses of smokers. Published risk estimates of lung cancer deaths among nonsmokers exposed to tobacco smoke conclude that ETS is responsible for 3000 deaths each year. It also seriously affects the respiratory health of hundreds of thousands of children. Very young children exposed of smoking at home are more likely to be hospitalized for bronchitis and pneumonia. Recent studies suggest that environmental tobacco smoke can also cause other diseases, including other cancers and heart disease in healthy nonsmokers.[27]

The best way to reduce exposure to cigarette smoke in the house is to quit smoking and discourage smoking indoors. Ventilation is the most common method of reducing exposure to these pollutants, but it will not eliminate it altogether. Smoking produces such large amounts of pollutants that neither natural nor mechanical methods can remove them from the air as quickly as they build up. In addition, ventilation practices sometimes lead to increased energy costs.

Respirable suspended particles (RSP) are particles or fibers in the air that are small enough to be inhaled. Particles can exist in either solid or liquid phase or in a combination. Where these particles are deposited and how long they are retained depends on their size, chemical composition, and density. Respirable suspended particles (generally less than 10 μm in diameter), can settle on the tissues of the upper respiratory tract, with the smallest particles (those less than 2.5 μm) penetrating the alveoli, the small air sacs in the lungs. Deposition rates, retention, and health and hazard risks of nanoparticles (10^{-3} μm, or 1.0 nm) are not known at this time.

Particulate matter is a broad class of chemically and physically diverse substances that present risks to health. These effects can be attributed to either the intrinsic toxic chemical or physical characteristics, as in the case of lead and asbestos, or to the particles acting as a carrier of adsorbed toxic substances, as in the case of attachment of radon daughters. Carbon particles, such as those created by combustion processes, are efficient adsorbers of many organic compounds and are able to carry toxic gases such as sulfur dioxide into the lungs.

Asbestos is a mineral fiber used mostly before the mid-1970s in a variety of construction materials. Home exposure to asbestos is usually due to aging, cracking, or physical disruption of insulated pipes or asbestos-containing ceiling tiles and spackling compounds. Apartments and school buildings may have an asbestos compound sprayed on certain structural components as a fire retardant. Exposure occurs when asbestos materials are disturbed and the fibers are released into the air and inhaled. Consumer exposure to asbestos has been reduced considerably since the mid-1970s, when the use of asbestos was either prohibited or stopped voluntarily in sprayed-on insulation, fire protection, soundproofing, artificial logs, patching compounds, and handheld hair dryers. Today, asbestos is most commonly found in older homes in pipe and furnace insulation materials, asbestos shingles, millboard, textured paints and other coating materials, and floor tiles. Elevated concentrations of airborne asbestos can occur after asbestos-containing materials are disturbed by cutting, sanding, or other remodeling activities. Improper attempts to remove these materials can release asbestos fibers into the air in homes, thereby increasing asbestos levels and endangering the people living in those homes. The most dangerous asbestos fibers are too small to be visible. After they are inhaled, they can remain and accumulate in the lungs. Asbestos can cause lung cancer, mesothelioma (a cancer of the chest and abdominal linings), and asbestosis (irreversible lung scarring that can be fatal). Symptoms of these diseases do not show up until many years after exposure began. A more detailed presentation on asbestos can be found in Chapter 8.

Lead has long been recognized as a harmful environmental pollutant. There are many ways in which humans are exposed to lead, including air, drinking water, food, and contaminated soil and dust. Airborne lead enters the body when an individual breathes lead particles or swallows lead dust once it has settled. Until recently, the most important airborne source of lead was automobile exhaust. Lead-based paint has long been recognized as a hazard to children who eat lead-contained paint chips. A 1988 National Institute of Building Sciences Task Force report found that harmful exposures to lead can be created when lead-based paint is removed from surfaces by sanding or open-flame burning. High concentrations of airborne lead particles in homes can also result from the lead dust from outdoor sources, contaminated soil tracked inside, and use of lead in activities such as soldering, electronics repair, and stained-glass artwork. Lead is toxic to many organs within the body at both low high concentrations. Lead is capable of causing serious damage to the brain, kidneys, peripheral nervous system (the sense organs and nerves controlling the body), and red blood cells. Even low levels of lead may increase blood pressure in adults. Fetuses, infants, and children are more vulnerable to lead exposure than are adults because lead is more easily absorbed into growing bodies, and the tissues of small children are more sensitive to the damaging effects to lead. The effects of lead exposure on fetuses and young children include delays in physical and mental development, lower IQ levels, shortened attention spans, and increased behavioral problems. Additional details on lead, as well as other metals, can be found in Chapter 8.

Particles present a risk to health out of proportion to their concentration in the atmosphere because they deliver a high-concentration package of potentially harmful substances. So, while few cells may be affected at any one time, those few that are can be badly damaged. Whereas larger particles deposited in the upper respiratory portion of the respiratory system are continuously cleared away, smaller particles deposited deep in the lung may cause adverse health effects. Particle sizes vary over a broad range, depending on source characteristics.

Major effects of concern attributed to particle exposure are impairment of respiratory mechanics, aggravation of existing respiratory and cardiovascular disease, and reduction in particle clearance and other host defense mechanisms. Respiratory effects can range from mild transient changes of little direct health significance to incapacitating impairment of breathing.

One method of reducing RSP concentrations is to properly design, install, and operate combustion sources. One should make sure there are no existing leaks or cracks in stovepipes, and that these appliances are always vented to the outdoors.

Also available are particulate air cleaners, which can be separated into mechanical filters and electrostatic filters. Mechanical filtration is generally accomplished by passing the air through a fibrous medium (wire, hemp, glass, etc.). These filters are capable of removing almost any sized particles. Electrostatic filtration operates on the principle of attraction between opposite electrical charges. Ion generators, electrostatic precipitators, and electric filters use this principle for removing particles from the air.

The ability of these various types of air-cleaning devices to remove respirable particles varies widely. High-efficiency particulate air (HEPA) filters can capture over 99 percent of particles and are advantageous in that filters only need changing every 3 to 5 years, but costs can reach $500 to $800. It is also important to note the location of air-cleaning device inlets in relation to the contaminant sources as an important factor influencing removal efficiencies.

Biological Contaminants

Heating, ventilation, and air conditioning systems and humidifiers can be breeding grounds for biological contaminants when they are not properly cleaned and maintained. They can also bring biological contaminants indoors and circulate them. Biological contaminants include bacteria, mold and mildew, viruses, animal dander and cat saliva, mites, cockroaches, and pollen. There are many sources for these pollutants. For example, pollens originate from plants; viruses are transmitted by people and animals; bacteria are carried by people, animals, and soil and plant debris; and household pets are sources of saliva, hair, and dead skin (known as dander).

Available evidence indicates that a number of viruses that infect humans can be transmitted via the air. Among them are the most common infections of humans. Airborne contagion is the mechanism of transmission of most acute respiratory infections, and these are the greatest of all causes of morbidity.

The primary source of bacteria indoors is the human body. Although the major source is the respiratory tract, it has been shown that 7 million skin scales are shed per minute per person, with an average of 4 viable bacteria per scale.[28] Airborne transmission of bacteria is facilitated by the prompt dispersion of particles. Infectious contact requires proximity in time and space between host and contact, and is also related to air filtration and air exchange rate.

Although many important allergens—such as pollen, fungi, insects, and algae—enter buildings from outdoors, several airborne allergens originate predominately in homes and office buildings. House dust mites, one of the most powerful biologicals in triggering allergic reactions, can grow in any damp, warm environment. Allergic reactions can occur on the skin, nose, airways, and alveoli.

The most common respiratory diseases attributable to these allergens are rhinitis, affecting about 15 percent of the population, and asthma, affecting about 3 to 5 percent.[28] These diseases are most common among children and young adults but can occur at any age. Research has shown that asthma occurs four times more often among poor, inner-city families than in other families. Among the suspected causes are mouse urine antigens, cockroach feces antigens, and a type of fungus called Alternia.

Hypersenitivity pneumonitis (HP), characterized by shortness of breath, fever, and cough, is a much less common disease, but is dangerous if not diagnosed and treated early. HP is most commonly caused by contaminated forced-air heating systems, humidifiers, and flooding disasters. It can also be caused by inhalation of microbial aerosols from saunas, home tap water, and even automobile air conditioners. Humidifiers with reservoirs containing stagnant water may be important sources of allergens in both residential and public buildings.

Some biological contaminants trigger allergic reactions, while others transmit infectious illnesses, such as influenza, measles, and chicken pox. Certain molds and mildews release disease-causing toxins. Symptoms of health problems caused by biologicals include sneezing, watery eyes, coughing, shortness of breath, dizziness, lethargy, fever, and digestive problems.

Attempts to control airborne viral disease have included quarantine, vaccination, and inactivation or removal of the viral aerosol. Infiltration and ventilation play a large role in the routes of transmission. Because many contaminants originate outdoors, attempts to reduce the ventilation rate might lower indoor pollutant concentrations. However, any reduction in fresh air exchange should be supplemented by a carefully filtered air source.

Central electrostatic filtration (as part of a home's forced-air system) has proven effective in reducing indoor mold problems. Careful cleaning, vacuuming, and air filtration are effective ways to reduce dust levels in a home. Ventilation of attic and crawl spaces help prevent moisture buildup, keeping humidity levels between 30 to 50 percent. Also, when using cool mist or ultrasonic humidifiers, one should remember to clean and refill water trays often, since these areas often become breeding grounds for biological contaminants.

Indoor Air Monitoring Methods

Methods and instrumentation for measuring indoor air quality vary in their levels of sensitivity (what levels of pollutant they can detect) and accuracy (how close they can come to measuring the true concentration). Instruments that can measure low levels of pollutant very accurately are likely to be expensive and require special expertise to use. Some level of sensitivity and accuracy is required, however, to ensure that data collected are useful in assessing levels of exposure and risk.

Methods to monitor indoor air fall into several broad categories. Sampling instruments may be fixed-location, portable, or small personal monitors designed to be carried by an individual. These samplers may act in an active or passive mode. Active samplers require a pump to draw in air. Passive samplers rely on diffusion or permeation.

Monitors may be either analytical instruments that provide a direct reading of pollutant concentration, or collectors that must be sent to a laboratory for analysis. Instruments may also be categorized according to the time period over which they sample. These include grab samplers, continuous samplers, and time-integrated samplers, each of which is briefly described below.

1. Grab sampler: Collects samples of air in a bag, tube, or bottle, providing a short-term average.
2. Continuous sampling: Allows sampling of real-time concentration of pollutants, providing data on peak short-term concentrations and average concentrations over the sampling period.

3. Time-integrated sampling: Measures an average air concentration over some period of time (active or passive), using collector monitors that must be sent out for analysis; cannot determine peak concentrations.

More details regarding monitoring methods for specific indoor air pollutants can be found in the IAQ Handbook[28] and in the next section.

3.6   MONITORING METHODS

Actual measurements of nanoparticles, and nanonemissions in particular, pose major problems for the technical community. Concern with micrometer (10^3-nm) particulates has existed since the mid-1970s. More recently, these concerns have been extended down to 5- and 2.5-μm particulates, with much of the effort being directed toward measurement. Nothing of industrial value was available for the measurement of nanoparticles at the time of the preparation of this manuscript. Notwithstanding this, an overall general review of present-day monitoring methods is warranted, and the material that follows addresses this issue.

Although this chapter addresses air issues, this section focuses on methods used to obtain data regarding contamination of air, water, and soil. Information is provided on how to set up a monitoring program and how to ensure that the data obtained are reliable. Much of this information is based on the American Society for Testing and Materials (ASTM) standards, the EPA's Contract Laboratory Program, which was established under Superfund, and the EPA's Environment Technology Verification (ETV) Program. A number of additional references are available for collection and preservation of samples.

Selecting a Methodology

The method used to sample a given area will directly affect the accuracy of the analysis. It is imperative that an appropriate methodology be selected in order to obtain the most reliable results possible. Several factors should be considered when selecting a method, including

1. The program objective (documenting exposures, determining regulatory compliance, locating a source)
2. The type of material to be sampled (soil, vegetation, air, water, sludge, etc.); in this section it is air
3. The physical and chemical properties of the contaminant
4. Other contaminants that affect the results
5. Regulatory requirements
6. Safety requirements
7. Difficulty of utilizing a particular method

8. Cost
9. Reliability
10. Scale of sample area (small-scale site related to individual persons versus a large-scale site)
11. Short- versus long-term sampling requirement

In addition to the above, basic professional judgment also plays a role in selecting a sampling method.

The extension of the above to nanoemissions to the atmosphere may test the validity of this methodology. The physical and chemical characterization, including the time variation of these properties, will be critical to the development and application of new monitoring methods. These ultrafine particles present an enormous challenge because of their small size, mass, and volume. Obtaining needed information on particle size distribution (PSD) and any real-time variation of this distribution will aid the practitioner in this field. It should be understood that some of the measurements will require instrumentation that presently does not exist.

Standard Practices for Sampling of Ambient Air

This section presents the broad concepts of sampling the ambient air for the concentrations of contaminants. Detailed procedures are not discussed. General principles in planning a sampling program are given, including guidelines for the selection of sites and the location of the air sampling inlet. The reader is referred to the reference materials for details, including background information, air quality modeling techniques, and special-purpose air sampling program.[30,31]

Investigations of atmospheric contaminants often involve the study of a heterogeneous mass under uncontrolled conditions. Interpretation of the data derived from the air sampling program must often be based on the statistical theory of probability. Extreme care must be observed to obtain measurements over a sufficient length of time to obtain results that may be considered representative.

The variables that may affect the contaminant concentrations are the atmospheric stability (temperature–height profile), turbulence, wind speed and direction, solar radiation, precipitation, topography, emission rates, chemical reaction rates for the formation and decomposition of contaminants, and the physical and chemical properties of the contaminant. The ambient temperature and atmospheric pressure at the location sampled must be known to obtain concentrations of gaseous contaminants in terms of mass per unit volume.

Because the analysis of the atmosphere is influenced by phenomena in which all factors except the method of sampling and composition are beyond the control of the investigator, statistical consideration must be given to determine the adequacy of the number of samples obtained, the length of time that the sampling program is carried out, and the number of sites sampled. The purpose of the sampling and the characteristics of the contaminant to be measured will have an influence in determining this adequacy. Regular or, if possible, continuous measurements of the contaminant with simultaneous pertinent meteorological observations should be obtained during all seasons of the year. Statistical techniques may then be applied to determine the influence of the meteorological variables on the concentrations measured.[32]

The choice of sampling techniques and measurement methodology, the characteristics of the sites, the number of sampling stations, and the amount of data collected all depend on the objectives of the monitoring program. These objectives may be one or more of the following:

1. Health and vegetation effects studies
2. Trend analysis
3. Evaluation of pollution abatement programs
4. Establishment of air quality criteria and standards relating to effects
5. Enforcement of control regulations
6. Development of air pollution control strategies
7. Activation of alert or emergency procedures
8. Land use, transportation, and energy systems planning
9. Background evaluations
10. Atmospheric chemistry studies

In order to cover all the variable meteorological conditions that may greatly affect the air quality in an area, air monitoring for lengthy periods of time may be necessary to meet most of the above objectives. The topography, demography, and micrometeorology of the area, as well as the contaminant measured, must be considered in determining the number of monitoring stations required in the area. Photographs and a map of the locations of the sampling stations is desirable in describing the sampling station.

Unless the purpose of the sampling program is site-specific, the sites monitored should, in general, be selected as to avoid undue influence by any local source that may cause local elevated concentrations that are not representative of the region to be characterized by the data. Monitoring sites for determining the impact on air quality by individual sources should be selected, if possible, so as to isolate the effect of the source being considered. When there are many sources of the contaminant in the area, the sites sampled should be strategically located so that with wind direction data obtained simultaneously near the sites, the monitoring results will provide evidence of the contributions of the individual sources. Multiple samplers or monitors operating simultaneously upwind and downwind from the source are often very valuable and efficient.

The meteorological parameters that are most important in an atmospheric sampling program are: wind direction and speed, the degree of persistence in direction, and gustiness; temperature and its changes with height above ground; the mixing height, that is, the height above ground that the pollutants will diffuse to during the afternoon; and solar radiation and hours of sunshine, humidity, precipitation, and barometric pressure. These parameters are important in assessing the pollution potential of an area and should be considered in the planning of a monitoring program and in the interpretation of the data. Pertinent meteorological and climatological information may be obtained from the local weather department. In many localities, however, the micrometeorology may be unique and meteorological investigations to provide data specific to the area may be needed.[33–35]

Topography can influence the contaminant concentrations in the atmosphere. For example, a valley will cause persistence in wind directions and intensify low-level nocturnal inversions that will limit the dispersion of pollutants emitted into it. Mountains or plateaus may act as barriers affecting the flow of air as well as the contaminant concentrations in their vicinity. Consideration should be given to the influence of these features as well as that of large lakes, the sea, and oceans.[32,33]

The choice of procedure for the air sampling is dependent on the contaminant to be measured. The reader is referred to ASTM Practice D3249 for recommendations for general ambient air analyzer procedures. ASTM-recommended methods have been published for most of the common contaminants that are sampled. Automatic instruments providing a continuous record of the concentrations of the contaminant should be utilized whenever possible to save manpower and increase efficiency. Very often, factors such as temperature, humidity, and vibrations, as well as the power line voltage, can influence the output of the air-monitoring instrument, and these should be controlled.

The monitors must be supplied with sample air that represents the ambient air under investigation. Careful consideration should be given to the sample conveying system. A duct system is often utilized for this purpose. There should be as few abrupt enlargements and elbows as possible, because these may affect the uniformity and hence the concentration of the contaminants measured. The material for the duct should be such that there will be little or no interaction between it and the contaminants in the air sampled. Temperature control should be employed to limit the condensation forming in the sampling lines. The samples are drawn from straight sections of the duct; also, the inlet lead to the monitoring instrument should be as short as possible.

The following guidelines are recommended for sampling locations unless site-specific measurements are desired. The height of the inlet to the sampling duct should normally be from 2.5 to 5 m above ground, whenever possible. The height of the inlet above the sampling station structure or vegetation of adjacent to the station should be greater than 1 m. Sampling should preferably be through a vertical inlet with an inverted cone over the opening. For a horizontal inlet, there should be a minimum of 2 m from the face of the structure. For access to representative ambient air in the area sampled, the elevation angle from the inlet to the top of nearby buildings should be less than 30°. To be representative of the area in which a large segment of the population is exposed to contaminants emitted by automobiles, the inlet should be at a distance greater than 15 m from the nearest high-volume traffic artery. Photochemical oxidants or ozone samplers should be located at distances greater than 50 m from high-volume traffic locations. Particulate matter samplers should be sited at locations that are greater than 200 m from unpaved streets or roads.[36–38]

Air sampling can be conducted for long or very short periods, depending on what type of information is needed. Instantaneous or grab sampling is the collection of an air sample over a short period, whereas longer period sampling is called integrated air sampling. There is no sharp dividing line between the two sampling methods; however, grab samples are usually taken in a period of less than 5 min.

Grab samples represent the environmental concentration at a particular point in time. It is ideal for following several phases of a cyclic process and for determining airborne concentrations of brief duration but is seldom used to estimate 8-h average concentrations. Grab samplers consist of various devices. An evacuated flask or plastic bag can be useful in gas and vapor concentration analysis. After introducing the sample of air into the container; it is sealed to prevent loss or further contamination and is then sent to a laboratory for analysis. There, trace analysis procedures such as gas chromotography, infrared spectrophotometry, or other methods are used to determine concentrations of gaseous contaminants.

In integrated air sampling, a known volume of air is passed through a collection medium to remove the contaminant from the sampled airstream. It is the preferred method of determining time-weighted average exposures. Integrated sampling consists of one or a series of samples taken for the full or partial duration of the time-averaging period. The time-averaging period can be from 15 min to 8 h, depending upon whether a ceiling, short-term, or 8-h exposure limit is being evaluated. Details of the apparatus or instruments employed in sampling the air or in carrying out associated meteorological investigations are discussed in the references.

The procedure for sampling should be undertaken in the following steps. First, conduct a general exploratory survey of the area including the topography, an inventory of sources for the contaminants, the height of their emissions, traffic, and land-use data. Next, complete a preliminary meteorology analysis to identify wind direction frequencies, wind velocity, and temperature–height profiles. Exploratory short-term temporary sampling requires a number of temporary sampling stations, to determine the need and the best sites for extensive long-term monitoring. Using air quality models, as well as the input of emission inventory and meteorological information for the area, an estimate for the levels of air quality over the area may be calculated. The model results will provide guidance in determining the locations for monitors that will measure the maximum levels and the number of monitors required to characterize the air quality in the area of concern.[39,40]

Sample Packaging and Shipment

This section provides general requirements for shipping soil and water samples. Similar methods are required for shipping air samples. These procedures are based on requirements for participation in the Federal Contract Laboratory Program.[41] The laboratory for analyzing a sample should be consulted regarding packaging requirements before the initiation of a sampling program. Samples must be packaged for shipment in compliance with current U.S. Department of Transportation (DOT) and commercial carrier regulations. All required government and commercial carrier shipping papers must be filled out and shipment classifications made according to current DOT regulations.

Traffic reports, shipment records, packing lists, chain-of-custody records, and any other shipping/sample documentation accompanying the shipment must be enclosed in a waterproof plastic bag and taped to the underside of the shipping cooler lid. Coolers must be sealed with custody seals in such a manner that the custody seal would be broken if the cooler were opened. Shipping coolers must have clearly visible return address labels on the outside.

Inside the cooler, sample containers must be enclosed in clear plastic bags through which sample tags and labels are visible. Table 3.2 describes the containers used for collection of various samples. Dioxin samples (as well as water and soil should samples be

TABLE 3.2   Types of Sample Collection Bottles

Description of Bottle Used	Sample Type[a]
80-oz amber glass bottle with Teflon-lined black phenolic cap	Extractable organics; low-concentration water samples
40-mL glass vial with Teflon-lined silicon septum and black phenolic cap	Volatile organics; low- and medium-concentration water samples
1-liter high-density polyethylene bottle with white poly cap	Metals, cyanide; low-concentration water samples
120-mL wide-mouth glass vial with white poly cap	Volatile organics; low- and medium-concentration soil samples
16-oz wide-mouth glass jar with Teflon-lined black phenolic cap	Metals, cyanide; medium-concentration water samples
8-oz wide-mouth glass jar with Teflon-lined black phenolic cap	Extractable organics; low- and medium-concentration soil samples
	Metals, cyanide; low- and medium-concentration soil samples
	Dioxin; soil samples
	Organics and inorganics; high-concentration liquid and solid samples
4-oz wide-mouth glass jar with Teflon-lined black phenolic cap	Extractable organics; low- and medium-concentration soil samples
	Metals, cyanide; low- and medium-concentration soil samples
	Dioxin; soil samples
	Organics and inorganics; high-concentration liquid and solid samples
1-liter amber glass bottle with Teflon-lined black phenolic cap	Extractable organics; low-concentration water samples
32-oz wide-mouth glass jar with Teflon-lined black phenolic cap	Extractable organics; medium-concentration water samples
4-liter amber glass bottle with Teflon-lined black phenolic cap	Extractable organics; low-concentration water samples

[a]This column specifies the *only* type(s) of samples that should be collected in each container.

suspected of having medium or high concentration, or of containing dioxin) must be enclosed in a metal can with a clipped or sealable lid (paint cans are normally used for this purpose) and surrounded by packing material such as vermiculite. The outer metal can must be labeled with the number of the sample contained inside.

Water samples for low- or medium-level organics analysis and low-level inorganics analysis must be shipped cooled with ice. No ice is to be used in shipping inorganic low-level soil samples or medium/high-level water samples, organic high-level water or soil samples, or dioxin samples. Ice is not required in shipping soil samples but may be utilized at the option of the sampler. All cyanide samples, however, must be shipped cooled to 4°C. Low- and medium-level water samples for inorganic analysis require chemical preservation.

Waterproof, metal ice chests or coolers are the only acceptable type of sample shipping container. Cardboard and Styrofoam containers should not be used. Shipping containers should be packed with noncombustible, absorbent packing material (vermiculite is recommended) surrounding the plastic-enclosed, labeled sample bottles (or labeled metal cans containing samples) to avoid sample breakage in transport. Sufficient packing material should be used so that sample containers will not make contact during shipment. Earth or ice should never be used to pack samples. Ice should be in sealed plastic bags to prevent melting ice from soaking packing material which, when soaked, makes handling of samples difficult in the lab.

Samples for organics analysis must be shipped "Priority One/Overnight." If shipment requires more than a 24-h period, sample holding times can be exceeded, thereby compromising the integrity of the sample analyses. Samples for inorganics analysis should be held until sampling for the entire area is complete and shipped for delivery within 2 days. Three days is the recommended period for collection of the samples. All samples should be shipped through a reliable commercial carrier.

Sample Documentation

Each sample must be properly documented to ensure timely, correct, and complete analysis for all parameters requested, and, most importantly, to support use of sample data in potential enforcement actions concerning a site. The documentation system provides the means to identify, track, and monitor each sample individually from the point of collection through final data reporting. As used herein, a sample is defined as a representative specimen that is collected at a specific location of a site at a particular point in time for a specific analysis, and may refer to field samples, duplicates, replicates, splits, spikes, or blanks that are shipped from the field to a laboratory. The reader should note that the material presented here also applies to the monitoring sections of Chapter 4 (water) and Chapter 5 (solid waste/land).

Sample Traffic Report The sample documentation system is usually based on the use of the EPA Sample Traffic Report, a four-part carbonless form printed with a unique sample identification number. One Traffic Report (TR) and its pre-printed identification number is assigned by the sampler to each sample collected. The two types of TRs currently in use are organic and inorganic.

To provide a permanent record for each sample collected, the sampler completes the appropriate TR, recording the case number, site name or code, location and site/spill ID, analysis laboratory, sampling office, dates of sample collection and shipment, and sample concentration and matrix. Numbers of sample containers and volumes are entered by the sampler beside the analytical parameter(s) requested for particular sample portions.

A strip of adhesive sample labels printed with the TR sample number comes attached to the TR for the sampler's use in labeling sample bottles. The sampler affixes one of these numbered labels to each container making up the sample. In order to protect the label from water and solvent attack, each label must be covered with clear waterproof tape. The sample labels, which bear the TR identification number, permanently identify each sample collected and link each sample component throughout the analytical process.

Sample Tag To render sample data valid for enforcement purposes, individual samples must be traceable continuously from the time of collection until the time of introduction as evidence during litigation. One mechanism utilized is the use of the "sample tag." Each sample removed from a waste site and transferred to a laboratory for analysis is identified by a sample tag containing specific information regarding the sample, as defined by the EPA National Enforcement Investigations Center (NEIC). Following sample analysis, sample tags are retained by the laboratory as physical evidence of sample receipt and analysis, and may later be introduced as evidence in EPA litigation proceedings.

The information recorded on an EPA sample tag includes:

1. Case number(s)—the unique number(s) assigned to identify the sampling event
2. Sample number—the unique sample identification number used to document that sample
3. Project code—the number assigned by EPA to the sampling project
4. Station number—a two-digit number assigned by the sampling team coordinator
5. Date—a six-digit number indicating the month, day, and year of collection
6. Time—a four-digit number indicating the military time of collection
7. Station location—the sampling station description as specified in the project plan
8. Samplers—signatures of samplers on the project team
9. Tag number—a unique serial number preprinted or stamped on the tag
10. Lab sample number—reserved for laboratory use

Additionally, the sample tag contains appropriate spaces for noting that the sample has been preserved and indicating the analytical parameter(s) for which the sample will be analyzed. Each sample tag is completed and securely attached to the sample container. Samples are then shipped under chain-of-custody procedures as described next.

Chain-of-Custody Record Official custody of samples must be maintained and documented from the time of sample collection up to introduction as evidence in court, in accordance with EPA enforcement requirements. The following custody documentation procedure was developed by NEIC and is used in conjunction with sample documentation.

A sample is considered to be in an individual's custody if the following criteria are met: It is in your possession or it is in your view after being in your possession; or it was in your possession and then was locked up or sealed to prevent tampering; or it is in a secured area. Under this definition, the team member actually performing the sampling is personally responsible for the care and custody of the samples collected until they are transferred or dispatched properly. In follow-up, the sampling team leader reviews all field activities to confirm that proper custody procedures were followed during the field work.

The Chain-of-Custody Record is employed as physical evidence of sample custody. The sampler completes a Chain-of-Custody Record to accompany each cooler shipped from the field to the laboratory.

Information similar to that entered on the sample tag is recorded on the Chain-of-Custody Record. Information includes the project number, sampler's signatures, and the case number. For each station number, the sampler indicates the following: date, time, whether the sample is a composite or grab, station location, number of containers, analytical parameters, sample number(s), and sample tag number(s). When relinquishing the samples for shipment, the sampler signs in the space indicated on the form, entering the date and time the samples are relinquished.

The top, original signature copy of the Chain-of-Custody Record is enclosed in plastic (with the sample documentation) and secured to the inside of the cooler lid. A copy of the custody record is retained for the sampler's files.

Shipping coolers are secured, and custody seals are placed across cooler openings. As long as custody forms are sealed inside the sample cooler and custody seals remain intact, commercial carriers are not required to sign off on the custody form.

The laboratory representative who accepts the incoming sample shipment signs and dates the Chain-of-Custody Record to acknowledge receipt of the samples, completing the sample transfer process. It is then the laboratory's responsibility to maintain internal log books and records that provide a custody record throughout sample preparation and analysis.

The EPA: Environmental Technology Verification Program

More recently, the EPA has put in place a verification program titled, Environmental Technology Verification (ETV) Program, which verifies the performance of innovative or improved technologies as an independent third party. Verification tests generate credible performance information with quality-assured data approved by EPA. The ETV Program addresses problems that threaten human health or the environment; it was designed to accelerate the entrance of new environmental technologies into the domestic and international marketplace.

The Air Pollution Control Technology Verification Center (APCTVC) is part of the EPA ETV Program and is operated as a partnership between Research Triangle

Institute (RTI) and EPA. The center verifies the environmental performance of commercial-ready air pollution control technologies. Verification provides potential purchasers and permitters with an independent and credible assessment of what they are buying and permitting. Verification tests use approved protocols, and verified performance is reported in verification statements signed by EPA. RTI contracts with Midwest Research Institute (MRI), ETS, Inc., and Southwest Research Institute (SRI) to perform verification tests.

Generally, the verification process described below follows similar steps, regardless of the technology area:

1. The vendor initiates the verification process by submitting an application to the APCTVC.
2. The APCTVC, the testing contractor, and the applicant discuss the intent of the test and develop a testing outline.
3. The APCTVC responds with contract outlining terms and conditions, statement of work, and cost.
4. The applicant approves and returns a signed copy of the terms and conditions with initial payment.
5. A test/quality assurance plan is prepared by the APCTVC and its testing organization with input from the applicant.
6. Testing is conducted.
7. A draft report is prepared and reviewed by the APCTVC, testing organization, EPA, and applicant.
8. The applicant submits a verification report and statement.
9. EPA approves and signs the verification report and statement.
10. The APCTVC distributes the verification report and statement to the applicant and EPA and posts them on the APCTVC and EPA ETV web sites so they are available to the general public.

The APCTVC Verification Center benefits technology developers and vendors, users and purchasers, permitters, and the public. Because the technologies are verified by RTI, which is an independent, objective third party, everyone is assured that the findings are based on high-quality, credible, consistent, useful, and widely accepted performance data and procedures. Technology verification provides many benefits, as detailed below.

For Developers and Vendors

1. Gives a sound science-based marketing tool.
2. Reduces costs for advertising and marketing.
3. Expands markets and business opportunities.
4. Enhances regulatory acceptance.
5. Accelerates new or improved technologies into the marketplace.

6. Provides access to expertise in performance measurement.

7. Adds confidence for investors and lenders.

For Users and Purchasers

1. Allows easier evaluation of competing technologies.

2. Facilities permitting process.

3. Reduces noncompliance risks.

For Permitters

1. Makes job easier.

2. Adds confidence in control systems performance.

3. Increases ability to make informed decisions.

More recently, Hartzell and Waits[42] provided an excellent detailed review of the Environmental Technology Verification Program.

3.7 SUMMARY

1. There are two classes of pollutants emitted into the atmosphere: gases and particulates. Traditional control devices for these pollutants are available and will probably be employed for many nanoemissions.

2. There are a number of factors to be considered prior to selecting a particular piece of air pollution control hardware. In general, they can be grouped into three categories: environmental, engineering, and economic. The final choice in the equipment selection is usually dictated by that equipment capable of achieving compliance with regulatory codes at the lowest uniform annual cost (amortized capital investment plus operation and maintenance costs). One area that has recently received some attention is hybrid systems, equipment that can in some cases operate at higher efficiency more economically than conventional devices.

3. The effective stack height is usually considered the sum of the actual stack height, the plume rise due to velocity of the issuing gases, and the buoyancy rise, which is a function of the temperature of the gases being emitted and the atmospheric conditions. The Pasquill–Gifford model has been the basis of most models developed and accepted today for atmospheric dispersion calculations.

4. Stacks discharging to the atmosphere have long been one of the methods available to industry for disposing of waste gases. The concentration to which humans, plants, animals, and structures are exposed at ground level can be reduced significantly by emitting the waste gases from a process at great heights.

5. Indoor air quality is rapidly becoming a major environmental concern because a significant number of people spend a substantial amount of time in a variety of different indoor environments. Health effects from indoor pollutants fall into two

categories: those that are experienced immediately after exposure and those that do not show up until years later.

6. Variables that affect air contaminant concentrations include atmospheric stability, turbulence, wind speed and direction, solar radiation, precipitation, topography, emission rates, chemical reaction rates for their formation and decomposition, and the physical and chemical properties of the contaminant. All of these factors must be accounted for when developing and analyzing an air sampling system.

REFERENCES

1. L. Theodore and R. Allen, *Air Pollution Control Equipment*, A Theodore Tutorial, Theodore Tutorials, East Williston, NY, 1994.

2. J. Mycock, J. McKenna, and L. Theodore, *Handbook of Air Pollution Control Engineering and Technology*, CRC/Lewis, Boca Raton, FL, 1995.

3. L. Theodore and A. Buonicore, *Air Pollution Control Equipment: Selection, Design, Operation and Maintenance*, ETS International, Roanoke, VA, 1982.

4. L. Theodore and M. Blenner, *Unit Operations for the Practicing Engineer*, John Wiley, Hoboken, NJ, 2005.

5. Adaped from M. K. Theodore and L. Theodore, *Major Environmental Issues Facing the 21st Century*, Theodore Tutorials, East Williston, NY, 1996.

6. Adapted from G. Burke, B. Singh, and L. Theodore, *Handbook of Environmental Management and Technology*, 2nd ed., Wiley, Hoboken, NJ, 2000.

7. E. Cunningham, *Proc. R. Soc. London Ser. A.*, **83**, 357 (1910).

8. A. Einstein, *Ann. Physik*, **19**, 289 (1906).

9. U.S. EPA. *Engineering Handbook for Hazardous Waste Incineration*, Monsanto Research Corporation, Dayton, OH, EPA Contract No. 68-03-3025, September 1982.

10. U.S. EPA. *Revised U.S. EPA Engineering Handbook for Hazardous Waste Incineration*, unpublished.

11. ETS International, assorted technical literature, author unknown, ETS International, Roanoke, VA, 1989.

12. L. Theodore, unpublished notes, 1984.

13. U.S. EPA, *Guideline on Air Quality Models*, Publication No. EPA-450/2-78-027, Research Triangle Park, NC (OAQPS No. 1.2-08), 1978.

14. U.S. EPA, *Industrial Source Complex (ISC) Dispersion Model Users Guide*, 2nd ed., Vols. 1 and 2, Publication Nos. EPA-450/4-86-005a, and EPA-450/4-86-005b, Research Triangle Park, NC, (NRTIS PB86 234259 and 23467), 1986.

15. L. Theodore et al., *Accident and Emergency Management Student Manual*, U.S. EPA APTI Course 503, 1989.

16. H. E. Hesketh, *Understanding and Controlling Air Pollution*, Ann Arbor Science, Ann Arbor, MI, 1972, pp. 33–70.

17. A. Gilpin, *Control of Air Pollution*. Butterworth, New York, 1963, pp. 326–333.

18. C. D. Cooper and F. C. Alley, *Air Pollution Control: A Design Approach*, Waveland Press, Prospect Heights, IL, 1986, pp. 493–515, 519–552.

19. R. M. Bethea, *Air Pollution Control Technology*, Van Nostrand Reinhold, New York, 1978, pp. 39–59.

20. L. Theodore and A. Buonicore, *Air Pollution Control Equipment*, Volume I: Particulates, Volume II: Gases, CRC Press, Boca Raton, FL, 1988.

21. A. M. Flynn and L. Theodore, *Health, Safety and Accident Management in the Chemical Process Industries*, Marcel Dekker, New York, 2002.

22. D. B. Turner, *Workbook of Atmospheric Dispersion Estimates* (EPA Publication No. AP-26). Research Triangle Park, NC, Environmental Protection Agency, 1970 (revised).

23. F. Pasquill, *Meteorology Magazine*, **90**(33), 1063 (1961).

24. F. A. Gifford, *Nuclear Safety*, **2**(4), 47 (1961).

25. H. Cota, *Journal of the Air Pollution Control Association*, **31**(8), 253 (1984).

26. L. Theodore, personal notes, 1989. Triangle Park. NC, author unknown, (NRTIS PB 86 234259 and 23467), 1986.

27. U.S. EPA. "Environmental Progress and Challenges." *EPA's Update*, August 1988.

28. Mueller Associates, Inc. *Indoor Air Quality Environmental Information Handbook: Building System Characteristics*, author unknown, Baltimore, MD, 1987.

29. J. Taylor, *Sampling and Calibration for Atmospheric Measurements*, ASTM Publications, Philadelphia, PA, 1987.

30. ASTM, 1991 *Annual Book of ASTM Standards*, Volume 11.03, ASTM D1357, New York, 1991.

31. S. Calvert and H. Englund, *Handbook of Air Pollution Technology*, Wiley, Hoboken, NJ, 1984.

32. L. Kornreich, *Proceedings of the Symposium on Statistical Aspects of Air Quality Data*. EPA-650/4-74-038, October 1974.

33. R. Munn, *Descriptive Micrometeorology*, Academic Press, New York, 1966.

34. D. Slade, *Meteorology and Atomic Energy*, TID-24180, U.S. Atomic Energy Commission, National Bureau of Standards, U.S. Department of Commerce, Springfield, VA, 1968.

35. A. Morris and R. Barras, *Air Quality Meteorology and Ozone*, ASTM Special Technical Publication, STP 653, Philadelphia, 1978.

36. U.S. EPA, *Guidance for Air Quality Monitoring Network Design and Instrument Siting*, QAQPS No. 1.2-012, September 1975.

37. U.S. EPA, *Selecting Sites for Carbon Monoxide Monitoring*. EPA-450/3-75-077, September 1975.

38. U.S. EPA, *Optimum Site Exposure Criteria SO_2 Monitoring*, EPA-450/3-77-0113, April 1977.

39. J. H. Seinfeld, "Optimal Locations of Pollutant Monitoring Stations in an Airshed," *Atmospheric Environment*, Vol. 6, Pergamon, Elmsford, NY, 1972, pp. 847–858.

40. M. M. Benairic, *Urban Air Pollution Modeling*, MIT Press, Cambridge, MA, 1980.

41. U.S. EPA, *User's Guide to the Contract Laboratory Program*, December 1986.

42. E. Hertzell and A. Waits, *EPA's Environmental Technology Verification Program: Raising Confidence in Innovation*, EM (AWHA), Pittsburgh, May 2004.

CHAPTER 4

WATER ISSUES

4.1 INTRODUCTION

Of primary concern in evaluating unknown "water" situations in nanoapplications are health and safety hazards. Concentrations (or potential concentrations) of emissions into water systems, and any potential time variation of these variables, also represent immediate concerns. Initial resolutions will unquestionably be based on experience, judgment, and professional knowledge.

As indicated in the previous chapter, extending standard procedures and design of equipment to control emissions from nanotechnology processes is subject to question at this time. The chemical and physical properties of these nanoemissions are not fully known, thus rendering judgments regarding equipment essentially mute. This information is critical to the advancement and development of new water pollution control technologies. This is especially true of ultrafine particles, that is, at the nanosize, since their mass, volume, and size is extremely small. However, it is anticipated that "new" emissions arising from nanoapplications will probably be "managed" initially by employing control technology as it exists today. Thus, it can be concluded that the equipment presently in place and available will see extensive use initially in the control of emissions from nanotechnology processes, and therefore deserved review.

Details in the description, design, prediction of performance, and operation and maintenance of traditional control equipment for these pollutants are available in the literature and have been routinely employed in practice for both control and

Nanotechnology: Environmental Implications and Solutions, L. Theodore and R. G. Kunz
ISBN 0-471-69976-4   Copyright © 2005 John Wiley & Sons, Inc.

recovery purposes. Qualitative information on these control devices is presented in this chapter.

Early History[1]

In a very real sense, many of the environmental problems faced by today's world are of relatively recent occurrence. For eons, effective nutrient and end-product cycling in natural systems eliminated any excess, useless residues. Industrial production, however, has evolved in a more linear mode, with the input of raw materials and output of both "useful" and seemingly "useless" (waste) products.

For much of human history, "the solution of pollution is dilution" was the general approach in dealing with wastes. In the case of wastewater, when the sewage of previous years was dumped into waterways, the natural processes of purification began. First, the sheer volume of clean water in the stream diluted the small amounts of wastes. Bacteria and other small organisms in the water consumed the sewage or other organic matter, turning it into new-bacterial cells, carbon dioxide, and other products. But the bacteria normally present in water must have oxygen in order to do their part in breaking down the sewage. Water acquires this all-important oxygen by absorbing it from the air and from plants that grow in the water itself. These plants use sunlight to turn the carbon dioxide present in water into oxygen.

The life and death of any body of water depends mainly upon its ability to maintain a certain amount of dissolved oxygen (DO). This DO is what fish breathe. If only a small amount of sewage is dumped into a stream, fish are not affected and the bacteria can do their work; the stream can quickly restore its oxygen loss from the atmosphere and from plants. Trouble begins when the sewage load is excessive. If carried to the extreme, the water could lose all of its oxygen, resulting in the death of fish and beneficial plant life.

It was not until the early 1800s that any sort of organized system for collection of wastewater was conceived. And it took nearly a century thereafter for systematic treatment to commence. Development of the germ theory by Koch and Pasteur in the latter half of the nineteenth century marked the beginning of a new era in sanitation. Before that time, the relationship of pollution to disease had been only faintly understood, and the science of bacteriology, then in its infancy, had not been applied to the subject of wastewater treatment.[2]

Both industrial and municipal wastewater treatment plants were originally designed in part to speed up the natural processes by which water purifies itself. However, these natural processes, even though they were accelerated in a waste treatment plant, were not sufficient to remove other contaminants such as disease-causing germs, excessive nutrients such as phosphates and nitrates, and chemicals and trace elements. New pollution problems of modern industry placed additional burdens on wastewater treatment systems, with the result that these pollutants have become even more difficult to remove from the wastewater.

In a very real sense, clean water is a resource that has been taken for granted. Pure water is often necessary for growing food, manufacturing goods, disposing of wastes, and for consumption. Water conservation is most frequently thought of as a measure to protect against water shortages. While protecting water supplies is an excellent reason to practice conservation, there is another important benefit of water conservation—improved water quality.

The link between water use and water pollution may not be immediately apparent, yet water use is a considerable source of pollution to waste systems. When water is used for household, industrial, agricultural, or other purposes, it is almost always degraded and polluted in the process. Called wastewater, this by-product of human and industrial activities may carry nutrients, biological and chemical contaminants, floating wastes, or other pollutants. Upon discharge, wastewater ultimately finds its way into groundwater or surface waters, contributing to their pollution.

Every day U.S. industry discharges billions of gallons of wastewater generated by industrial processes. This liquid waste stream often contains many toxic metals and organic pollutants. Unfortunately, the discharge point for a large portion of these industries is frequently a municipal sewer system that leads to a publicly owned treatment works (POTW). It is estimated that roughly 60 percent of the total toxic metals and organics discharged by industry winds up at municipal treatment plants.

This flood of toxic wastewater varies from day to day, and from region to region. As indicated above, its principal pollutants are toxic metals and organic chemicals. Some important toxic metals are lead, zinc, copper, chromium, cadmium, mercury, and nickel. Toxic organics include benzene, toluene, and trichloroethylene. Each of these substances, to a greater or lesser degree, is known to be harmful to human health. Many are toxic to aquatic life as well.

The consequences of these wastewater discharges have been severe. It is estimated that 14,000 miles of stream in 39 states have been polluted by toxic substances. It is also estimated that over a half million acres of lakes in 16 states and nearly 1000 square miles of estuaries in 8 states have been adversely affected.

Industries that send their wastes to POTWs are known as *indirect dischargers*; that is because their discharges enter America's surface waters by an indirect route via municipal sewage treatment works. *Direct dischargers*, on the other hand, are industries that release their treated wastewater directly to surface waters.

Another area of concern is groundwater pollution. A few years ago this was almost an unknown problem. Today, groundwater is one of the major environmental areas of concern. Some of the reasons are detailed in the next three paragraphs.

Groundwater is that part of the underground water that is below the water table. Groundwater is in the zone of saturation within which all the pore spaces of rock materials are filled with water. The United States has approximately 15 quadrillion gallons of water stored in its groundwater systems within one-half mile of the surface.

Annual groundwater withdrawals in the United States are on the order of 90 billion gallons per day, which is only a fraction of the total estimated water in storage. This represents about a threefold increase in American groundwater usage since 1950. Most of this is replenished through rainfall and offsets the hydraulic effects of pumpage, except in some heavily pumped, arid regions of the Southwest. American groundwater use is expected to rise in the future. Public drinking water accounts for

14 percent of groundwater use. Agricultural uses, such as irrigation (67 percent) and water for rural household and livestock (6 percent), account for 73 percent of groundwater usage. Self-supplied industrial water accounts for the remaining U.S. groundwater use. Approximately 50 percent of all Americans obtain all or part of their drinking water from groundwater sources.

The richest reserves of American groundwater are in the mid-Atlantic coastal region, the Gulf Coast states, the Great Plains, and the Great Valley of California. The Ogallala aquifer, which extends from the southern edge of North Dakota southwestward to the Texas and New Mexico border, is the single largest American aquifer in terms of geographical area. The most important American aquifer in agricultural terms is the large unconsolidated aquifer underlying the Great Valley of California. The most important groundwater sources of public drinking water are the aquifers of Long Island, New York, which have the highest per capita usage concentration in the United States.

This Chapter serves as an introduction to water pollution. The next two sections are concerned primarily with industrial water pollution control equipment and municipal water pollution control equipment. Some overlap exists in these two sections because these two topics are interrelated. The general subject of dispersion modeling in water systems is also provided. The chapter concludes with a section devoted to monitoring methods. Much of this material has been drawn from two literature sources.[1,3]

Before proceeding to the heart of this chapter, a review of certain water-related terms would enhance the textual materials to follow. Biochemical oxygen demand (BOD) is defined as the amount of oxygen required by a living organism engaged in the utilization and stabilization of the organic matter present. Standard tests are conducted at $20°C$ with a 5-day incubation period. BOD is usually exerted by dissolved and colloidal organic matter and imposes a load on the biological units of a treatment plant. Oxygen must be provided so that bacteria can grow and oxidize the organic matter. An added BOD load, caused by an increase in organic waste, requires more bacterial activity, more oxygen, and greater biological-unit capacity for its treatment.

Two other tests are generally used to estimate waste organic content: total organic carbon (TOC) and chemical oxygen demand (COD). TOC and COD are primary measures of total organic content, a portion of which may not be removed by biological treatment means. Other terms that have appeared in the literature include dissolved organic carbon (DOC), dissolved inorganic carbon (DIC), suspended organic carbon (SOC), suspended inorganic carbon (SIC), and total oxygen demand (TOD).

4.2 INDUSTRIAL WASTEWATER MANAGEMENT[3]

As indicated in the introductory section, this and the next section are concerned with industrial and municipal wastewater management, respectively. Because of the unknowns associated with future nanotechnology applications, and their accompanying unknown emissions, only the traditional approaches to control/ recovery are reviewed. It should be noted that the EPA is currently supporting nanotechnology research that will develop improved treatment of nanoparticles in water.

Types of Pollutants

The various types of pollutants found in industrial wastewaters can be classified into eight categories: common sewage and other oxygen-demanding wastes, disease-causing agents, plant nutrients, synthetic organic chemicals, inorganic chemicals and other mineral substances, sediments, radioactive substances, and "heat."

Oxygen-demanding wastes are the traditional organic wastes, ammonia, iron, or any other oxidizable compound contributed by industrial wastes. Such wastes result from food processing, paper mill production, tanning, and other manufacturing processes. These wastes are usually destroyed by bacteria if there is sufficient oxygen present in the water. Because fish and other aquatic life depend on oxygen for life, the oxygen-demanding wastes must be controlled—otherwise the fish will die.

The disease-causing agents include infectious organisms that are carried into surface water and groundwater by certain kinds of industrial wastes, such as from tanning and meat-packing plants. Humans or animals may come in contact with these microbes either by drinking the water or through swimming, fishing, or other activities. Although modern disinfection techniques have greatly reduced the danger from this type of pollutant, the problem must be carefully monitored.

Plant nutrients are the substances in the food chain of aquatic life (such as algae and water weeds) that support and stimulate plant growth. Carbon, nitrogen, and phosphorus are the three chief nutrients present in natural waters. Large amounts of these nutrients are present in sewage, certain industrial wastes, and drainage from fertilized land. Biological waste treatment processes do not remove the phosphorus and nitrogen to any substantial extent—in fact, they convert the organic forms of these substances into mineral form, making them more usable by plant life. The problem starts when an excess of these nutrients overstimulates the growth of water plants (particularly algae), a condition called *eutrophication.* When the plants and algae eventually die, the decomposition of their organic matter can lead to a severe depletion of dissolved oxygen in the body of water.

The synthetic organic chemicals include pesticides, synthetic industrial chemicals, and wastes from their manufacture. Many of these substances are toxic to aquatic life and are possibly harmful to humans. They cause taste and odor problems in drinking water, and they resist waste treatment. Some are known to be poisonous at very low concentrations.

A vast array of metal salts, acids, solid matter, and many other chemical compounds are included in the category of inorganic pollutants. Their sources include mining and manufacturing processes, oil field operations, and agricultural practices. While a wide variety of acids are discharged as waste by industry, the largest single source comes from mining operations and mines that have been abandoned.

Many of these types of chemicals are being created each year. If untreated, they can interfere with natural stream purification, destroy fish and other aquatic life, and cause excessive hardness of water supplies. Moreover, such chemicals can corrode expensive water treatment equipment and increase the cost of waste control. The EPA has developed a list of 129 priority pollutants that include both toxic organic and inorganic chemicals. The heightened danger of these chemicals has necessitated the drafting of specific limitations in most treatment plant permits.

While they are not as dangerous as some other types of pollutants, sediments are a major problem because of the sheer magnitude of the amount reaching our waterways. Sediments fill streams channels and harbors, requiring expensive dredging; they also fill reservoirs, reducing their capacity and useful life. They erode power turbines and pumping equipment, and they reduce fish and shellfish populations by blanketing fish nests and food supplies.

Radioactive pollution results from the mining and processing of radioactive ores, from the use of refined radioactive materials in power reactors, and from their use in industrial, medical, and research purposes. Because radiation bioaccumulates, control of this type of pollution must take into consideration total exposure in the environment.

While not directly toxic, "heat" reduces the capacity of water to absorb oxygen. Tremendous volumes of water are used by power plants and industry for cooling. Most of the water, with the added heat, is returned to natural bodies of water, raising their temperatures. Having less oxygen, the water is not as efficient in assimilating oxygen-consuming wastes and in supporting fish and aquatic life. Unchecked discharges of waste heat can seriously alter the ecology of a lake, a stream, or even part of the sea.

To complicate matters, most industrial wastes are a mixture of the eight types of pollution. Such combinations described above make the problems of treatment and control that much more difficult and costly.

Characterization of Wastewater

The volume and strength of industrial wastewaters are usually defined in terms of units of production. For example, for a pulp-and-paper mill, waste is often measured in gallons per ton of pulp. However, because of variations in the production process as well as differences in housekeeping and water reuse, considerable variation may occur in the flow and characteristics of wastewaters of similar industries. In fact, very few industries utilize identical process operations. Thus, in the development of an effective strategy for wastewater control, an industrial waste survey is usually required to determine the character of a particular industry's waste load.[4]

The industrial waste survey involves a definite procedure designed to develop a flow and material balance of all processes using water and producing wastes and to establish the variation in waste characteristics from specific process operations as well as from the plant as a whole. The results of the survey should establish possibilities for water conservation and reuse, and the variation in flow and strength of the effluents that need to undergo wastewater treatment.

The general procedure to be followed in developing the necessary information with a minimum of effort can be summarized in four steps:

1. Develop a sewer map from consultation with the plant engineer and an inspection of the various process operations. This map should indicate possible sampling stations and a rough order of magnitude of the anticipated flow.
2. Establish sampling and analysis schedules. Continuous samples with composites weighted according to flow are the most desirable, but these either are not always possible or do not lend themselves to the physical sampling location. The period of sample composite and the frequency of sampling must be established according to the nature of the process being investigated.
3. Develop a flow and material balance diagram. After the survey data are collected and the samples analyzed, a flow and material balance diagram should be developed that considers all significant sources of waste discharge. How closely the summation of the individual sources checks the measured total effluent provides a check on the accuracy of the survey.
4. Establish statistical variation in significant waste characteristics. The variability of certain waste characteristics is significant for waste treatment plant design. These data should be prepared as a probability plot showing frequency of occurrence.

Data from industrial waste surveys are highly variable and are usually susceptible to statistical analysis. Statistical analysis of variable data provides the basis for treatment plant process design.

Determination of Wastewater Composition

As implied in the industrial waste survey discussed above, the objective of water quality management is to control the discharge of pollutants so that water quality is not degraded to an unacceptable extent below the natural background level. Direct measurement of the pollutants must be made. The impact of the pollutant on water quality must be predicted and compared to the background water quality that would be present without human intervention. Based on this, a decision can be made on the levels acceptable for the intended uses of the water.[5]

The *biochemical oxygen demand (BOD) test* is used to determine the amount of biodegradable contents in a sample of wastewater. As discussed in the introductory section, the BOD test measures the amount of oxygen consumed by living organisms (mainly bacteria) as they metabolize the organic matter present in a waste. The test simulates conditions as close as possible to those that occur naturally.[6] As is true for any bioassay, the success of a BOD test depends on the control of such environmental and nutritional factors as pH and osmotic conditions, essential nutrients, constant temperature, and population of organisms representative of natural conditions.

Although it theoretically takes an infinite amount of time for all the oxidizable material in a sample of water to be consumed, it has been empirically determined that a period of 20 days is required for near completion. Because the 20-day waiting period is too long to wait in most cases, a 5-day incubation period (the time in which 70 to 80 percent of available material is usually oxidized) has been adopted as standard procedure. This shorter test is referred to as BOD_5. The BOD_5 test also serves the purpose of avoiding the contribution of nitrifying bacteria to the overall DO measurement, since these bacteria populations do not become large enough to make a significant oxygen demand until about 8 to 10 days from the start of the BOD test.[7]

While the BOD test is the only test presently available that gives a measure of the amount of biologically oxidizable organic material present in a body of water, it does not provide an accurate picture of the total amount of overall oxidizable material and thus the total oxygen demand. For such measurements, the *chemical oxygen demand (COD) test* is used. In this test, a very strong oxidizing chemical, usually potassium dichromate, is added to samples of different dilution. To ensure full oxidation of the various compounds found in the samples, a strong acid and a chemical catalyst are added. One of the products of this oxidation–reduction reaction is the chromate ion, which gives a very sharp color change that can be easily detected in a spectrophotometer. Thus, the greater the absorbance measured (inverse of transmittance), the greater the amount of oxidation that has taken place, the more the oxidizable material originally present, and the greater the oxygen demand of that material. To find the corresponding concentration of DO in the sample, the absorbance measured is related to an absorbance–concentration graph of a standard whose oxygen concentration is known.

A third method for measuring the organic matter present in a wastewater is the *total organic carbon (TOC) test*. The test is performed by placing a sample into a high-temperature furnace or chemically oxidizing environment. The organic carbon is oxidized to carbon dioxide. The carbon dioxide that is produced can then be measured. While the TOC test does directly measure the concentration of organic compounds, it does not provide a direct measurement of the rate of reaction nor the degree of biodegradability. For this reason, the TOC test has been accepted as a monitoring technique but has not been utilized in the establishment of treatment regulations.

The BOD, COD, and TOC tests provide estimates of the general organic content of a wastewater. However, because the particular composition of the organics remains unknown, these tests do not reflect the response of the wastewater to various types of biological treatment technologies. It is therefore necessary to separate the wastewater into its specific components.

The *solids analysis* focuses on the quantitative investigation and measurement of the specific content of solid materials present in a wastewater. The *total solids* are the materials that remain after the water from the solution has evaporated. For any solution, the total solids consist of *suspended* and *dissolved* particles. Dissolved particles are classified according to size as either *soluble* or *colloidal*.

When a solution is poured through a filter (usually 0.45 μm in size), those solids that are too large to pass through the filter openings will be retained. These solids are called *suspended solids*. The filter containing the suspended solids is heated to remove the water and to produce the *total suspended solids (TSS)*. The TSS are further heated in a muffle furnace at 550°C. At such high temperatures, the *volatile suspended solids* (VSS), mostly organics, will be vaporized, leaving behind the remaining *fixed suspended solids.* Those solids that are small enough to pass through the filter are called *dissolved (soluble) solids.* By heating the solution to remove the water, the *total dissolved solids* are obtained. As with the suspended solids, these solids are further heated at extreme temperatures to remove the *volatile dissolved solids* (VDS), leaving the *fixed dissolved solids.*

The assumption is made that all volatile solids (VSS and VDS) are organic compounds. A distinction is made as to whether such solids may or may not be adsorbed by a specific treatment medium. The soluble organics that are nonsorbable are further separated into degradable and nondegradable constituents.

Additional measurements of pH, nitrogen, and phosphorus levels in wastewaters are often required when complex pollution conditions exist. Various measuring devices are available for direct measurement of these inorganic components.

Wastewater Treatment Processes

Numerous technologies exist for treating industrial wastewater. These technologies range from simple clarification in a settling pond to a complex system of advanced technologies requiring sophisticated equipment and skilled operators. Finding the proper technology or combination of technologies adequate to treat a particular wastewater to meet federal and local requirements and yet still be cost effective can be a challenging task.

Treatment technologies can be divided into three broad categories: physical, chemical, and biological. Many treatment processes combine two or all three categories to provide the most economical treatment. There are a multitude of treatment technologies for each of these categories. Therefore, although the technologies selected for discussion in this chapter are among the most widespread in the industrial field, they represent only a fraction of the available technologies. Figure 4.1 matches various types of industry with candidate treatment technologies.[8-10]

Physical Treatment Processes

Clarification (Sedimentation)   When an industrial wastewater containing a suspension of solid particles that have a higher specific gravity than the transporting liquid is in a relatively calm state, the particles will settle out because of the effects of gravity. This process of separating the settleable solids from the liquid is called *clarification* or *sedimentation.* In some treatment systems employing two or more stages of treatment and clarification, the terms *primary, secondary*, and *final* clarification are used. *Primary clarification* is the term normally used for the first clarification process in the system. This process is used to remove the readily settleable solids prior to subsequent treatment processes, particularly biological treatment. This treatment step results in significantly lower pollutant loadings to downstream processes and is appropriate for industrial wastewaters containing a high suspended solids content.

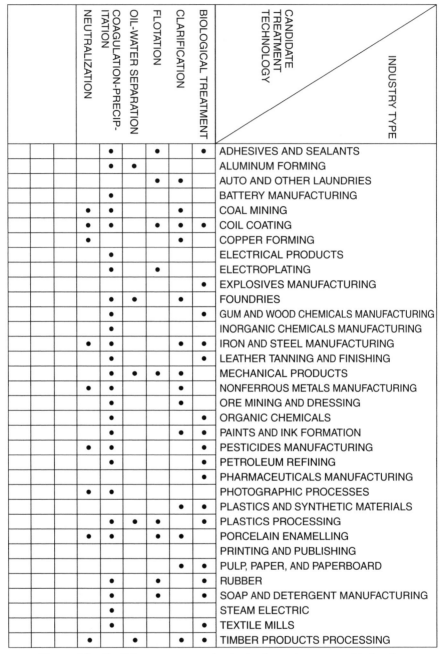

Figure 4.1   Candidate treatment technologies for various types of industries.

The actual physical sizing of the clarifier (depth, surface area, inlet structure, etc.) is highly dependent upon the quantity and composition of the wastewater flow to be treated. Because these criteria will vary substantially among industries, sizing must be performed following a detailed study conducted on a site-by-site basis.

Clarification units can be either circular or rectangular and are normally designed to operate on a continuous, flow-through basis. Circular units are generally called *clarifiers*, whereas rectangular units are commonly referred to as *sedimentation tanks*.

In general, clarifiers are designed on the basis of the type of suspended solids to be removed from the waste stream. There are three types of suspended solids. Class I solids are discrete particles that will not readily flocculate and are typical of raw influents. Class II particles are characterized by a relatively low solids concentration of flocculated material. These types of particles are usually found in wastewaters that have been subjected to chemical addition. Either of the above two particles may be found in industrial wastewaters. Class I particles are typical of physically manufactured operations, whereas Class II particles are more common to chemically manufactured operations. The last type are Class III particles. These are normally the solids generated from a biological treatment process. Because of their poor settling properties, special care must be exercised in the design for their removal. Clarification is effective in removing a substantial portion of the suspended solids and BOD in an industrial wastewater. Not only does this aid in the performance of other, downstream processes, but it also protects them from negative effects, such as clogging. This technology is relatively receptive to modifications, such as chemical addition, which allows for alterations to compensate for changes in wastewater characteristics. Finally, clarification is a relatively simple operation, which requires neither sophisticated operation nor highly skilled operators. The disadvantages of clarification as a component of an industrial wastewater treatment system include high capital cost and limitations on specific pollutant loadings and flows.

Flotation In contrast to clarification, which separates suspended particles from liquids by gravitational forces, flotation separates these particles by their density, through the introduction of air into the system. Fine bubbles adhere to, or are absorbed by, the solids, which are then lifted to the surface. There are several methods of achieving flotation. In one method—dissolved air flotation—small gas bubbles are generated as a result of the precipitation of a gas from a solution supersaturated with that gas. Supersaturation occurs when air is dispersed through the sludge in a closed pressure tank. When the sludge is removed from the tank and exposed to atmospheric pressure, the previously dissolved air leaves the solution in the form of fine bubbles.

Flotation separator tanks can be either rectangular or circular in shape and constructed of either concrete or steel. Tanks are constructed with equipment to provide uniform flow distribution at the inlet, to provide pressurized gas, and to skim off float material. In designing a dissolved air flotation separator system, the following variables are typically considered: full, partial, or recycled pressurization;

feed characteristics; surface area; float characteristics; hydraulic loading; chemical usage; type of pressurization equipment; and operating pressure.

Flotation separation is an appropriate technology for treating suspended solids and oil and grease in industrial waters. Generally, the process will achieve 40 to 65 percent suspended solids removal and 60 to 80 percent of oil and grease removal. The reduction of BOD levels will also be realized through the removal of suspended solids. The addition of chemicals prior to actual flotation can significantly improve the performance of this technology.

In addition to the pollutant removals mentioned, flotation has other advantages. The resulting float material may be used as an auxiliary fuel source. Minimized construction costs may result because the higher overflow rates and shorter retention times require smaller tanks. Also, because of the short detention periods and the presence of DO in the wastewater, odor problems are kept at a minimum. Some disadvantages associated with this type of treatment process are high operating and maintenance costs, inability to remove "heavy" suspended solids, and the requirement of more skilled personnel than for conventional settling processes.

Oil–Water Separation   In practically all manufacturing industries, oil and grease can be found in a plant's wastewater. This generally results from equipment lubrication, accidental spills, and similar incidence. However, for some industries, such as petroleum and oil refining, oil compounds can represent a significant constituent of the flow because of the use of these oils as the raw materials in production. Not only can the oil have a detrimental effect on the performance of a wastewater treatment system, but it also represents a valuable raw material being wasted. Thus, specific treatment processes have been developed to separate the major portion of the oil from the wastewater stream. These are called oil–water separators.

The configuration of the separator is that a flow-through tank. The basic principle by which oil–water separators work is the differential between the specific gravities of water versus the oils to be removed. Since the oils generally have lower specific gravities, they will rise to the top of the unit while the heavier water sinks to the bottom. An important consideration in the separator design is the oil globule size, since the terminal velocity of spheres in a liquid medium will determine the rate at which the oil rises. Retaining baffles and skimmers capture the oil compounds as the separated water leaves the tank. Sludge collectors can be used to scrape the bottom of the tank to remove any deposited solids.

The major advantage of oil–water separators is their ability to treat wastewater that is heavily laden with oil compounds. Because of the design of these separators, they represent a very simple treatment operation that minimizes personnel requirements. Since this technology relies on natural forces (gravity) rather than on chemicals or aeration, its operating costs are minimized. Also, it results in a more "pure" oil, which can make recycling much easier.

The disadvantages of oil–water separators are as follows. The ability to remove only specific size oil globules can result in very small particles passing through the system. Also, if the wastewater has a high solids content and the oil and solids become mixed, the resultant globules may be too heavy to float to the surface and thus remain suspended. Finally, improperly maintained units are subject to odor problems.

Chemical Treatment Processes

Coagulation—Precipitation The nature of an industrial wastewater is often such that conventional physical treatment methods will not provide an adequate level of treatment. Particularly, ordinary settling or flotation processes will not remove ultrafine colloidal particles and metal ions. In these instances, natural stabilizing forces (such as electrostatic repulsion and physical separation) predominate over the natural aggregating forces and mechanisms, namely, van der Waals forces and Brownian motion, which tend to cause particle contact. Therefore, for adequate treatment of these particles in industrial wastewaters, coagulation–precipitation may be warranted.

The first and most important part of this technology is coagulation, which involves two discrete steps. Rapid mixing is employed to ensure that the chemicals are thoroughly dispersed throughout the wastewater flow for uniform reaction. Next, the wastewater undergoes flocculation, which provides for particle contact so that the particles can agglomerate to a size large enough for removal. The final part of this technology involves precipitation. This is really the same as settling and thus can be performed in a unit similar to a clarifier.

Coagulation–precipitation is capable of removing industrial wastewater pollutants such as BOD, COD, and TSS. In addition, depending upon the specifics of the wastewater being treated, coagulation–precipitation can remove additional pollutants such as phosphorus, nitrogen compounds, and metals. This technology is attractive to industry because a high degree of clarifiable and toxic pollutants removal can be combined in one treatment process. A disadvantage of this process is the substantial quantity of sludge generated, which presents a sludge disposal problem.

Neutralization In virtually every type of manufacturing industry, chemicals play a major role. Whether they result from the raw materials or from the various processing agents used in the production operation, some residual compounds will ultimately end up in a process wastewater. Thus, it can generally be expected that most industrial waste streams will deviate from the neutral state (i.e., will be acidic or basic in nature).

Highly acidic or basic wastewaters are undesirable for two reasons. First, they can adversely impact the aquatic life in receiving waters; second, they might significantly affect the performance of downstream treatment processes at the plant site or at a publicly owned treatment works (POTW). Therefore, in order to rectify these potential problems, one of the most fundamental treatment technologies, neutralization, may be employed at industrial facilities. Neutralization involves adding an acid or a base to a wastewater to offset or neutralize the effects of its counterpart in the wastewater flow, namely, adding acids to alkaline wastewaters and bases to acidic wastewaters.

The most important considerations in neutralization treatment are a thorough understanding of the wastewater constituents so that the proper neutralizing chemicals are used, and proper monitoring to ensure that the required quantities of these chemicals are used and that the effluent is in fact neutralized. For acid waste streams, lime, soda ash, and caustic soda are the most common base chemicals used in neutralization. In the case of alkaline waste streams, sulfuric, hydrochloric, and nitric acid are generally used for neutralization. Some industries have operations that separate acid and alkaline waste streams. If properly controlled, these waste streams can be combined to produce a neutralized effluent without the need for adding neutralizing chemicals.

Neutralizing treats the pH level of a wastewater flow. Although most people do not think of pH as a pollutant, it is in fact designated by the EPA as such. Since many subsequent treatment processes are pH-dependent, neutralization can be considered as a preparatory step in the treatment of all pollutants.

Eliminating the adverse impacts on water quality and wastewater treatment system performance is not the only benefit of neutralization. Acidic or alkaline wastewaters can be very corrosive. By neutralizing its wastewaters, a plant can protect its treatment units and associated piping. Thus, the major disadvantage of neutralization is that the chemicals used in the treatment process are themselves corrosive and can be dangerous.

Biological Treatment Processes For many industrial wastewaters, one of the most important concerns has to do with those constituents which can exert an oxygen demand and have an impact on receiving waters. Although most industries are discussed in terms of the toxics they discharge, they are also significant sources of BOD and COD. In many instances, the most appropriate industrial treatment technology for removing this oxygen-demanding pollutant is biological treatment. Discussion in this section will be limited to those processes frequently used in the industrial field: aerobic suspended growth processes (activated sludge), aerobic contact processes, aerated lagoons (stabilization ponds), and anaerobic lagoons.

Aerobic Suspended Growth Processes (Activated Sludge) An aerobic suspended growth process is one in which the biological growth products (microorganisms) are kept in suspension in a liquid medium consisting of entrapped and suspended colloidal and dissolved organic and inorganic materials. This biological process uses the metabolic reactions of the microorganisms to attain an acceptable effluent quality by removing those substances exerting an oxygen demand. Depending on the type of material in the raw wastewater stream, this process may be preceded by one or more other treatment technologies (i.e., clarification, oil and grease removal, etc.) to improve removal efficiencies.

In the suspended growth processes, wastewater enters a reactor basin, concrete–steel–earthen tank(s) where microorganisms are brought into contact with the organic components of the wastewater by some type of mixing device. This mixing device not only maintains all material in suspension but also promotes transfer of oxygen to the wastewater, thus providing the oxygen necessary for sustaining the biological activities in the reactor basin. The organic matter in the wastewater serves as a carbon and energy source for microbial growth and is converted into microbial cell tissue and oxidized end products, mainly carbon dioxide. Contents of the reactor basin are referred to as *mixed liquor suspended solids* (MLSS) and consist mainly of microorganisms and inert and nonbiodegradable matter.

When the MLSS are discharged from the reactor basin, a means of separating them is normally provided. Concentrated microbial solids are recycled back to the reactor basin to maintain a concentrated microbial population for degradation of the wastewater. Because microorganisms are usually synthesized in the process, a means must be provided for wasting some of the microbial solids. Wasting of the solids is usually done from the settling basin, although wasting from the reactor basin is an alternative.

The first suspended growth process, now called the *conventional activated sludge process*, was developed to achieve carbonaceous BOD removal. However, since its inception, many modifications to the basic process have taken place. The variations in the activated sludge process are too numerous to be discussed in detail here. However, some of these are step aeration, contact stabilization, aerated lagoon, and deep tank aeration.

Aerobic suspended growth systems can be adapted to treat a wide range of industrial wastewaters. The process can be easily expanded to accommodate increased flows. However, these systems do have some drawbacks. Suspended growth systems generally perform best under uniform hydraulic and pollutant loadings. For some industries it is extremely difficult to maintain these conditions because of their manufacturing operations. A common event is a "shock" loading of high strength entering the treatment process, resulting in poor pollutant treatment. Also, certain toxic pollutants can kill microorganisms in the reactor basin, causing a loss of treatment from this part of the overall system. Operational skills and controls required to operate an aerobic suspended growth system effectively are higher than for most other biological processes. Finally, high energy costs can be expended in providing mixing and oxygen in the process to sustain microbial growth.

Excluding those waste streams with very high concentration of toxic pollutants, aerobic suspended growth systems can be used to treat any wastewater containing biodegradable matter. Removal efficiencies in excess of 90 percent for carbonaceous BOD have been achieved through this process.

Aerobic Attached Growth Processes   An aerobic attached growth process is one in which the biological growth products (microorganisms) are attached to some type of medium (i.e., rock, plastic sheets, plastic rings, etc.) and where either the wastewater trickles over the surface or the medium is rotated through the wastewater. The process is related to the aerobic suspended growth process in that both depend upon the biochemical oxidation of organic matter in the wastewater to carbon dioxide, with a portion oxidized for energy to sustain and promote the growth of microorganisms. There are three general types of aerobic attached growth systems: conventional trickling filters, roughing filters, and rotating biological contractors (RBCs).

There are several advantages for attached growth processes over other biological processes. First, microorganism growth can be easily reinstituted in the case of an accidental "kill." Secondly, since oxygen is supplied naturally, the need for air- or oxygen-generating equipment is eliminated. This, along with a much simpler operation, lowers the requirements for highly skilled operational personnel. Both can result in substantial cost savings. The disadvantages are that attached growth treatment processes experience operating difficulties in cold climates. Enclosing the units for temperature protection can lead to other problems such as condensation. These units are also susceptible to clogging if dense media are used and/or high solids loadings are applied.

Excluding those waste streams with very high concentrations of toxic pollutants, aerobic attached growth processes can be used to treat any wastewater containing biodegradable matter. In general, the aerobic attached growth processes are not quite as efficient as the aerobic suspended growth processes in removing BOD suspended solids, and toxic pollutants.

Aerobic Lagoons (Stabilization Ponds)   Aerobic lagoons are large, shallow earthen basins that are used for wastewater treatment by utilizing natural processes involving both algae and bacteria. The objective is microbial conversion of organic wastes into algae. Aerobic conditions prevail throughout the process.

In aerobic photosynthesis, the oxygen produced by the algae through the process of photosynthesis is used by the bacteria in the biochemical oxidation and degradation of organic waste. Carbon dioxide, ammonia, phosphate, and other nutrients released in the biochemical oxidation reactions are, in turn, used by the algae, forming a cyclic–symbiotic relationship. Aerobic lagoons are used for treatment of weak industrial wastewater containing negligible amounts of toxic and/or nonbiodegradable substances.

Anaerobic Lagoons   Anaerobic lagoons are earthen ponds built with a small surface area and a deep liquid depth of 8 to 20 ft. Usually these lagoons are anaerobic throughout their depth, except for an extremely shallow surface zone. Once greases form an impervious layer, completely anaerobic conditions develop. In a typical anaerobic lagoon, raw wastewater enters near the bottom of the lagoon (often at the center) and mixes with the active microbial mass in the sludge blanket, which is usually about 6 ft deep. The discharge is located near one of the sides of the lagoon, submerged below the liquid surface. Excess undigested grease floats to the top, forming a heat-retaining and airtight cover. Excess undigested grease floats to the top, forming a heat-retaining and airtight cover. Excess sludge is washed out with the effluent. Anaerobic lagoons are effective prior to aerobic treatment of high-strength organic wastewater that also contains a high concentration of solids. Under optimal operating conditions, BOD removal efficiencies of up to 85 percent are possible.

The advantage in using either aerobic or anaerobic lagoons are low cost (excluding land if not readily available), simplicity of operation, low operating and maintenance cost, and when designed properly, high reliability. The disadvantages in using any lagoon process are high land requirements, possible odor emissions, and the potential for seepage of wastewater into groundwater unless the lagoon is adequately lined. In addition, in most locales of the United States there are seasonal changes in both available light and temperature. Typically in the winter, biological activity decreases because of a reduction in temperature.

4.3 MUNICIPAL WASTEWATER TREATMENT[1]

In the United States, the treatment and disposal of wastewater did not receive much attention in the late 1800s because the extent of the nuisance caused by the discharge of untreated wastewater into the relatively large bodies of water was not severe, and large areas of land were available for disposal purposes. During the early 1900s, nuisance and declining health conditions created the need for more effective means of wastewater management. Land space was no longer readily available, especially in larger cities. This led to the planning, design, construction, and operation of a higher level of pollution technology in wastewater treatment facilities.

If untreated wastewater is allowed to accumulate, the decomposition of the organic materials it contains can lead to the production of offensive odors and gases. In addition, untreated wastewater contains numerous pathogenic (i.e., disease-causing) microorganisms, released from the human intestinal system. It contains nutrients that can stimulate the growth of aquatic life, and it may also contain toxic compounds. For these reasons, the immediate removal of wastewater from its sources of generation, followed by treatment and disposal, is imperative.

Today, not only must a wastewater treatment plant satisfy effluent quality requirements, it must also satisfy many other environmental conditions. Some of these conditions are designed to meet requirements for aesthetics and the minimization of obnoxious odors at treatment and disposal sites; to prevent contamination of water supplies from physical, chemical, and biological agents; to prevent destruction of fish, shellfish, and other aquatic life; to prevent degradation of water quality of receiving waters from overfertilization; to prevent impairment of beneficial uses of natural water (recreation, agriculture, commerce, or industry); to protect against the spread of disease from crops grown on sewage irrigation or sludge disposal; to prevent decline in land values and, therefore, not to restrict the level of community growth and development; and to encourage other beneficial uses of effluent.

The purpose of any wastewater treatment plant is to convert the components in raw wastewater, with its inherently harmful characteristics, into a relatively harmless final effluent for discharge to a receiving body of water and to dispose of safely the solids (sludge) produced in the process. The planning, design, construction, and operation of wastewater treatment facilities is a complex problem. This section presents a brief review of the various factors that lead to the development and operation of a wastewater treatment plant.

Wastewater Characteristics and Solids Production

Municipal wastewater is composed of a mixture of dissolved, colloidal, and particulate organic and inorganic materials. The total amount of the substances accumulated in a body of wastewater is referred to as the *mass loading*. The concentration of any individual component is constantly changing as a result of sedimentation, hydrolysis, and microbial transformation and degradation of organic compounds.

Wastewater characteristics are described in terms of water flow conditions and chemical quality. The characteristics depend largely on the types of water usage in the community and industrial and commercial contributions. During wet weather, a significant quantity of infiltration or inflow may also enter the municipal collection system. This will change the characteristics of wastewater significantly.

The characteristics of a wastewater may be obtained from plant flow records and laboratory data at the municipal wastewater treatment plant. The data describing the wastewater characteristics should include minimum, average, and maximum dry-weather flows, peak wet-weather flows, sustained maximum flows, and chemical parameters such as biological oxygen demand (BOD), total suspended solids, total dissolved solids, pH, total nitrogen, phosphorus, and toxic chemicals. It is important that reliable estimates of the wastewater characteristics be made since this is what the municipal wastewater plant will be treating.

Quality of Wastewater Municipal wastewater contains 99.9 percent water. The remaining materials include suspended and dissolved organic and inorganic matter as well as microorganisms. These materials make up the physical, chemical, and biological qualities that are characteristic of residential and industrial waters.

The physical quality of municipal wastewater is generally reported in terms of its temperature, color, odor, and turbidity. The temperature of wastewater is slightly higher than that of the water supply. The latter is an important parameter because of its effect upon aquatic life and the solubility of gases. The temperature varies slightly with the seasons, normally remaining higher than the air temperature during most of the year and falling lower only during the hot summer months. The color of a wastewater is usually indicative of age. Fresh water is usually gray; septic wastes impart a black appearance. Odors in wastewater are caused by the decomposition of organic matter that produces offensive-smelling gases such as hydrogen sulfide. Wastewater odor generally can provide a relative indication of its condition.

Turbidity in wastewater is caused by a wide variety of suspended solids. Suspended solids are defined as the matter that can be removed from water by filtration through prepared membranes. Volatile suspended solids for the most part represent organics and may affect oxygen resources of the stream; however, they are not a direct measure of total organics. Suspended solids may cause the undesirable conditions of increased turbidity and silt load in the receiving water.

Chemical characteristics of wastewater are expressed in terms of organic and inorganic constituents. Different chemical analyses furnish useful and specific information with respect to the quality and strength of wastewater.

Organic components in wastewater are the most significant factor in the pollution of many natural waters. The principal groups of organic substances found in municipal wastewater are proteins (40 to 60 percent), carbohydrates (25 to 50 percent), and fats and oils (10 percent). Carbohydrates and proteins are easily biodegradable.

Fats and oils are more stable but are decomposed by microorganisms. In addition, wastewater may also contain small fractions of synthetic detergents, phenolic compounds, and pesticides and herbicides. These compounds, depending on their concentration, may create problems such as nonbiodegradability, foaming, or carcinogenicity. The concentrations of these toxic organic compounds in wastewaters are very small. Their sources are usually industrial wastes and surface runoff.

The inorganic compounds most often found in wastewater are chloride, hydrogen ions (pH), alkalinity-causing compounds, nitrogen, phosphorus, sulfur, and heavy metals. Trace concentrations of these compounds can significantly affect organisms in the receiving water through their growth-limiting characteristics. Algae and macroscopic plant forms are capable of using inorganics as substrate in their metabolism. The major elements that serve as inorganic metabolites are carbon, ammonia–nitrogen, and phosphorus.

Gases commonly found in raw wastewater include nitrogen, oxygen, carbon dioxide, hydrogen sulfide, ammonia, and methane. Of all these gases mentioned, the ones that are considered most in the design of a treatment facility are oxygen, hydrogen sulfide, and methane concentrations. Oxygen is required for all aerobic life forms either within the treatment facility or in the receiving water. During the absence of aerobic conditions, oxidation is brought about by the reduction of inorganic salts such as sulfates or through the action of methane-forming bacteria. The end products are often very obnoxious. To avoid such conditions, it is important that an aerobic state be maintained.

The quality and species of micro- and macroscopic plants and animals that make up the biological characteristics in a receiving body of water may be considered as the final test of wastewater treatment effectiveness. Within the treatment facility, the wastewater provides the perfect medium for good microbial growth, whether it be aerobic or anaerobic. Bacteria and protozoa are the keys to the biological treatment process used at most treatment facilities, and to the natural biological cycle in receiving waters. In the presence of sufficient dissolved oxygen, bacteria convert the soluble organic matter into new cells and inorganic elements. This causes a reduction of organics loading through the buildup of more complex organisms and/ or removal.

Within a wastewater treatment plant handling domestic wastes, bacteria will be the dominant plant/animal species, with other organisms achieving varying degrees of importance depending on the plant design and the specific processes used for treatment. Consequently, wastewater treatment is directed toward using and removing the common bacteria along with organic and inorganic components.

Water quality in a receiving body of water is strongly influenced by the biological interactions that take place. The discharge (effluent) to the receiving waters becomes a normal part of the biological cycle, and its effect on aquatic organisms is the ultimate consideration of treatment plant operation. Typically, the species and organisms found in the biological examination of the receiving waters include zooplankton and phytoplankton, peryphyton, macroinvertebrates, macrophytes, and fish.

Because of the increasing awareness that enteric viruses can be waterborne, attempts have been made to identify and quantify virus contributions to receiving waters from wastewater treatment plants. Different physical and chemical analytical techniques have been used in an effort to identify such viruses. Virus removal is mandated in connection with reclaimed wastewater discharged for indirect reuse in groundwater, and into lakes and rivers used for body-contact purposes.

Sludge Characteristics Most wastewater treatment plants use primary sedimentation to remove readily settleable solids from raw wastewater. In a typical plant, the dry weight of primary sludge solids (those removed by filtration, settling, or other physical means) is roughly 50 percent of that for the total sludge solids. Primary sludge is usually easier to manage than biological and chemical sludges—which are produced in the advanced or secondary stages of treatment—for several reasons. First, primary sludge is readily thickened by gravity, either within a primary sedimentation tank or within a separate gravity thickener. In comparison with biological and many chemical sludges, primary sludge with low conditioning requirements can rapidly be mechanically dewatered. Furthermore, the dewatering device will produce a drier cake and give better solids capture than it would for most biological and chemical sludges.

Primary sludge always contains some grit, even when the wastewater has been processed through degritting. Typically, it also contains different anaerobic and facultative species of bacteria, such as sulfate-reducing and oxidizing bacteria. Primary sludge production is typically within the range of 800 to 2500 lb per million gallons (100 to 300 mg/L) of wastewater. A basic approach to estimating primary sludge production for a particular plant is to compute the quantity of TSS entering the primary sedimentation tanks, assuming an efficiency of removal.

Biological sludges are produced by secondary treatment processes such as activated sludge, trickling filters, and rotating biological contactors. Quantities and characteristics of biological sludges vary with the metabolic and growth rates of the various microorganisms present in the sludge. Biological sludge that contains debris such as grit, plastics, paper, and fibers is produced at plants lacking primary treatment. Plants with primary sedimentation normally produce a fairly pure biological sludge. Biological sludges are generally more difficult to thicken and dewater than are primary sludge and most chemical sludges. Additional details are provided in the next chapter.

Wastewater Plant Design[11-16]

A significant amount of times is involved in the planning and design of a wastewater treatment facility. The initial phase consists of a facility plan, which must be prepared before grants from the federal and state governments can be obtained.

A facility plan is prepared to identify the water pollution problems in a specific area, develop design data, evaluate alternatives, and recommend a solution. Most of the data developed in a facility plan are used in the preparation of design plans, specifications, and cost estimates of the wastewater treatment facility.

Steps in the facility plan include:

1. *List effluent limitations:* The plan should list the effluent limitations applicable to the facility being planned. Publicly owned wastewater treatment plants built after June 30, 1974, must achieve "best practicable waste treatment technology" (equivalent of secondary treatment).
2. *Assess current situation:* Describe briefly the existing conditions to be considered when examining alternatives during planning. The following conditions should be described:
   a. Planning area
   b. Organizational context
   c. Demographic data
   d. Water quality
   e. Other existing environmental conditions
   f. Existing wastewater flows and treatment systems, including system performance
   g. Infiltration and inflow
3. *Assess future conditions:*
   a. Twenty-year planning period beyond startup of the facility. Phased construction should be considered.
   b. Land use; must be carefully coordinated with the state, municipal, and regional regulations, policies, and plans.
   c. Demographic and economic projections.
   d. Forecasts of flow and wasteloads; includes projections of economic and population growth, infiltration/inflow estimates, analysis of pollutant content and flows in the existing system, sewer overflow data, industrial wasteload data projections, and pollution-reducing possibilities.
   e. Future environment of the planning area without the project.
4. *Develop and evaluate alternatives:* Alternative treatment systems and their impact on the environment, long-range sewer plans for the planning area, sludge utilization and/or disposal, and facility location.
5. *Select plan:* The public is provided with alternative proposals and hearings are held to explain each proposal.
6. *Preliminary design of treatment works:* The following items are included: a schematic flow diagram; unit processes; plant site plans; sewer pipe plans and profiles; design data regarding detention items, flow rates, and sizing of units; operation and maintenance summary; cost estimates; and a completion schedule.
7. *Arrangements for implementation:* Following selection of plan and design, existing institutional arrangements should be reviewed and a financial program developed, including preliminary allocation of costs among various classes of users of the system.

Wastewater treatment plants utilize a number of unit operations and processes to achieve the desired degree of treatment. The collective treatment schematic is called a *flowsheet*. Many different flow schemes can be developed from various unit operations and processes for the desired degree of treatment. However, the most desirable flow scheme is the one that is the most cost-effective.

Wastewater treatment plants are designed to process the liquid and solid portions of the wastewater. Treatment systems and solids disposal systems must be put together so as to assure the most efficient utilization of resources such as money, materials, energy, and work force in meeting treatment requirements. Logic dictates what the process elements must be and the order in which they go together.

A methodical process of selection must be followed in choosing a resource-efficient and environmentally sound system from the numerous treatment and disposal options available. The basic selection mechanism used is the "principle of successive elimination," an iterative procedure in which less effective options are progressively eliminated until only the most suitable system or systems for the particular site remain.

The concept of a "treatment train" is a result of a systems approach to problem solving. However, this concept is useful only if all components of the train are considered. This includes not only sludge treatment and disposal components, but also wastewater treatment options and other critical linkages such as sludge transportation, storage, and side stream treatment.

The general sequence of events in system selection is:

1. Selecting relevant criteria
2. Identifying options
3. Narrowing the list of candidate systems
4. Selecting a system

The basis of unit process design is the initial and future volume and characteristics of the wastewater, anticipated variations, and statutory requirements of regulatory agencies. Data acquisition required will be, in part, determined by the treatment processes considered and largely determined by the treatment requirements. The plant process design is usually a function of peak and minimum loading conditions and not a function of average or median conditions. Wastewater flows for design purposes are estimated by two methods:

1. Gauging flow in existing systems and making corrections appropriate to increased future requirements
2. Estimating and totaling the various components of the flow

For updating an existing wastewater facility, the first method of estimating flow is more reliable. In the case of a new facility, the second method is preferred. A commonly used basis for wastewater treatment plant design is the adoption of a maximum flow of two to four times the average dry-weather flow. The value depends on local factors, including population.

Several flow rates are used for design of various elements in a wastewater treatment plant. The average daily flow rate for the period of design is determined by totaling the 24-h average of all components. This rate is used to determine such items as pumping and chemical costs, sludge solids, and organic loading. The design average flow rate is generally used for mass loading of treatment units. Peak design rate, usually 2 to 2.25 times the design average flow rate, is used for hydraulic sizing.

The method of solids disposal usually controls the selection of solids treatment systems and not vice versa (see Chapter 5). Thus, the system selection procedure normally begins when the solids disposal option is specified. The process selection procedure consists of developing treatment/disposal systems that are compatible with one another and appear to satisfy local relevant criteria, and choosing the best system or systems by progressive elimination. Proper selection of the sludge processing equipment is important for trouble-free operation of a wastewater treatment plant. Sludge is quite odorous and may cause environmental concerns. Therefore such factors as solids capture, chemical quality of return flows, ability to handle variable quality of sludge, ease of operation, and odors are often given serious consideration.

Wastewater treatment plants utilize a number of individual or unit operations and processes to achieve the desired degree of treatment. The collective treatment schematic is called a *flow scheme*, *a flow diagram*, *a flowsheet*, *a process train*, or a *flow schematic*. Many different flow schemes can be developed from various unit operations and processes for the desired level of treatment. Unit operations and processes are grouped together to provide what is known as primary, secondary, and tertiary (or advanced) treatment. The term *primary* refers to physical unit operations, *secondary* refers to chemical and biological unit processes, and *tertiary* refers to combinations of all three.

Treatment methods in which the application of physical forces predominate are known as *physical unit operations*.[17] These were the first methods to be used for wastewater treatment. Screening, mixing, flocculation, sedimentation, flotation, and filtration are typical physical unit operations.

Treatment methods in which the removal or conversion of contaminants is brought about by the addition of chemicals or by other chemical reactions are known as *chemical unit processes*. Precipitation, gas transfer, adsorption, and disinfection are the most common examples used in wastewater treatment.

Treatment methods in which the removal of contaminants is brought about by biological activity are known as *biological unit processes*. Biological treatment is used primarily to remove the biodegradable organic substances (colloidal or dissolved) in wastewater. Basically these substances are converted into gases that can escape to the atmosphere or into biological cell tissue that can be removed by settling. Biological treatment is also used to remove the nitrogen in wastewater. With proper environmental control, wastewater can be treated biologically in most cases.

Municipal wastewater treatment plants are mandated by the federal government to provide secondary-level treatment in order to comply with the final effluent discharge to the receiving waters. As a result of this mandate, secondary treatment is considered the minimum acceptable design criterion for a wastewater treatment plant. Secondary treatment implies that chemical and biological processes are utilized in the overall treatment process in addition to physical treatment. An overview of a typical secondary wastewater treatment plant may be described as being grouped in the following sections:

1. Primary treatment and handling facilities
2. Activated sludge treatment/secondary facilities
3. Sludge treatment, storage, and disposal facilities

As influent plant flow, primary raw sewage enters the plant by means of interceptors; then the flow enters the wet wells, which are located at the lowest elevation in the plant. The purpose of the low elevation is to provide gravity flow and to create sufficient pressure differential for the use of sewage pumps leading into the collection systems. Upstream of the pumps, the raw sewage first enters screens such as a bar screen rack mechanism. This device physically separates large objects such as paper cups, rags, and sticks from the raw sewage flow. Grinders and shredders are also frequently incorporated into the screening facilities. The next phase of processing involves grit removal; here, materials such as sand, coffee grounds, and cigarette filter tips are separated in grit tanks or chambers. Typically a grit tank is rectangular in shape, equipped with velocity control devices, raking devices, and aeration piping. Grit tanks are designed to perform the following functions:

1. Protect moving mechanical equipment from abrasion and abnormal wear.
2. Reduce clogging in pipes and sludge hoppers.
3. Prevent accumulations in aeration tanks and sludge digesters and the consequent loss of usable volume.

From the grit tanks, the raw sewage enters the primary settling tanks. The purpose of the primary settling tanks is to provide the detention time needed to separate the settleable and floatable solids from the wastewater for appropriate handling. The objectives of primary settling are:

1. Removal of finely dispersed solids by floc formation with larger particles
2. Removal of colloidal material via adsorption to larger particles

At this stage of the processing and handling of the primary effluent, further treatment will now employ biological and chemical processes (secondary treatment). The most common secondary treatment process is the activated sludge process. The aeration tank is the heart of the activated sludge process. Here, oxygen is introduced into the system, along with the effluent from the primary treatment process, to satisfy the requirements of the organisms in the sludge and to keep the sludge dispersed in the aeration liquor. These organisms break down the waste material remaining in the primary effluent.

The activated sludge process is used to convert nonsettleable substances from a finely divided, colloidal, and dissolved form to biological floc. This newly formed biological floc, known as *sludge*, is removed from the system through sedimentation, thereby providing a high degree of secondary treatment. The biological floc is developed in aeration tanks and settled out in final settling tanks.

There are different types of aeration systems; the two most common types are subsurface diffusion and mechanical aeration. In the diffused air system, compressed air is introduced at the bottom of the tank near one side. This causes the tank's contents to be circulated by the air-lift effect. The floating or fixed bridge aerators are common mechanical aeration devices. Some mechanical aerators use a blade to agitate the surface of the tank and disperse air bubbles into the aeration liquor. Others circulate the aeration liquor by an updraft or downdraft pump or turbine. This action produces surface and subsurface turbulence, while at the same time diffusing air through the liquid mass.

The basic function of separating the newly formed activated sludge from the treated wastewater cannot be accomplished without proper operation of the final settling tanks. The final settling tanks' function is liquid–solid separation, solids concentration, and solids return.

Secondary treatment as described above can remove up to 99 percent of the bacteria from raw wastewater. Tertiary treatment is capable of an even greater removal. In some instances, however, regulatory agencies may require removal or inactivation of pathogenic organisms in excess of that achieved by secondary or tertiary treatment. In those cases, disinfection is required. Chlorine is the disinfectant used at the vast majority of wastewater treatment plants. It is effective and reliable and may be the least costly disinfection alternative. Disinfection usually must be performed year round if the receiving stream is used as a source of drinking water by downstream communities or if it flows into shellfish areas. Seasonal disinfection may be required if the main usage is bathing or other primary contact recreation. After disinfection, the treated water is now ready for use by the receiving community.

The problem of sludge treatment still needs to be addressed. The principal sources of sludge at a municipal wastewater treatment plant are the primary sedimentation tanks and the final settling tanks. Additional sludge may come from chemical precipitation, nitrification facilities, screening, and grinder devices. Sludge contains large volumes of water. The small fraction of solids in the sludge is highly offensive, thus requiring treatment such as conditioning. Common sludge management processes include thickening, stabilization, dewatering, and disposal.

Sludge thickening is used to concentrate solids and reduce volume. Thickened sludge requires less tank capacity and chemical dosage for stabilization, and smaller piping and pumping equipment for transport. Common methods of sludge thickening used at medium-to-large plants are gravity thickening, dissolved air flotation, and centrifugation.

The main purpose of sludge stabilization is to reduce pathogens, eliminate offensive odors, and control the potential for putrification of organic matter. Sludge stabilization can be accomplished by biological, chemical, or physical means. Selection of any method depends largely on the ultimate sludge disposal method. Various methods of sludge stabilization are: anaerobic or aerobic digestion (biological), chemical oxidation or lime stabilization (chemical), and thermal conditioning (physical). The anaerobic digestion process is the most widely selected stabilization process used in most plants.

The principal purpose of sludge digestion is twofold:

1. To reduce in volume the solids content from the treatment process
2. To decompose highly putrescible organic matter to relatively stable or inert organic and inorganic compounds

The digestion process is carried out in the digester or digestion tanks. In an anaerobic digestion process, the tanks are covered to exclude air and oxygen, to prevent release of offensive odors, and to collect digester gases. In the process of decomposition, gases are formed by bacterial action. These gases are comprised mostly of carbon dioxide and methane. Anaerobic digesters are most commonly operated at 35 to 42°C.

Anaerobic sludge digesters are of two types, namely, standard rate and high rate. In the standard-rate digestion process the digester contents are usually unheated and unmixed. The digester period may vary from 30 to 60 days. In a high-rate digestion process, the digester contents are heated and completely mixed. The required detention period is 10 to 20 days. Often a combination of standard- and high-rate digestion is achieved in two-stage digestion. The second-stage digester mainly separates the digested solids from the supernatant liquor, although additional digestion and gas recovery may also be achieved.

Overview of Advanced Wastewater Treatment Technology (Tertiary Treatment)

Advanced wastewater treatment technology is designed to remove those constituents that are not adequately removed in the secondary treatment plants. These include nitrogen, phosphorus, and other soluble organic and inorganic compounds. Nitrogen and phosphorus are nutrients that accelerate the growth of plants in the receiving waters. Ammonia is toxic to fish, exerts nitrogenous oxygen demand, and increases chlorine demand. Heavy metals, hydrogenated hydrocarbons, and phenolic compounds are toxic to fish and other aquatic life, concentrate in the food chain, and may create taste and odor problems in water supplies. Many of these constituents must be removed to meet stringent water quality standards and to allow reuse of the effluent for municipal, industrial, irrigation, recreation, and other water needs.

The most commonly used advanced wastewater treatment processes are chemical precipitation of phosphorus, nitrification, denitrification, ammonia stripping, breakpoint chlorination, filtration, carbon adsorption, ion exchange, reverse osmosis, and electrodialysis. Some of these unit operations and processes are discussed below.

Coagulation and Flocculation   Coagulation involves the reduction of surface charges and the formation of complex hydrous oxides. Flocculation involves combining the coagulated particles to form settleable floc. The coagulant (alum, ferric chloride, ferrous sulfate, ferric sulfate, etc.) is mixed rapidly and then stirred in order to encourage formation of floc piror to settling. The objective is to improve the removal of BOD, TSS, and phosphorus. To accomplish this, chemicals are added prior to primary treatment, in biological treatment processes, or in separate facilities following the biological treatment processes. Polymers are also used in conjunction with these chemicals. The chemical dosage is adjusted to give the desired amount of floc formation and BOD, TSS, or phosphorus removal. The approximate average alum and ferric chloride dosages in municipal wastewater are 170 and 80 mg/L, respectively. Lime may be needed in conjunction with iron salts.

Major equipment for a coagulation–flocculation system includes: chemical storage; chemical feeders, piping, and control systems; flash mixer; flocculator; and sedimentation basin. Proper mixing of chemicals at the point of addition, along with proper flocculation prior to clarification, is essential for maximum effectiveness. Flocculation may be accomplished in a few minutes to half an hour in basins equipped with mixers, paddles, or baffles. The coagulants react with alkalinity to produce insoluble metal hydroxides for floc formation. If sufficient alkalinity is not present, lime or soda ash (sodium carbonate) is added in the desired dosages.

In biological treatment units the addition of coagulants has a marked influence on biota. Population levels of protozoans and higher animals are adversely affected. However, the BOD, TSS, and phosphorus removal is significantly improved. Overdosing of chemicals may cause toxicity to microorganisms necessary to the treatment process. Finally, large quantities of sludge are produced from chemical precipitation, the chemical sludges may cause serious handling and disposal problems.

Lime Precipitation   Lime reacts with bicarbonate alkalinity and orthophosphate, causing flocculation. The objectives of lime addition are to increase removal of BOD, TSS, and phosphorus. Lime is added prior to primary sedimentation, in biological treatment, or in a separate facility after secondary treatment. Lime addition may be in a single stage or in two stages. Major equipment for a lime precipitation process includes: lime storage; lime feeders, piping, and control systems; flash mixer; flocculator; and sedimentation basin. For a two-stage system, a CO_2 source such as an incinerator or an internal combustion engine is required.

Lime in a single stage is used in primary biological treatment or after secondary treatment. The procedure is the same as in coagulation. The pH of the wastewater is raised to about 10, and the wastewater is flocculated and then settled. Normal lime dosage is about 180 to 250 mg/L as CaO. The actual dosage depends primarily on phosphorus concentration, hardness, and alkalinity. The biological system is not adversely affected by lime addition in moderate amounts (80 to 120 mg/L as CaO). The microbial production of carbon dioxide is sufficient to maintain a pH near neutral. High dosage may upset the biological process.

Lime addition prior to primary and after secondary treatment may be achieved in two stages. First, the pH of wastewater is raised to greater than 11, flocculated, and settled. Lime dosage up to 450 mg/L as CaO may be necessary. Next, the effluent is carbonated by adding CO_2 to lower the pH, and then it is flocculated and settled. Higher BOD, TSS, and phosphorus removal is achieved in a two-stage lime process. However, use of excess lime causes scale formation in tanks, pipes, and other equipment. Also, handling and disposal of large quantities of lime sludge is a problem.

Nitrification   Nitrification converts ammonia to nitrate form, thus eliminating toxicity to fish and other aquatic life and reducing the nitrogenous oxygen demand. Ammonia oxidation to nitrite and then to nitrate is performed by auto-trophic bacteria. The reactions are shown by Equations (4.1) and (4.2):

$$NH_3 + \frac{2}{3}O_2 \longrightarrow NO_2^- + H^+ + H_2O + \text{biomass} \tag{4.1}$$

$$NO_2^- + \frac{1}{2}O_2 \longrightarrow NO_3^- + \text{biomass} \tag{4.2}$$

Temperature, pH, dissolved oxygen, and the ratio of BOD to total Kjeldahl nitrogen (TKN) are important factors in nitrification.

Denitrification   Nitrite and nitrate are reduced to gaseous nitrogen by a variety of facultative heterotrophs in an anaerobic environment. An organic source, such as acetic acid, acetone, ethanol, methanol, or sugar, is needed to act as hydrogen donor (oxygen acceptor) and to supply carbon for synthesis. Methanol is preferred because it is least expensive. Equations (4.3) through (4.5) express the basic reactions:

$$3O_2 + 2CH_3OH \longrightarrow 2CO_2 + 4H_2O \tag{4.3}$$
$$6NO_3^- + 5CH_3OH \longrightarrow 3N_2 + 5CO_2 + 7H_2O + 6OH^- \tag{4.4}$$
$$2NO_2^- + CH_3OH \longrightarrow N_2 + CO_2 + H_2O + 2OH^- \tag{4.5}$$

Biological Nutrient Removal   Biological phosphorus and nitrogen removal has received considerable attention in recent years. Among other benefits reported, bio-logical nutrient removal can reduce aeration capacity and eliminate the need for chemical treatment, thereby saving money. Biological nutrient removal involves anaerobic and anoxic treatment of return sludge prior to discharge into the aeration basin. Based on the anaerobic, anoxic, and aerobic treatment sequence and internal recycling, several processes have been developed. Over 90 percent phosphorus and high nitrogen removal (by nitrification and denitrification) has been reported by biological means.

Ammonia Stripping   Ammonia gas can be removed from an alkaline solution by air stripping as expressed by Equation (4.6):

$$NH_4^+ + OH^- \longrightarrow NH_3 \uparrow + H_2O \tag{4.6}$$

The basic equipment for an ammonia-stripping system includes chemical feed, a stripping tower, a pump and liquid spray system, a forced-air draft, and a recarbonation system. This process requires raising the pH of the wastewater to about 11, formation of droplets in the stripping tower, and providing air–water contact and droplet agitation by countercurrent circulation of large quantities of air through the tower. Ammonia-stripping towers are simple to operate and can be very effective in ammonia removal, but the extent of their efficiency is highly dependent on the air temperature. As the air temperature decreases, the ammonia removal efficiency drops significantly. This process, therefore, is not recommended for use in a cold climate. A major operational disadvantage of stripping is the need for neutralization and prevention of calcium carbonate scaling on the tower. Also, there is some concern over the discharge of ammonia into the atmosphere.

4.4 DISPERSION MODELING IN WATER SYSTEMS[3]

Four distinct periods can be distinguished in the development of mathematical models:[18]

1. *The precomputer age (1900–1950).* During this time, the focus was entirely on water quality with little concern for the environmental aspects.
2. *The transition period of the 1950s.* Data collection was accelerated but the analysis was slow and costly.
3. *The early years of computer use.* During the 1960s, the first computer models were developed, and many models were developed during the 1970s owing to greater computer access.
4. *The mid-1970s to date.* Because of the development of inexpensive microcomputers, models can now be used for routine evaluations.

Models are classified by the number of dimensions modeled and by the type of model employed. These can be described as follows:

1. *One-dimensional models.* The only direction modeled is the direction of flow. This is a valid model for flowing streams, where the concentration of pollutants is taken to be constant with stream cross section.
2. *Two-dimensional models.* This is used in wide rivers where concentration may not be uniform across the entire width. The model is a function of both width and flow direction. For deep, narrow rivers, lakes, or estuaries, the horizontal and vertical dimensions are modeled, while the lateral dimension is held constant.
3. *Three-dimensional models.* The assumption here is that concentration can vary with length, width, and depth. Of the three models, this is the most accurate. It is also, however, both tedious and time consuming. The potential accuracy is greater, but the development and running costs are also greater.

There is also a zero-dimensional model that takes none of the lateral, vertical, or longitudinal motion into consideration. In this model, a segment of stream is treated as a completely mixed reactor.[19] The chemical engineer often refers to this type of model as a CSTR, that is, a continuous stirred tank reactor. These models are an assembly of concepts in the form of one or more mathematical equations that approximate the behavior of a natural system or phenomenon.[20]

Rather than focus on the water systems themselves, however, this section will examine the individual pollutants and components that are modeled. These include microorganisms, dissolved oxygen, eutrophication, and toxic chemicals. Within each of these areas, specific types of contaminants as well as the different water systems that are affected will be explored.

Mathematical Models

There are mass (componential) and flow (overall mass) balance equations for rivers and streams. They are as follows:

Mass Balance

$$\begin{array}{l}\text{Mass rate of} \\ \text{substance upstream}\end{array} + \begin{array}{l}\text{Mass rate of} \\ \text{substance added} \\ \text{by outfall}\end{array} = \begin{array}{l}\text{Mass rate of substance} \\ \text{immediately downstream} \\ \text{from outfall assuming} \\ \text{complete mixing}\end{array} \quad (4.7)$$

Flow Balance

$$\begin{array}{l}\text{Flow rate} \\ \text{upstream}\end{array} + \begin{array}{l}\text{Flow rate added} \\ \text{by outfall}\end{array} = \begin{array}{l}\text{Flow rate immediately} \\ \text{downstream from outfall}\end{array} \quad (4.8)$$

The physical characteristics of lakes set them apart from other water systems in modeling for microorganisms, as well as for the other contributors to water quality. These characteristics include evaporation due to a large surface area, and temperature stratification due to poor mixing within the lake. Therefore, these differences must be taken into account in the model.[21]

$$\begin{array}{l}\text{Net flow into and} \\ \text{out of the lake due to} \\ \text{river and/or} \\ \text{groundwater flow}\end{array} + \begin{array}{l}\text{Precipitation} \\ \text{directly onto} \\ \text{the lake}\end{array} - \text{Evaporation} = \begin{array}{l}\text{Change in the} \\ \text{lake volume} \\ \text{with time}\end{array} \quad (4.9)$$

The last, and most complicated water system that will be examined is the estuary. Unlike lakes and rivers, there are no simple balance equations that can be written for estuaries. Estuaries are coastal water bodies where freshwater meets the sea. They are traditionally defined as semienclosed bodies of water having a free connection with the open sea and within which seawater is measurably diluted with freshwater entering from land drainage.[22]

The seaward end of an estuary is easily defined because it is connected to the sea. The landward end, however, is not that well defined. Generally, tidal influence in a river system extends further inward than salt intrusion. That is, the water close to the fall line of the estuary may not be saline, but it may still be tidal. Thus, the estuary is limited by the requirement that both salt and freshwater be measurably present. The exact location of the salt intrusion depends on the freshwater flow rate, which can vary substantially from one season to another.[23]

The variations in an estuary throughout the year, together with the fact that each estuary is different, makes modeling rather difficult. Some simplifications can, however, be made that provide some remarkably useful results in estimating the distribution of estuarine water quality. The simplifications can be summarized through the following assumptions:

1. The estuary is one-dimensional.
2. Water quality is described as a type of average condition over a number of tidal cycles.
3. Area, flow, and reaction rates are constant with distance.
4. The estuary is in a steady-state condition.

A water body is considered to be a one-dimensional estuary when it is subjected to tidal reversals (i.e., reversals in direction of the water velocity) and where only the longitudinal gradient of a particular water quality parameter is dominant.[21]

Microorganisms

The transmission of waterborne diseases (e.g., gastroenteritis, amoebic dysentery, cholera, and typhoid) has been a matter of concern for many years. The impact of high concentrations of disease-producing organisms on water uses can be significant. Bathing beaches may be closed permanently or intermittently during rainfall conditions when high concentrations of pathogenic bacteria are discharged from urban runoff and combined sewer overflows. Diseases associated with drinking water continue to occur.[21]

There are four types of organisms that can affect water quality. The first are indicator bacteria, which may reflect the presence of pathogens. In the past they were used as a measure of health hazard. Pathogenic bacteria are the cause of such diseases as salmonella, cholera, and dysentery and continue to be a problem worldwide. Viruses are submicroscopic, inert particles that are unable to replicate or adapt to environmental conditions outside a living host,[24] and, if ingested, they can cause hepatitis. And finally, pathogenic protozoa are parasitic, but able to reproduce, and are responsible for amoebic dysentery. Table 4.1 lists examples of communicable disease indicators and organisms.

The factors that can affect the survival or extinction of these microorganisms are:[21]

1. Sunlight
2. Temperature
3. Salinity
4. Predation

TABLE 4.1 Examples of Communicable Disease Indicators and Organisms

Indicators	Viruses
Bacteria	Hepatitis A
Total coliform	Enteroviruses
Fecal coliform	Polioviruses
Fecal streptococci	Echoviruses
Obligate anaerobes	Coxsackieviruses
Bacteriophages (bacterial viruses)	
Pathogenic Bacteria	Pathogenic Protozoa and Helminths
Vibrio cholerae	*Giardia lambia*
Salmonella	*Entamoeba hystolytica*
Shigella	Facultatively parasitic ameobae
	Nematodes

Source: From Ref. 21.

5. Nutrient deficiencies
6. Toxic substances
7. Settling of organism population after discharge
8. Resuspension of particulates
9. Growth of organisms within the body of water

The overall decay rate equation for microorganisms is given as:

$$K_B = K_{B1} + K_{Bl} \pm K_{Bs} - K_a \qquad (4.10)$$

where K_B is the overall rate of decay, K_{B1} is the death rate due to temperature, salinity, and predation, K_{Bl} is the death rate due to sunlight, K_{Bs} is the net loss or gain due to settling or resuspension, and K_a is the aftergrowth rate.[21]

One describing equation for the downstream distribution of bacteria in rivers and streams is given as

$$N = N_0 \exp(-K_B t^*) \qquad (4.11)$$

where N is the concentration of an organism, N_0 is the concentration of the organism at the outfall, K_B is the overall net rate of decay as given previously, and t^* is the time it takes to travel a downstream distance x at a water velocity U.[21] As the equation states, the organism will decay exponentially with time (thus, the form of a classical first-order chemical reaction).

Dissolved Oxygen

The problems of dissolved oxygen (DO) in surface waters have been recognized for over a century. The impact of low DO concentrations or of anaerobic conditions was reflected in an unbalanced ecosystem, fish mortality, odors, and other aesthetic nuisances. While coliform was a surrogate variable for communicable disease and public health, DO is a surrogate variable for the general health of the aquatic ecosystem.[21]

The variations in dissolved oxygen levels are caused by sources and sinks. The sources include: reaeration from the atmosphere, which is dependent upon turbulence, temperature, and surface films; photosynthetic oxygen production, where plants react CO_2 and H_2O to form glucose and oxygen; and incoming DO from tributaries (streams that feed a larger stream or a lake) or effluents. The sinks of DO are: oxidation of carbonaceous (CBOD) and nitrogenous (NBOD) waste materials, oxygen demand of the sediments of the water body, and the use of oxygen for respiration by aquatic plants.[21]

With these inputs, sources, and sinks, the following general mass balance equation for DO in a segmented volume can be written as:

$$
\begin{array}{l}
\text{Rearation} + (\text{Photosynthesis} - \text{Respiration}) - \begin{array}{l}\text{Oxidation}\\ \text{of CBOD}\\ \text{\& NBOD}\end{array} \begin{array}{l}\text{Sediment}\\ -\ \text{oxygen}\\ \text{demand}\end{array} \\[2em]
+ \text{Oxygen input} \pm \begin{array}{l}\text{Oxygen transport}\\ \text{(into or out}\\ \text{of segment)}\end{array} = \begin{array}{l}\text{Change with time of}\\ \text{dissolved oxygen in a}\\ \text{specific volume of water}\end{array}
\end{array}
\tag{4.12}
$$

This equation can be applied to a specific water body where the transport, sources, and sinks are unique to that aquatic system.[1]

The discharge of municipal and industrial waste, and urban and other non-point-source runoff will necessitate a continuing effort in understanding the DO resources of surface waters. The DO problem can thus be summarized as the discharge of organic and inorganic oxidizable residues into a body of water, which, during the processes of ultimate stabilization of the oxidizable material (in the water or sediments), and through interaction of aquatic plant life, results in the decrease of DO to concentrations that interfere with desirable water uses.[21] The balance equations that can be applied to the various water systems, namely rivers and lakes, are the same as described in the previous section.

Eutrophication

Even the most casual observer of water quality has probably had the dubious opportunity of walking along the shores of a lake that has turned into a sickly green pea soup. Or perhaps, one has walked the shores of a slow-moving estuary or bay and had to step gingerly to avoid rows of rotting, matted, stringy aquatic plants. These problems have been grouped under a general term called *eutrophication*. The unraveling of the causes of eutrophication, the analysis of the impact of human activities on the problem, and the potential engineering controls that can be exercised to alleviate the condition have been a matter of special interest for the past several decades.

Eutrophication is the excessive growth of aquatic plants, both attached and planktonic (those that are free-swimming), to levels that are considered to be an interference with desirable water uses. One of the principal stimulants is an excessive level of nutrients such as nitrogen and phosphorus. In recent years, this problem has been increasingly acute due to the discharge of such nutrients by municipal and industrial sources, as well as agricultural and urban runoff. It has often been observed that there is an increasing tendency for some water bodies to exhibit increases in the severity and frequency of phytoplankton blooms and growth of aquatic weeds apparently as a result of elevated levels of nutrients.[21]

The principal variables of importance in the analysis of eutrophication are:[21]

1. Solar radiation at the surface and with depth
2. Geometry of water body; surface area, bottom area, depth, volume
3. Flow, velocity, and dispersion
4. Water temperature
5. Nutrients
   a. Phosphorus
   b. Nitrogen
   c. Silica
6. Phytoplankton

The nonorganic products that result from oxidation are referred to as nutrients. They include nitrogen found in the form of ammonia, nitrite, and nitrate and phosphorus, which occurs in the form of phosphates. A third nutrient, silicon in the form of silicate, enters the system through the weathering of soils and rocks. These nutrients, along with carbon dioxide, "feed" the process of photosynthesis, which creates the beginning components of the biological cycle—phytoplankton (the microscopic plants that drift around in the water), diatoms (which need silicon for their shells), flagellates (organisms possessing one or more whiplike appendages often used for locomotion), and green and blue-green algae among them. Some of these nutrients will go into producing a complementary pool of rooted aquatic plants. Figure 4.2 shows the basic biological cycle in lakes and estuaries.[25] A critical portion of the cycle is the phytoplankton pool. Increased nutrient availability can lead to unsightly plankton blooms and to anoxic conditions as the available oxygen is used up in the plankton decay.[25]

Other processes related to algal growth and nutrient recycling are sorption and desorption of inorganic material, settling and deposition of phytoplankton, uptake of nutrients and growth of phytoplankton, death of phytoplankton, mineralization of organic nutrients, and nutrient generation from the sediment.[26]

Toxic Substances

The issue of the release of chemicals into the environment at a level of toxic concentration is an area of intense concern in water quality and ecosystem analyses.

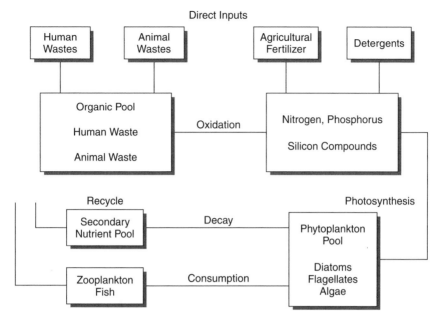

Figure 4.2 Anthropogenic inputs.

Passage of the Toxic Substances Control Act (TSCA) of 1976 in the United States, unprecedented fines, and continual development of data on lethal and sublethal effects attest to the expansion on control on the production and discharge of such substances. However, as illustrated by pesticides, the ever-present potential for insect and pest infestations with attendant effects on humans and livestock results in a continuing demand for product development. As a result of these competing goals, considerable effort has been devoted in recent years to the development of predictive schemes that would permit an a priori judgment of the fate and effects of a chemical in the environment.[21]

Table 4.2 summarizes a few specific chemicals that are of special interest in evaluating water quality, with brief descriptions of the problems that they may cause. Note especially the differences in the effects on water quality caused by various chemical forms of the same element, e.g., sulfur. This illustrates the importance, sometimes, of assaying specific ions or molecules instead of merely total content of the element itself.[27]

The uniqueness of the toxic substances problem lies in the potential transfer of a chemical to humans with possible attendant public health impacts. This transfer occurs primarily through two principal routes:

1. Ingestion of the chemical from the drinking water supply
2. Ingestion of the chemical from contaminated aquatic foodstuffs (e.g., fish and shellfish) or from food sources that utilize aquatic foodstuffs as a feed.

TABLE 4.2   Potential Water Quality Problems That May Be Caused by a Few Selected Chemicals

Chemical	Potential Problems
Arsenic	Toxicity to humans
	Toxicity to aquatic life
Chlorine	Organic reactions form trihalomethanes
	Toxicity to fish and other aquatic life
Calcium	Causes "hardness" in water
	May result in scale formation in pipes
Iron	Causes stains in laundry and on fixtures
	May kill fish by clogging their gills
Nitrogen: ammonia	May accelerate eutrophication in lakes
	May improve productivity of the water
	May be toxic to aquatic life
Nitrogen: nitrates	May be toxic to babies
	May accelerate eutrophication in lakes
	May improve productivity of the water
Oxygen, dissolved	Low concentrations harmful to fish
	Low concentrations may cause odor problems
	High concentrations accelerate metal corrosion
	Low or zero concentration may allow sulfide formation and concrete corrosion
Phenolics	Tastes and odors in drinking water
	Can cause tainting of fish flesh
	May be toxic to aquatic life
Sulfur: sulfides	Objectionable odors in and near water
	May be toxic to aquatic life
	May corrode concrete through acid formation
	Oxidation of sulfide to sulfate exerts an oxygen demand
Sulfur: sulfites	React with DO and exert oxygen demand
Sulfur: sulfates	Increase water corrosiveness to metals
	Decompose anaerobically to form sulfides
	Salty taste and laxative effects

The toxic substances water quality problem can therefore be summarized as the discharge of chemicals into the aquatic environment. This results in concentrations in the water or aquatic food chain at levels that are determined to be toxic, in a public health sense or to the aquatic ecosystem itself, and thus may interfere with the use of the water body for water supply or fishing or contribute to ecosystem instability.[21]

4.5   MONITORING METHODS[1]

This section focuses on methods used to obtain data regarding contamination of water. In Section 3.6 information was provided on how to set up a monitoring program and how to ensure that the data obtained are reliable. Additional references and material were presented on the collection and preservation of samples as well as details on the EPA Environmental Technology Verification Program, which also applies to water applications.[28]

Many argue that appropriate monitoring equipment and procedures are a precursor to the protection of human health and the ecosystem from nanoemissions. Although traditional monitoring methods will continue to have their place, units that provide rapid, precise, simultaneous pollutant quantifications will be required in the future for detection at the nanolevel. The material to follow will be based primarily on present standard procedures.

Standard Practices for Sampling of Water

This section includes three procedures for water sample collection. The first procedure is for the collection of a grab sample of water at a specific site representing conditions only at the time of sampling. It is the only procedure suitable for bacteriological analyses and some radiological test procedures. The second practice is for collection of a composite sample at a specific site, portions of which are collected at various time intervals. Alternatively, the composite may consist of portions collected at various sites or may consist of a combination of both site and time variables. The third procedure provides a continuous flowing sample, from one or more sampling sites, suitable for on-stream analyzers.[29]

The goal of sampling is to obtain for analysis a portion of the main body of water that is truly representative. The most critical factors necessary to achieve this are points of sampling, time of sampling, frequency of sampling, and maintenance of integrity of the sample prior to analysis. Homogeneity is frequently lacking, necessitating multiple-point sampling. If it is impractical to utilize a most-representative sampling point, it may be practical to determine and understand interrelationships so that results obtained at a minimum number of points may be extrapolated. A totally representative sample should not be an absolute prerequisite to the selection of a sampling point. With adequate interpretation, a nonrepresentative sample can yield valuable data about trends and can indicate areas where more representative data would be available. Most samples collected from a single point in a system must be recognized as being nonrepresentative to some degree. Therefore, it becomes important to recognize the degree of representation in the sample and to make it a part of the permanent record. Otherwise, an artificial degree of precision is assigned to the data when it is recorded.

The following general rules are applicable to all sampling procedures. First, the samples must represent the conditions existing at the point taken. Secondly, the samples must be of sufficient volume and must be taken frequently enough to permit reproducibility of testing requisite for the desired objective, as conditioned by the method of analysis to be employed. Thirdly, the samples must be collected, packed, shipped, and manipulated prior to analysis in a manner that safeguards against change in the particular constituents or properties to be examined.

Grab Samples The procedure for collecting grab samples is applicable to sampling water from sources such as wells, rivers, streams, lakes, oceans, and reservoirs for chemical, physical, bacteriological, or radiological analysis. A grab sample represents the conditions existing only at the point and time of sampling.

A reasonably accurate estimate of the composition of a raw water piped from a large body of water (such as the Great Lakes) far enough from the shoreline to avoid variation from inflowing tributaries and waste discharges may be made by taking individual samples at infrequent intervals (such as biweekly or monthly) sufficient to cover seasonal changes. If samples are taken from near the shoreline of such a body of water or from a river, they should be taken at shorter intervals (e.g., daily) to provide more exact knowledge of the variations in composition where these are of importance in the use to which the water is to be put. If greater variations or cycles of pollution occur or close surveillance of plant intake water is required, more frequent samples should be collected (e.g., at hourly intervals).

Water undergoing continuous or intermittent treatment must be sampled with such frequency that adequate control is ensured. The interval between samples is directly related to the rate at which critical characteristics can reach intolerable limits.

Normally, samples are taken without separation of particulate matter. If constituents are present in colloidal or flocculent suspension, the samples should be taken so that they are present in representative proportion.

The specific method of analysis for any given constituent to determine the volume of sample required needs to be consulted. Frequently, the required volume will vary with concentration level of any given constituent. The minimum volume collected should be three to four times the amount required. When sampling highly radioactive water, smaller sizes may be desirable to reduce the radiation hazard.

The point of sampling for open bodies of water should be chosen with extreme care so that a representative sample of the water to be tested is obtained. Surface scum is to be avoided. Because of a wide variety of conditions found in streams, lakes, reservoirs, and other bodies of water, it is not possible to prescribe the exact point of sampling. Where the water in a stream is mixed so as to approach uniformity, a sample taken at any point in the cross section is satisfactory. For large rivers or for streams not likely to be uniformly mixed, more samples are desirable and are usually taken at a number of points at the surface across the entire width and at a number of depths at each point. When boats are used, care should be exercised to avoid collecting samples where the turbulence caused by a propeller or by oars has disturbed the characteristics of the water. Ordinarily, samples are taken at these points and then combined to obtain an integrated sample of such a stream of water. Alternatively, the single grab samples may be tested—for example, to determine the point of highest bacterial density.

The location of the sampling point should be chosen with the information desired and in conformity to local conditions. Sufficient distance downstream (with respect to steam flow at the time of sampling) from a tributary or source of pollution must be allowed in order to permit thorough mixing. If this is not possible, it is better to sample the stream above the tributary or source of pollution and, in addition, to sample the tributary source of pollution. In general, a distance of 1 to 3 miles below the tributary is sufficient. Samples are collected at least one-half mile below dams or waterfalls to allow time for the escape of entrained air. When lakes, reservoirs, or other bodies of water are sampled, it is necessary to avoid nonrepresenative areas such as those created by inlet streams, more stagnant areas, or abrupt changes in shorelines, unless determining the effect of such local conditions is a part of the sampling program. It is desirable to take a series of samples from any source of water to determine whether differences in composition are likely to exist before final selection of the sampling point.

For sampling to determine chemical and physical analyses of unconfined water at any specific depth in ponds, lagoons, reservoirs, and so on, during which contact with air or agitation of the water would cause a change in concentration of characteristics of a constituent to be determined, one should use a sampling apparatus so constructed that the solution at the depth to be sampled flows through a tube to the bottom of the container, and that a volume of sample equal to 4 to 10 times the volume of the receiving container passes through it. When no determinations of dissolved gases are to be made, any less complicated apparatus may be used that will permit the collection of a sample at a desired depth, or of an integrated sample containing water from all points in a vertical section.

When samples are to be shipped, the bottle should not be filled entirely, in order to allow some room for expansion when the contents are subjected to a change in temperature. An air space of 10 percent usually suffices for this purpose, although this does not protect against bursting of the container due to freezing. However, if contact with air would cause a change in the concentration or characteristics of a constituent to be determined, the sample must be secured without contact with air and completely fill the container.

Chemical preservatives are added to samples for chemical or physical examination only as specified in a given test method. Quick freezing has been found to be beneficial in preserving some organic constituents. Any preservatives added should be noted on the label.

Samples collected for biological examination should be iced or refrigerated immediately after collection. They are to be held or transported at a temperature of not more than 4°C. Samples for microbiological analyses should be held no longer than 6 h from time of collection to analysis. Field examination should be considered if these time limits cannot be met.

Drinking water samples delivered to the laboratory by the collector are to be analyzed on the same day. A specific exception can be made for drinking water samples mailed or sent by public transportation to control laboratories since they are permitted to be held up to 30 h.

In general, one should allow as short a time as possible to elapse between the collection of a sample and its analysis. Under some conditions, analysis in the field is necessary to secure reliable results. The actual time that may be allowed to intervene between the collection and analysis of a sample varies with the type of examination to be conducted, the character of the sample, and the time interval allowable for applying corrective treatment.

Composite Samples Composite sampling is applicable for subsequent chemical and physical analyses. It may not be suitable for sample collection for radiological examination, particularly for short-lived radionuclides. Composite samples are not suited for bacteriological examination, for constituents that change with contact to air, or for purgable organics.

Composite samples may be made by mutual agreement of the interested parties by combining individual (grab) samples taken at frequent intervals or by means of automatic samplers. Individual test methods should be consulted for the effect of time interval and temperature prior to analysis.[30-32] Whether or not the volume of sample is proportional to the rate of flow should be indicated. At the end of a definite period, the composite sample should be mixed thoroughly so that determinations on a portion of the composite sample will represent the average for the stable constituents. Variations of unstable constituents may be determined by analysis on the individual samples.

In sampling process waters, composite samples are collected in at least one 24-h period. If the process is cyclic in nature, the sample is collected during at least one complete process cycle. Increments for composite samples are collected at regular intervals from 15 min to 1 h, and in proportion to the rate of flow of the water. This may be conveniently done by taking a simple multiple in milliliters per minute, in gallons per minute, or in some other unit of flow. A suitable factor is chosen to give the proper volume (about 4 liters) for the composite sample. When samples are taken from a stream, composite samples for analysis normally consist of equal quantities of daily samples for a suitable number of consecutive days—for example, 7 days. The point of sampling should be the same as described for collection of grab samples.

Continual Sampling Continual sampling is applicable to sampling water from sources such as wells, rivers, streams, lakes, oceans, and reservoirs on a continual basis for use in chemical, physical, or radiological analyses. The apparatus employed consists of the following:

1. Delivery valve or pump
2. Piping system
3. Flow regulation system
4. Waste disposal system

Sampling is essentially on a continuous basis. Intermittent operation is possible through use of sample bypass equipment, although this is seldom used except in measuring variables with a time relationship, such as rate of oxygen uptake. In these cases, deviations from this method are handled under descriptions of the specific measurement involved.

The size, quantity, and, in some cases, type of particulate matter often account for one or more of the variables to be measured and, in other cases, introduce errors in the analysis if the variables are distributed. The water delivery system should flow fast enough to keep the heavier particles in suspension, and the system volume should be large enough to prevent undesirable filter action through restrictions.

When simultaneous samples from several locations are required, water is drawn continually from each individual source proportionate to flow and is then mixed into a single sample. The selection of sampling points is determined in a manner similar to that of selection of grab sample sites.

Because pumps employing suction principles disturb the gas–liquid balance, a submersible-type pump is used for pumping samples from open bodies of water whenever the measurements to be made concern dissolved gases such as oxygen or carbon dioxide. Pumps, screens, valves, and piping must be selected of corrosion-resistant material to prevent sample contamination from corrosion products and to prevent need for undue maintenance. The debris screens employed around the pump intake should be of sufficient size to preclude a significant pressure drop developing across the screen in the event of partial clogging.

Manufacturers of continual analyzers and samplers will generally specify minimum volume and pressure requirements for proper operation. Sample pump selection must be based on these minimums and the configuration of sample piping. The piping system between the pump and the sample container should be designed so that the pump is operating against the lowest practical head. The piping system should be constructed with a continual rise in elevation from the pump to the point of delivery without reverse bends in which sediment and algae can accumulate. To prevent freezing in outdoor installations, the check valve is removed from the pump in order that the piping will drain in the event of power failure.

In a continually operating sample system, the time lag between the system intake and the point of sample delivery is a function of the flow rate of the water and the dimensions of the intervening pipe. Usually, the system dimensions make this time so short that its effect on the accuracy of the determination is negligible. Wherever special precautions should be observed, they will be described in the particular method covering that analysis.

Groundwater Monitoring Wells   The quality of groundwater has become an issue of national concern. Groundwater monitoring wells are one of the more important tools for evaluating the quality of groundwater, delineating contamination plumes, and establishing the integrity of hazardous material management facilities. The goal in sampling groundwater monitoring wells is to obtain samples that are truly representative of the aquifer or groundwater in question.

Water that stands within a monitoring well for a long period of time may become unrepresentative of formation water because chemical or biochemical change may cause water quality alternations; and even if it is unchanged from the time it entered the well, the stored water may not be representative of formation water at the time of sampling. Because the representativeness of stored water is questionable, it should be excluded from samples collected from a monitoring well.

There is a fairly large choice of equipment presently available for groundwater sampling from single-screened wells and well clusters. The sampling devices can be categorized into eight basic types, as described in detail in ASTM D4448. The type of equipment selected will depend upon the sampling methodology.[29]

4.6 SUMMARY

1. To meet wastewater management objectives, four major areas of concern must be addressed: (a) Sources and characteristics of wastes and wastewaters must be ascertained, (b) the particular type of wastewater treatment process, or combination of processes, must be determined, (c) wastewater effluent must be effectively controlled, and (d) sludge and other solid wastes must be properly managed. Controlling waste discharges must be a quantitative endeavor. Direct measurement must be made of the pollutants, along with (a) a prediction of the impact of the pollutant on water quality and (b) a determination of acceptable levels for the intended uses of the water.

2. At present, municipal wastewater plants are being upgraded and expanded to meet all government regulations. Primary treatment, which includes filtration and settling tanks, removes paper, rags, sand, coffee grounds, and other solid waste materials from the wastewater. Secondary treatment typically utilizes a biological process for further removal of substances from the wastewater.

3. Advanced wastewater treatment, known as *tertiary treatment*, is designed to remove those constituents that may not be adequately removed by secondary treatment. This includes removal of nitrogen, phosphorus, and heavy metals.

4. A model is an assembly of concepts in the form of one or more mathematical equations that approximate the behavior of a natural system or phenomena. There exist mass balance equations for both rivers and lakes, but a simple equation for estuaries does not exist because of their complexity.

5. Three types of water sampling are grab samples, composite samples, and continuous flowing samples. Under site-specific conditions, these types of sampling can be used for almost any body of water.

REFERENCES

1. G. Burke, B. Singh, and L. Theodore, *Handbook of Environmental Management and Technology*, 2nd ed., Wiley, Hoboken, NJ, 2000.
2. Metcalf & Eddy, Inc., *Wastewater Engineering: Treatment, Disposal, and Reuse*, 3rd ed., McGraw-Hill, New York, 1991.
3. M. K. Theodore and L. Theodore, *Major Environmental Issues Facing the 21st Century*, Theodore Tutorials, East Williston, NY, 1996.
4. W. W. Eckenfelder, Jr., *Industrial Water Pollution Control*, 2nd ed., McGraw-Hill, New York, 1989.

5. D. A. Cornwell and D. L. Mackenzie, *Introduction to Environmental Engineering*, 2nd ed., McGraw-Hill, New York, 1991, pp. 265–266.

6. C. N. Sawyer and P. L. McCarty, *Chemistry for Environmental Engineering*, 3rd ed., McGraw-Hill, New York, 1978, p. 416.

7. Metcalf & Eddy. Inc., *Wastewater Engineering: Treatment, Disposal, and Reuse*, 3rd ed., McGraw-Hill, New York, p. 82.

8. U.S. EPA, *Pretreatment of Industrial Wastes*, EPA Publication, Washington, DC, 1978.

9. U.S. EPA, *Innovative and Alternative Technology Assessment Manual*, EPA Publication, Washington, DC, 1980.

10. R. H. Perry and D. W. Green (editors) *Chemical Engineering Handbook*, 7th ed., "Industrial Wastewater Treatment," McGraw-Hill, New York, 1997, p. 25-58–25-80.

11. Malcolm Pirnie. Inc., and M. Baker. Jr., of New York, Inc., J.V., 201 Facilities Plan for WP-287, the Coney Island WPCP, December 1980.

12. New England Interstate Water Pollution Control Commission. *Guides for the Design of Wastewater Treatment Works*. 1980.

13. R. H. Perry and D. W. Green (editors) *Chemical Engineering Handbook*. 6th ed., Industrial Wastewater Management Section, McGraw-Hill Publishers, New York, 1985.

14. S. Qasim, *Wastewater Treatment Plants*. CBS College Publishing, New York, 1985.

15. US EPA. *Design Guides for Biological Wastewater Treatment Process*. Project No. 11010 ESQ. Austin, TX, August 1971.

16. Water Pollution Control Federation. *Wastewater Treatment Plant Design*. Lancaster Press. Lancaster, PA, 1977.

17. L. Theodore and M. Blenner, *Unit Operations for the Practicing Engineer*, Wiley, Hoboken, NJ, 2005.

18. V. Novotny, "Agricultural Nonpoint Source Pollution, Model Selection and Application," *Dev. Environ. Model*, 10, 1986.

19. M. Dortch and J. Martin, *Alternatives in Regulated Flow Management*, CRC Press, Boca Raton, FL, 1988.

20. American Society for Testing and Materials (ASTM). *Standard Practice for Evaluating Environmental Fate Models for Chemicals*, proposed standard, Subcommittee E-47.06 on Environmental Fate. Committee E-47 on Biological Effects and Environmental Fate, ASTM, Philadelphia, 1983.

21. R. Thomann and J. Mueller, *Principles of Surface Water Quality Modeling and Control*, Harper & Row, New York, 1987.

22. Pritchard. D. "What Is an Estuary?" *Estuaries*, American Association for the Advancement of Sciences. **83**(2) (1967).

23. W. Mills, D. Porcella, M. Ungs, S. Gherini, K. Summers, L. Mok, G. Rupp, and G. Bowie, "Water Quality Assessment: A Screening Procedure for Toxic and Conventional Pollutants," Part II. EPA/600/6-85/002b. 1985.

24. National Academy of Sciences, *Drinking Water and Health*. Safe Drinking Water Committee. Natural Resources Council, Washington, DC, 1977.

25. C. Officer and J. Page *Tales of the Earth*, Oxford University Press, New York, 1993.

26. W. Lung, "Application to Estuaries." *Water Quality Modeling, Vol. III*, CRC Press, Boca Raton, FL, 1993.

27. J. Lamb, III. *Water Quality and Its Control*, Wiley, New York, 1989.

28. E. Hartzell and A. Waits, "EPA's Environmental Technology Verification Program: Raising Confidence in Innovation," EM (AWMA), Pittsburgh, May 2004.

29. ASTM, *Annual Book of ASTM Standards*, Volume 11.04, ASTM D3370, New York, 1990.

30. G. Friedlander, J. Kennedy, and J. Miller, *Nuclear and Radiochemistry*, 2nd ed., Wiley, Hoboken, NJ, 1964.

31. R. T. Overman and H. M. Clark, *Radioisotope Techniques*, McGraw-Hill, New York, 1960.

32. *Hazardous Material Regulations of the D.O.T.*, Title 49, Parts 170 to 190. Consult current edition. Contact the U.S. Government Printing Office, Superintendent of Documents, Washington, DC, 20401.

CHAPTER 5

SOLID WASTE ISSUES

5.1 INTRODUCTION[1]

Details on the description, design, prediction of performance, operation, and maintenance of traditional control equipment for solid wastes are available in the literature.[2,3] They have been routinely employed in practice for both control and recovery purposes. Information on these solid waste management practices is presented in this chapter.

Of primary concern in evaluating unknown "solid waste" situations are health and safety hazards. Concentrations (or potential concentrations) of pollutants and any potential time variation of these variables, all represent immediate concerns. Initial resolutions will unquestionably be based on experience, judgment, and professional knowledge.

As with air and water emissions, extending standard procedures and designs of equipment to control pollutant problems arising from nanotechnology processes is subject to question at this time. The chemical and physical properties of these nano-emissions are not fully known, thus rendering judgments regarding equipment essentially mute. However, this information is critical to the advancement and development of new management/control practices. This is especially true of ultra-fine particles, that is, at the nanosize, since their mass, volume, and size is extremely small. Notwithstanding the above, it is anticipated that "new" control technologies and management practices arising from nanoapplications will probably be "managed" by employing procedures in place today.

Nanotechnology: Environmental Implications and Solutions, L. Theodore and R. G. Kunz.
ISBN 0-471-69976-4   Copyright © 2005 John Wiley & Sons, Inc.

Regarding the chapter title, solid wastes encompass all of the wastes arising from human and animal activities that are normally solid and that are discarded as unwanted or useless. The term *solid waste* is all-inclusive of the heterogeneous mass of throwaways from urban areas, as well as the more homogeneous accumulation of agricultural, industrial, and mineral wastes. In urban communities especially, the accumulation of solid wastes is a direct and primary consequence of life. This accumulation adds up to approximately 4 lb of solid waste per person every day in the United States. That is equal to one billion pounds of waste requiring disposal every day. Properly dealing with this huge amount of waste is what solid waste management is all about. This chapter is concerned with a variety of topics in this area, including past and future trends, legislative impacts, and current practices.

History

From the days of primitive society, humans and animals have used this planet's resources to live and dispose of wastes. In early times, the disposal of these wastes did not cause a significant problem because populations were small and the amount of land available for assimilation of wastes was large. Problems with solid waste disposal began developing, though, when humans first began to congregate in communities and the accumulation of wastes became a consequence of life.

An example of the early consequences of improper, or more appropriately, nonexistent solid waste management can be seen in fourteenth-century Europe. The practice of throwing solid wastes into the unpaved roads and vacant land in medieval towns led to the breeding of rats, with their fleas carrying the bubonic plague. Thus, the lack of any plan for the management of solid wastes led to the Black Death, which killed half of the population of Europe.

It was not until the last century that public health control measures became a vital issue to public officials, who finally realized that solid wastes had to be collected and disposed of in a sanitary manner in order to control the spread of disease.[4] In the United States, at the turn of the century, the most common methods for the final disposal of solid wastes were dumping on land, dumping in water, plowing into the soil, feeding to hogs, reduction, and incineration. These methods, though a good start, were for the most part unregulated and inefficient. In many cases they wound up causing other types of pollution problems. Recent solid waste management practices, with emphasis on sanitary landfilling, began in the 1940s in this country. Standards and guidelines were established for municipal sanitary landfills. However, municipalities did not follow these programs with consistency. By 1965, there were still major problems with solid waste disposal. As a result, the government, in that year, passed the first in a series of solid waste legislation designed to ensure proper solid wastes practices, as well as to protect the environment. This event thus marked the beginning of solid waste management as it is known today. In addition to sanitary landfilling, modern-day practices include pollution prevention, that is, source reduction and recycling/reuse, incineration, and so forth.[4]

Elements of a Solid Waste Management System

There are six main elements associated with the management of solid wastes from the point of generation to final disposal:[4]

1. Waste generation
2. Waste handling and separation
3. Collection
4. Separation, processing, and transformation of solid wastes
5. Transfer and transport
6. Disposal

Waste generation includes activities in which materials are identified as no longer being of value and are either thrown away or gathered together for disposal. In a typical community, solid waste is generated by many sources including residential, commercial, industrial, agricultural, and municipal services.

Waste handling and separation involves the management of wastes until they are placed in storage containers for collection. This includes the separation of waste materials for reuse and recycling, as well as the movement of loaded containers to the point of collection.

The collection of solid wastes includes the gathering of solid wastes and recyclable materials, as well as the transport of these materials to their destination. This destination may be a materials-processing facility, a transfer station, or a landfill site. In addition, transport of the waste becomes a problem in large cities where long hauling distances can make disposal costly.

Next comes the separation, processing, and transformation of solid waste materials. The activities included in this element occur in locations away from the source of waste generation. Processing includes: the separation of waste components by size using screens, manual separation of waste components, size reduction by shredding, separation of ferrous metals using magnets, compaction, and combustion. Transformation of solid wastes to reduce volume and weight is achieved using one of two types of processes: chemical or biological. The most common form of chemical transformation is combustion, while the most common form of biological transformation is aerobic composting. The transfer and transport of solid wastes involves two steps: (1) the transfer of wastes from the smaller collection vehicle to larger transport equipment, and (2) the subsequent transport of the wastes over long distances to a disposal site. The transfer usually takes place at a transfer station.

The last element in the solid waste management system is disposal. The disposal of solid wastes by landfilling or land spreading is the ultimate fate for most solid wastes. This includes wastes transported directly to the site, processed wastes, residue(s) from the combustion of solid waste, and compost.

Hierarchy of Solid Waste Management

A hierarchy is often employed in waste management to rank actions in implementing programs in the community. This hierarchy was adopted by the EPA (often referred to as the pollution prevention hierarchy) and is composed of four elements: source reduction, recycling, waste transformation, and ultimate disposal.[4]

The highest ranking element in the solid waste management hierarchy, source reduction, involves reducing the amount and/or toxicity of the wastes that are generated. It is obviously the most effective way to reduce the quantity of waste, the cost of handling it, and its impact on the environment. Waste reduction may occur through the design, manufacture, and packaging of products with minimum toxic content, minimum volume of material, and a longer useful life. It can also occur at the residential or commercial level through selective buying patterns and the reuse of materials and products.

The second element in the hierarchy is recycling. It involves:

1. The separation and collection of waste materials
2. The preparation of these materials for reuse, reprocessing, and remanufacture
3. The actual reuse, reprocessing, and remanufacture of these materials

Recycling plays an important part in helping to reduce the demand on resources and the amount of waste that ends up in landfills.

The third ranking element is waste transformation. It deals with the physical, chemical, or biological alteration of wastes. This is performed to improve the efficiency of solid waste management operation, to convert waste(s) to reusable and recyclable materials, and to recover heat and useful combustible gases. In addition, waste transformation ultimately results in the reduced use of landfill capacity.

Ultimate disposal, which usually involves landfilling, is the least desirable means of dealing with society's wastes. It is used for the solid wastes that cannot be recycled and are of no further use, the residual matter left after separation of solid wastes at a materials recovery facility, and the residual matter remaining after the recovery of conversion products or energy. Landfilling involves the controlled disposal of wastes into the land and is the most common method of ultimate disposal for waste residuals.

Implementation and Operation of Options

In implementing a solid waste management plan, several areas need to be considered: the proper mix of alternatives and technologies, flexibility in meeting future changes, and the need for monitoring and evaluation.[4]

A wide variety of alternative programs and technologies now exists for the management of solid wastes. This has led to confusion as to what is the proper mix between the amount of waste separated for reuse and recycling, the amount of waste that is composted, the amount of waste that is combusted, and the amount of waste to be placed in landfills. This decision is further complicated by the wide range of people involved in the decision-making process (including regulatory individuals) for the implementation of solid waste management options.

The ability to be flexible in meeting future changes is of great importance in the development of a solid waste management system. Some important factors to consider include changes in the quantities in composition of the waste stream, changes in the specifications and markets for recyclable materials, and rapid developments in technology. By studying the possible outcomes related to these factors, it is possible for a local community to be protected from unexpected changes in local, regional, and national conditions.

Solid waste management is an ongoing process that requires continual monitoring and evaluation to determine if program objectives and goals are being met. Thus, timely adjustments can be made to the system that reflect changes in waste characteristics, changing specifications and markets for recovered materials, and new and improved waste management technologies.

In the actual operation of a solid waste management system, a number of other management issues must also be considered. Workable but protective regulatory standards must be set. They can neither be so lax nor so strict that they cause the failure of the whole process. Next, there must be more improved scientific methods for interpretation of data as well as in the way in which these data are presented to the public. Third, household hazardous wastes should be removed from the garbage can for disposal in smaller, highly controlled waste management units. The reasoning behind this is that the accumulation of household hazardous wastes at a municipal landfill could eventually contaminate it. Another issue is whether land disposal units should be placed near or at large urban centers. Concern here focuses on the transportation cost to a faraway landfill versus the "not in my back yard" (NIMBY) syndrome of the local urban residents. Compounding this is the "not in my election year" (NIMEY) syndrome. Finally, because of the increasing quantity and complexity of solid waste management units, more qualified managers must be trained and maintained in order to operate them.[4]

This chapter serves as an introduction to solid waste management. The general subject of solid waste management is first examined, followed by material on industrial waste management and municipal waste management. The section concludes with discussions on hospital waste, the highly sensitive issue of nuclear wastes, metals, Superfund with all of its ramifications, and monitoring methods. Some of this material has been drawn from the literature referenced herein.

5.2 INDUSTRIAL WASTE MANAGEMENT[1]

As indicated in the introduction to this chapter, industrial and municipal waste management practices are reviewed in this and the next section, respectively. For reasons discussed repeatedly earlier in the text, only information on traditional waste management approaches is presented. Material on nanotechnology solid waste practices, particularly as they apply to land and soils, is understandably absent.

Pollution has grown to proportions where a reasonable solution to the total problem is almost unfathomable, but not necessarily unattainable. One of the human problems has always been the proper disposal of refuse. Demand for sanitary systems grew upon the discovery that diseases and illnesses develop as the result of inadequate and unsanitary disposal of wastes. Industrial pollution then started to become intolerable to society.

Large processing plants may have a multitude of different individual waste problems, each requiring a separate solution to meet air, water, and solid waste pollution standards. The solutions to these problems may range from the very simple to the very complex. In developing solutions, it is of primary importance to know the characteristics of the waste. Learning as much as possible about the various management tools that are available is secondary.

The composition of industrial wastes varies not only with the type of industry but with the processes used within the same industry. They may be classified according to composition in many ways, but in general, the wastes may be classified as wastes that may be utilized, and as wastes that require treatment. Industry continues to recognize the importance of saving and making use of all available resources from certain wastes. The wastes that require treatment should attract more attention. These wastes are usually of a form that contains waste materials in a more or less dilute solution or suspension. Since many materials of value are present in small quantities in a dilute solution, they cannot be economically recovered by known processes. In turn, these wastes that require treatment can be divided into three classes of wastes. There are those where organic compounds predominate and constitute the undesirable components, those that contain poisonous substances, and those that contain certain inert materials in such concentrations as to have undesirable features.

This chapter will review a wide range of industries and their respective wastes. In discussing these wastes, both solid and liquid wastes are treated together since it is difficult, if not impossible, to compartmentalize each phase/class of waste in any presentation. Each particular waste requires a different method of handling and treatment. In general, the predominating compounds in a waste usually determine the treatment process that will be required for that distinctive waste.

Food Processing

Food-processing industries are industries whose main concern is the production of edible goods for human or animal consumption. The production processes usually consist of the cleaning, the removal of inedible portions, the preparation, and the packaging of the raw material. The generated wastes are the spoiled raw material or the spoiled manufactured product, the liquid or water used (rinsing, washing, condensing, cooling, transporting, and processing), the cleaning liquids for the equipment, the drainage of the product, the overflow from tanks, and the unused portions of the product.

The wastes that result from food processing usually contain varying degrees of concentrations of organic matter. To provide the proper environmental conditions for the microorganisms upon which biological treatment depends, additional adjustments such as continuous feeding, temperature control, pH adjustment, mixing, supplementary nutrients, and microorganism population adaptation are necessary.

The major and more effective methods of aerobic or anaerobic biological treatments make use of activated sludge, biological filtration, anaerobic digestion, oxidation ponds, and spray irrigation. The loadings of the biological units must be maintained with care since many of the wastes contain high concentrations of organic matter. Most often, long periods of aeration or high-rate two-stage biofiltration (a biological control process) is required to produce an acceptable effluent.

The selection of the type of treatment depends on the degree of treatment required, the nature and phase of the organic waste, the concentration of organic matter, the variation in waste flow (if applicable), the volume of the waste, and the capital and operating costs.

Cannery Wastes

Cannery wastes are classified according to the product being processed, its growth season, and its geographic location. Many canneries are designed to process more than one product because vegetables, fruits, and citrus fruits have short harvesting and processing periods. The wastes from these plants are primarily organic. These wastes are the result of the trimming, juicing, blanching, and pasteurizing of raw materials; the cleaning of the processing equipment; and the cooling of the finished product. The most common and effective methods of treatment for the bulk of these wastes are discharging to a municipal treatment plant, lagooning with the addition of chemical stabilizers, soil absorption or spray irrigation, and anaerobic digestion.

The vegetables that produce strong wastes when processed for canning are peas, beets, carrots, corn, squash, pumpkins, and beans. The origin of all vegetable wastes is analogous since the canning procedures are alike even though the processing preparations differ for each vegetable. The wastes that result from food processing consist of the wash liquid; the solids from sorting, peeling, and coring operations; the spillage from filling and sealing the machines; and the wash liquid from cleaning the facilities.

The fruits that present the most common problems in the discharge of waste after processing are peaches, tomatoes, cherries, apples, pears, and grapes. Their wastes come from lye peeling, spray washing, sorting, grading, slicing and canning, removing condensates, cooling of cans, and plant cleanup.

The main citrus fruits (oranges, lemons, and grapefruit) are usually processed in one plant to make canned citrus juices, concentrates, citrus oils, dried meal, molasses, and other by-products. The wastes come from cooling waters, pectin wastes, pulp-press liquors, processing-plant wastes, and floor washings. The canning solid waste is a mixture of peel, rag, and seeds of the fruits, surplus juices, and blemished fruits.

The selection of the most suitable type of treatment of cannery wastes involves the review of the volume, phase, and treatment involved in the process and the unique conditions of the packaging periods. Cannery wastes are most efficiently treated by screening, chemical precipitation, lagooning, and spray irrigation (digestion and biological filtration are also used, but to a lesser extent).

The preliminary step of screening is designed to remove large solids prior to the final treatment or discharge of the waste to a receiving stream or municipal wastewater system. Only slight reductions in BOD (biological oxygen demand) are accomplished by screening. The machines either rotate or vibrate and have loads ranging from 40 to 50 lb per 1000 gal of waste water. The wastes retained on the screens are disposed of by being spread on the ground, used as sanitary fill, dried and burned, or used as animal food supplement.

Chemical precipitation is used to adjust the pH to reduce the concentration of solids in the wastes. This method is quite effective for treating apple, tomato, and cherry wastes. Ferric salts or aluminate and lime have produced 40 to 50 percent BOD reductions.[5] The product of this procedure is normally dried on sand beds for a week without producing an odor.

Treatment in lagoons involves biological action, sedimentation, soil absorption, evaporation, and dilution. When adequate land is available, lagooning may be the only practical and economical treatment of cannery wastes. $NaNO_3$ (sodium nitrite) is used to eliminate odors produced by lagoons with unmaintained aerobic conditions. However, the use of these treated lagoons for complete treatment may be costly because of the large volumes of wastes involved. Surface sprays are used to reduce the flies and other insect nuisances that breed around these lagoons.

Whenever the cannery waste is nonpathogenic and nontoxic to plants, spray irrigation is the preferred, economical method to use. Ridge-and-furrow irrigation beads are used on soils of relatively high water-absorbing capacity. In general, wastes should be screened before spraying, although comminution alone has been used successfully in conjunction with spray irrigation.

Oxygen-demanding materials in cannery wastes can be removed by biological oxidation. When the operation is limited by seasonal conditions, it is difficult to justify capital investment for bio-oxidation facilities. However, in many instances cannery wastes can be combined with domestic sewage, and then bio-oxidation processes provide a practical and economic solution.

Dairy Wastes

Most dairy wastes are made up of various dilutions of whole milk, separated milk, buttermilk, and whey. They result from accidental or intentional spills, drippings, and washings. Dairy wastes are largely neutral or slightly alkaline but have a tendency to become acid quite rapidly because of the fermentation of milk sugar to lactic acid. Lactose in milk wastes may be converted to lactic acid when streams become deficient in oxygen, and the resulting lowered pH may cause precipitation of casein. Because of the presence of whey, cheese-plant waste is decidedly acid. Milk wastes have very little suspended material, and their pollution effects are almost entirely due to the oxygen demand that they impose on the receiving stream. Decomposing casein causes heavy black sludge and strong butyric acid (rancid) odors that characterize milk-waste pollution.

There is a considerable variation in the size of the dairy plants and in the type of products they manufacture. The disposal or treatment of milk waste may be accomplished through irrigation on land, hauling, biological filtration on either the standard or the recirculating filter, biochemical treatment, or the oxidized sludge process. Milk-plant wastes have a tendency to ferment and become anaerobic and odorous because they are composed mostly of soluble organic materials. This characteristic enables them to respond ideally to treatment by biological methods. The selection of a treatment method hinges on the location and size of the plant. The most effective conventional methods of treatment are aeration, trickling filtration, activated sludge, irrigation, lagooning, and anaerobic digestion.

There is a wide variation in the flow rates and strength of milk wastes, and through holding and equalization, a desirable uniform waste could be achieved. Aeration for one day often results in 50 percent BOD reduction and eliminates odors during conversion of the lactose to lactic acid. Some two-stage filters yield greater than 90 percent BOD reduction, while single-stage filters yield about 75 to 80 percent BOD reduction.[5]

A successful method for the complete treatment of milk wastes is the activated sludge process. It uses aeration to cause the accumulation of an adapted sludge. When supplied with sufficient air, the flora and fauna in the active sludge culture oxidize the dissolved organic solids in the waste. Excess sludge is settled out and subsequently returned to the aeration units. Properly designed plants that provide ample air for handling the raw waste and returned sludge are not easily upset, nor is the control procedure difficult.

The amount of milk and milk products lost in wastewater from factories depends significantly on the degree of control and attention to detail in the operation of the plants. The first and most important step in reducing pollution from milk factories is to make sure that whole whey and buttermilk are never discharged with the wastewater. Also, churns in which the milk is delivered should be adequately drained. The effects of whey and buttermilk on the environment are intense if neglected; in addition, they have high food values and can be used as food or in the preparation of foods.

Fermentation and Pharmaceutical Industries

The fermentation industries range from breweries and distilleries to some parts of the pharmaceutical industry (the producers of antibiotics); the pharmaceutical industry is treated later in this section. To produce alcohol or alcoholic products, starchy materials (barley, oats, rye, wheat, corn, rice, potatoes) and materials containing sugars (blackstrap and high-sugar molasses, fruits, sugar beets) are used. The process of converting these raw materials to alcohol depends upon the desired alcoholic product. Beer manufacturers focus on taste, while distillers are concerned about alcohol yield.

The brewing of beer has two stages. The first stage involves the malting of the barley and the second involves brewing the beer from the malt. Both these operations occur at the same plant. The two major wastes produced by the malting process come from the steep tank after grain has been removed, and those remaining in the germinating drum after the green malt has been removed. A considerable amount of water is required for cooling purposes in the actual brewing process.

Brewery wastes are composed mainly of liquor pressed from the wet grain, liquor from yeast recovery, and wash water from the various departments. The residue remaining after the distillation process is referred to as "distillery slops," "beer slops," or "still bottoms."

In a distillery, there are several sources of wastes. The dealcoholized still residue and evaporator condensate are major concerns. Minor wastes include redistillation residue and equipment washes. In the manufacture of compressed yeast seed, yeast is planted in a nutrient solution and allowed to grow under aerobic conditions until maximum cell multiplication is attained. The yeast is then separated from the spent nutrient solution, compressed, and finally packaged. The yeast-plant effluent consists of filter residues resulting from the preparation of the nutrient solutions, spent nutrients, wash water, filter press effluent, and cooling and condenser waters or liquid.

Pharmaceutical wastes arise primarily from spent liquors from the fermentation process, with the addition of the floor washings and laboratory wastes. Wastes from pharmaceutical plants producing antibiotics and biologicals can be categorized as strong fermentation beers, inorganic solids, washing of floors and equipment, chemical waste, and barometric condenser water from evaporation. The wastes from pharmaceutical plants that produce penicillin and similar antibiotics are strong and generally should not be treated along with domestic sewage unless the extra load is considered in the design and operation of the treatment plant.

Stillage is the principal pollution load from a distillery; it is the residual grain mash from distillation columns. Industry attempts to recover as much of this as possible as a by-product to manufacture animal feed or for conversion to chemical products. Centrifuging has also been used to concentrate distillery slops.

Meat Industry

The three main sources of waste in the meat industry are stockyards, slaughter-houses, and packinghouses. The stockyard is where the animals are kept until they are killed. The actual killing, dressing, and some by-product processing are carried out in the slaughterhouse. Packinghouse operations include the manufacture of sausages, canning of meat, rendering of edible fats into lard and edible tallow, cleaning of casings, drying of hog's hair, and some rendering of inedible fats into grease and inedible tallow. Packinghouse wastes are generated from various operations on the killing floor, during carcass dressing, rendering, bag-hair removal and processing, casing, and cleaning. Stockyard wastes contain both liquid and solid excretions. The amount and strength of the wastes vary widely, depending on the presence or absence of cattle horns, the thoroughness and frequency of manure removal, the frequency of washing, and so on.

Blood should be recovered as completely as possible, even in small plants. Blood is a rich source of protein and is more economical to recover for large plants. Small plants do not have the equipment nor the conditions necessary to profit from the sale of the blood. Paunch manure should be recovered and used for fertilizer purposes. There is little reason for this material to enter the waste system except as washings from the floor. Grease recovery or removal should be common practice in all packing houses and even in smaller slaughterhouses. Grease removal is accomplished through the use of baffled tanks or grease traps. Cleanup by water from high-pressure hoses has been and continues to be the general practice in the meat-packing industry. The use of dry cleanup prior to wet cleanup reduces pollution loads substantially; although it reduces wastewater volume, it does increase solid waste volume.

Slaughterhouse processes are centered about the killing floor. Meat plant wastes are similar to domestic sewage in regard to their composition and effects on receiving bodies of water. The total organic contents of these wastes are considerably higher than those of domestic sewage. Without adequate dilution, the principal detrimental effects of meat plant wastes are oxygen depletion, sludge deposits, discoloration, and general nuisance conditions. The total liquid waste from the poultry-dressing process contains varying amounts of blood, feathers, fleshings, fats, washings from evisceration, digested and undigested foods, manure, and dirt. The largest amount of pollution from the process is contributed by the manure from receiving and feeding stations and blood from the killing and sticking operations.

The treatment processes adapted to slaughterhouse and packing plant wastes depend on the size of the industry. The most common methods used for treatment are fine screening, sedimentation, chemical precipitation, trickling filters, and activated sludge. Biological filtration is perhaps the most dependable process for the medium- and larger-sized plants.

Poultry-plant wastes should and do respond readily to biological treatment; it is attainable if troublesome materials such as feathers, feet, heads, and so on are removed beforehand. Treatment facilities include stationary screens in pits, septic tanks, and lagoons.

The small packinghouse or slaughterhouse requires a process of treatment that is dependable and simple to operate. Small plants run sporadically resulting in undesirable operating conditions for biological processes. Biological processes are much more easily upset by careless treatment or large variations in waste content than are chemical processes.

Textile Industry

The textile industry has been one of the largest of users and polluters of water, and unfortunately, there has been little success in the development of the low-cost treatment methods needed by the industry to lessen the pollution that is discharged into streams. The operations of textile mills consist of weaving, dyeing, printing, and finishing. Many processes involve several steps, each contributing a particular type of waste, like sizing of the fibers, kiering (alkaline cooking at elevated temperature), desizing the woven cloth, bleaching, mercerizing, dyeing, and printing. Textile wastes are generally colored, highly alkaline, high in BOD and suspended solids, and high in temperature. Manufacturing synthetic fiber generates wastes that resemble chemical manufacturing wastes, and their treatment depends on the chemical process used. Equalization and holding are generally preliminary steps to the treatment of those wastes because of their varying compositions. Additional methods are chemical precipitation, trickling filtration, and, more recently, biological treatment and aeration.

5.3 MUNICIPAL SOLID WASTE MANAGEMENT[1,6,7]

Introduction

It is not news that many communities in America are faced with a garbage disposal problem. In 1990, Americans generated over 195 million tons of municipal solid waste, with ever increasing amounts expected in years to come. Since over two-thirds of municipal solid waste is sent to landfills, this section will key primarily on landfilling—the solid waste management option that simply will not go away. However, some landfills are closing and the siting of new landfills has become increasingly difficult because of public opposition. Past problems sometimes associated with older landfills might have contributed to this situation. Landfills that were poorly designed, that were located in geologically unsound areas, or that might have accepted toxic materials without proper safeguards have contaminated some groundwater sources. Many communities use groundwater for drinking, and people living where contamination has occurred understandably worry about its threat to their health and the cost of cleaning it up. Communities where new landfills are needed share these concerns. Consequently, at a time when more are needed, there is increasing resistance to building new landfills.

To ease these worries and to make waste management work better, federal, state, Native American tribal, and local governments have adopted an integrated approach to waste management. This approach involves a mix of three waste management techniques:

1. Decreasing the amount and/or toxicity of waste that must be disposed of by producing less waste to begin with (source reduction)
2. Increasing recycling of materials such as paper, glass, steel, plastics, and aluminum, thus recovering these materials rather than discarding them
3. Providing safer disposal capacity by improving the design and management of incinerators and landfills

The remainder of the section addresses the following topic areas: source reduction and recycle/reuse, incineration, and landfilling.

Source Reduction and Recycle/Reuse

To increase recycling nationwide, the EPA has undertaken a number of efforts to stimulate markets for secondary materials and to promote increased separation, collection, processing, and recycling of waste. The EPA also funded the establishment of a National Recycling Institute, composed of high-level representatives from business and industry, to identify and resolve issues in recycling.

Composting is another process commonly associated with recycling. Composting is the microbiological decay of organic materials in an aerobic environment. Materials that could potentially be composted include agricultural waste, grass clippings, leaves and other yard waste, food waste, and paper products. Many municipalities have implemented leaf composting programs.

One of the problems with the implementation of any recycling program is the public perception of associated costs. Many people believe that recycling is free or, at the very least, inexpensive. However, in most instances, that is not the case. Costs are associated with every aspect of the program, including collection of the materials, their processing, and disposing of any residues. Purchase of new equipment or the retrofitting of existing equipment that is used to separate or recycle materials, or incorporating recycled materials into a process, is often very expensive. Direct operational costs include labor and utilities. With the exception of glass and aluminum, it is usually more cost effective, unfortunately, to use virgin materials rather than recycled materials in manufacturing processes. Many times, markets are not available for sorted materials. The plastics industry is representative of this problem. Although much research has gone into plastics recycling in the past few years, markets for both the segregated material and the end products are very limited. The public needs to realize that recycling is not cheap; and, many times, the cost of recycling is offset only by the avoided cost of disposal rather than by any profits generated.[7,8]

Incineration[8,9]

Incineration is not a new technology and has been commonly used for treating wastes for many years in Europe and the United States. The major benefits of incineration are that the process actually destroys most of the waste rather than just disposing of or storing it; it can be used for a variety of specific wastes; and it is reasonably competitive in cost compared to other disposal methods.

Municipal solid waste incineration involves the application of combustion processes under controlled conditions to convert wastes containing hazardous materials to inert mineral residues and gases. Four parameters influence the mechanisms of incineration:

1. Adequate free oxygen must always be available in the combustion zone.
2. Turbulence, the constant mixing of waste and oxygen, must exist.
3. Combustion temperatures must be maintained; the combustion process must provide enough heat to raise the burning mixture to a sufficient temperature to destroy all organic components.

4. Elapsed time of exposure to combustion temperatures must be adequately long in duration to ensure that even the slowest combustion reaction has gone to completion. In other words, transport of the burning mixture through the high-temperature region must occur over a sufficient period of time.

Municipal solid waste can be combusted in bulk form or in reduced form. Shredding, pulverizing, or any other size reduction method that can be used before incineration decreases the amount of residual ash as a result of better contact of the waste material with oxygen during the combustion process.[10] Shredded waste used as fuel is generally referred to as refuse-derived fuel (RDF) and is sometimes combined with other fuel types. Table 5.1 lists the American Society for Testing and Materials (ASTM) classification for RDF.

The types of incinerators used in municipal waste combustion include fluidized-bed incinerators, rotary waterwall combustors, reciprocating grate systems, and modular incinerators. The basic variations in the design of these systems are related to the waste feed system, the air delivery system, and the movement of the material through the process. Specific details are available in the literature.

TABLE 5.1   ASTM Classifications for Refuse-Derived Fuel (RDF)

ASTM RDF	RDF Classification	Nomenclature Description
RDF1	Raw	Solid waste used as a fuel in discarded form, without oversize bulky waste.
RDF2	Coarse	Solid waste processed to a coarse particle size, with or without ferrous metal extraction, such that 95% by weight passes through a 6-inch, 2-mesh screen.
RDF3	Fine or fluff	Solid waste processed to a particle size such that 95% by weight passes through a 2-inch, 2-mesh screen, and from which the majority of metals, glass, and other inorganics have been extracted.
RDF4	Powder	Solid waste processed into a powdered form such that 95% by weight passes through a 10-mesh screen and from which most metals, glass, and other inorganics have been extracted.
RDF5	Densified	Solid waste that has been processed and densified into the form of pellets, slugs, cubettes, or briquettes.
RDF6	Liquefied	Solid waste that has been processed into a liquid fuel.
RDF7	Gaseous	Solid waste that has been processed into a gaseous fuel.

Landfilling[6,8]

As indicated earlier, approximately two-thirds of the nation's municipal solid waste is landfilled. The reason is that it is not possible to reuse, recycle, or incinerate the entire solid waste stream; therefore, a significant portion of the waste must be land-filled. Landfills have been a common means of waste disposal for centuries. A process that originally was nothing more than open piles of waste has now evolved into sophisticated facilities. Perhaps the best approach both to describe and discuss this solid waste management option is to examine the federal regulations pertaining to landfills. The federal regulations for municipal solid waste landfills coyer the following six basic areas:

1. Location
2. Operation
3. Design
4. Groundwater monitoring and corrective action
5. Closure and postclosure care
6. Financial assurance

The following material presents the applicable regulations in some detail. However, states and tribes with EPA-approved programs have the opportunity to exercise flexibility in implementing these regulations. Some of the exceptions described below are available only in states and tribes with EPA-approved programs.

Location　Because landfills can attract birds that can interfere with aircraft oper-ation, owners/operators of sites near airports must show that birds are not a danger to aircraft. This restriction applies to new, existing, and laterally expanding landfills. Landfills may not be located in areas that are prone to flooding unless the owner/operator can prove the landfill is designed to withstand flooding and prevent the waste from washing out. This restriction also applies to new, existing, and laterally expanding landfills. Since wetlands are important ecological resources, new landfills and laterally expanding ones may not be built in wetlands unless the landfill is in a state or on tribal lands with an EPA-approved program and the owner/operator can show that it will not pollute the area. The owner/operator must also show that no alternative site is available. This restriction does not apply to existing landfills. To prevent pollution that could be caused by earthquakes or other kinds of earth movement, new and laterally expanding landfills may not be built in areas prone to them. This restriction does not apply to existing landfills. Finally, landfills cannot be located in areas that are subject to landslides, mudslides, or sinkholes; this restriction applies to new, existing, and laterally expanding landfills.

Operation　The EPA and the states have developed regulations specifically cover-ing the disposal of hazardous wastes in special landfills. Owners/operators of municipal landfills must develop programs to keep these regulated hazardous wastes out of their units. In general, each day's waste must be covered to prevent the spread of disease by rats, flies, mosquitoes, birds, and other animals that are naturally attracted to landfills. Methane gas, which occurs naturally at landfills, must be monitored routinely. If emission levels at the landfill exceed a certain limit, the proper authorities must be notified and a plan must be developed to solve the problem. Owners/operators must restrict access to their landfills to prevent illegal dumping and other unauthorized activities. So that no pollutants are swept into lakes, rivers, or streams, landfills must be built with ditches and levees to keep storm water from flooding their active areas, and to collect and control stormwater runoff. Landfills cannot accept liquid waste from tank trucks or in 55-gal drums. This restriction helps reduce both the amount of leachate (liquids that have passed through the landfill) and the concentrations of contaminants in the leachate. Finally, landfills must be operated so they do not violate state and federal clean air laws and regulations. This means, among other things, that the burning of waste is prohibited at landfills, except under certain conditions.

Design   New and expanding landfills must be designed for groundwater protection by making sure that levels of contaminants do not exceed federal limits for safe drinking water. In states and tribes with EPA-approved programs, landfill owners/operators have flexibility in designing their units to suit local circumstances, providing the state or tribal program director approves the design. This allows owners/operators to ensure environmental protection at the lowest possible cost to citizens served by the landfill. This flexibility means, for example, that the use of a liner, and the nature and thickness of the liner system, may vary from state to state, and perhaps from site to site. In states and tribal areas without EPA-approved programs, owners/operators must build their landfills according to a design developed by EPA or seek a waiver. The EPA design lays out specific requirements for liners and leachate collection systems. Liners must be a composite, that is, a synthetic material over a 2-ft layer of clay. This system forms a barrier that prevents the escape of leachate from the landfill into groundwater. The design also requires leachate collection systems that allow the leachate to be captured and treated.

Groundwater Monitoring and Corrective Action   Generally, landfill owners/operators must install monitoring systems to detect groundwater contamination. Sampling and analysis must be conducted twice a year. States and tribes with EPA-approved programs have the flexibility to tailor facility requirements to specific local conditions. For example, they may specify different frequencies for sampling groundwater for contaminants, or phase in the deadline for complying with the federal groundwater monitoring requirements.

If the groundwater becomes contaminated, owners/operators in approved states and tribal areas must clean it up to levels specified by the state or tribal director. In states and tribes without EPA-approved programs, the federal regulations specify that contaminants must be reduced below the federal limits for safe drinking water.

Closure and Postclosure Care   When a landfill owner/operator stops accepting waste, the landfill must be closed in a way that will prevent problems later. The final cover must be designed to keep liquid away from the buried waste. For 30 years after closure, the owner/operator must continue to maintain the final cover, monitor groundwater to ensure the unit is not leaking, collect and monitor landfill gas, and perform other maintenance activities. (States and tribes with approved programs may vary this period based on local conditions.)

Financial Assurance   To ensure that monies are available to correct possible environmental problems, landfill owners/operators are now required to show that they have the financial means to cover expenses for site closure, postclosure maintenance, and cleanups. The regulations spell out ways to meet this requirement, including (but not limited to) surety bonds, insurance, and letters of credit.

5.4 HOSPITAL WASTE MANAGEMENT[1,11]

Virtually all of the 6600 hospitals in the United States house X-ray equipment, laboratories, kitchens, pharmacies, and waste disposal stations. More than half also have diagnostic radioisotope facilities, CT scanners, and ultrasound equipment. The environmental impact of the waste generated is considerable. All of these substances are subject to the EPA and/or the Occupational Safety and Health Administration (OSHA) rules and regulations on the environment and worker exposure. OSHA has numerous regulations pertaining specifically to workers in health care settings.

Medical wastes are not only generated by hospitals but also by laboratories, animal research facilities, and by other institutional sources. The term *biomedical waste* is coming into usage to replace what had been referred to as *pathological waste* or *infectious wastes*. Hospitals, however, are generating more and more medical waste with their increasing use of disposable products as well as their increasing service to the community. Hence, the focus of this section will be on the issue of hospital waste management.

Progress has been made in methods and equipment for the care of hospital patients. Hundreds of single-service items have been marketed to reduce the possibility of hospital-acquired infections. Yet hospitals generally have been slow to improve their techniques for the handling and disposing of the waste materials, which are increasing in quantity as a result of more patients and higher per-patient waste loads.

Medical waste comes in a wide variety of forms. These forms include packaging, such as wrappers from bandages and catheters; disposable items, such as tongue depressors and thermometer covers; and infectious wastes, such as blood, tissue, sharps, cultures, and stocks of infectious agents.

The location of these wastes include laboratories, X-ray facilities, surgical departments, pharmacies, emergency rooms, offices, and service areas.

There is an equally wide variety of sources. While hospitals, clinics, and health care facilities may generate the vast majority of medical waste, both infectious and noninfectious waste is also generated by private practices, home health care, veterinary clinics, and blood banks. In New York and New Jersey alone, there are approximately 150,000 sources producing nearly 250 million pounds of waste a year.

The beach closures along costal New Jersey in 1987 and along the south shore of Long Island in 1988 have focused attention on medical wastes. Their volume is relatively small (probably less than 1 percent of the total), but as with sewage wastes, concern centers around the issue of public health. Why these wastes are appearing more frequently is not certain. However, there are several possible contributing factors. The three major factors include:

1. A marked increase in disposable medical care materials
2. An increase in the use of medically associated equipment on the streets as drug paraphernalia
3. An increase in illegal disposal of medical wastes as a consequence of the increased costs of disposal.

Medical Waste Regulations and Definitions

On March 24, 1989, the U.S. EPA published regulations in the *Federal Register* as required under the Medical Waste Tracking Act of 1988. The term *medical waste* was defined as any solid waste that is generated in the diagnosis, treatment, or immunization of human beings or animals, in research pertaining thereto, or in the production or testing of biologicals. Medical waste can be either infectious or noninfectious. The term medical waste does not include any hazardous or household waste, defined in regulations under Subtitle C of the act.

Infectious waste is waste that contains pathogenic microorganisms. In order for a disease to be transmitted, the waste must contain sufficient quantity of the pathogen that causes the disease. There must also be a method of transmitting the disease from the waste material to the recipient.

Medical waste that has not been specifically excluded in the EPA provisions (e.g., household waste) and is either a listed medical waste or a mixture of a listed medical waste and a solid under the demonstration program of the act is known as *regulated medical waste*. Seven classes of listed wastes are defined by the EPA as regulated medical waste. Details on these seven classes are provided below.

1. *Cultures and stocks.* Cultures and stocks of infectious agents and associated biologicals, including: cultures and stocks of infectious agents from research and industrial laboratories; wastes from the production of biologicals; discarded live and attenuated vaccines; and culture dishes and devices used to transfer, inoculate, and mix cultures.

2. *Pathological waste.* Human pathological wastes, including (a) tissues, organs, body parts, and body fluids that are removed during surgery, autopsy, or other medical procedures and (b) specimens of body fluids and their containers.

3. *Human blood and blood products.* Products here include: liquid waste human blood; products of blood; items saturated and/or dripping with human blood that are now caked with dried human blood including serum, plasma, and other components; and containers that were used or intended for use in either patient care, testing, laboratory analysis, or the development of pharmaceuticals (intravenous bags are also included in this category).

4. *Sharps.* The category includes sharps that have been used in animal or human patient care or treatment, in medical research, or in industrial laboratories, including hypodermic needles, syringes (with or without the attached needles), pasteur pipettes, scalpel blades, blood vials, needles with attached tubing, and culture dishes (regardless of presence of infectious agents).

5. *Animal waste.* Contaminated animal carcasses, body parts, and bedding of animals that were known to have been exposed to infectious agents during research, production of biologicals, or testing of pharmaceuticals.

6. *Isolation wastes.* Biological waste and discarded materials contaminated with blood, excretions, or secretions from humans known to be infected with certain highly communicable diseases.

7. *Unused sharps.* These include hypodermic needles, suture needles, syringes, and scalpel blades.

Waste Storage and Handling

Hospital wastes are stored in many kinds of receptacles: wastepaper baskets, garbage cans, empty oil drums, laundry hampers, carts, buckets, and even on the floor. Plastic containers are coming into widespread use. They are easier to lift and clean than metal containers, and the bases and sides are impermeable to insects, since they do not rust, bend, or dent.

Most hospitals segregate their medical wastes prior to treatment and disposal. However, most hospitals do not segregate all medical waste categories from one another, although certain wastes, most often sharps, cultures, and stock, are segregated from other medical wastes prior to treatment or disposal. Most hospitals carefully segregate sharps in rigid plastic sharps containers. Medical wastes are usually segregated from the general trash (e.g., office and cafeteria wastes). Medical waste is almost always separated into red, orange, or biohazard marked bags, and general waste is usually placed in clear, white, or brown bags.

In some hospitals a sharps container is mounted on the wall of every patient room. In other hospitals, the sharps containers are placed in central collection areas on the patient floors, in other areas as necessary (operating room, emergency room, laboratory), and on the carts themselves.

Red bags are almost always used in the laboratory, operating rooms (OR), emergency rooms (ER), and isolation rooms. In some cases, red bags also appear in patient rooms. As an alternative to red-bagging all waste from patient rooms, some hospitals place red bags and sharps containers on patient care carts. In this fashion, medical wastes are segregated from other discarded wastes.

Medical waste from the "floors" (as patient wings are called) are sometimes stored in "soiled utility rooms" on the patient floors until carried to central storage rooms for incineration or transport. The housekeeping staff is often responsible for gathering the waste and carting it to the storage area. In some cases, general trash is collected in the same cart along with red bag wastes.

Suctioned fluids are more commonly discharged into a sanitary sewer rather than containerized or incinerated. However, some hospitals use disposable suction containers and place the entire container in red bags. Other fluids—from the laboratory, for example—are often contained and then red-bagged. In some cases, fluids are poured into red bags.

Sharps, other than needles and syringes (such as discarded slides and test tubes), often are placed in the red plastic bags without first being placed in a punctureproof container. Some of these are then placed in cardboard boxes to avoid punctures. The boxes also provide support for heavier sharps such as glassware, slides, and tubes of blood.

An unusual feature of hospital waste management is that wastes are generated continuously around the clock, but they are collected at fixed intervals during the day shift. The housekeeping department usually has the primary responsibility for collection within the hospitals, although a number of other departments have regular responsibility for other facets of waste collection. Generally, only minimal qualifications are required for individuals collecting wastes.

Hospitals often use manually propelled carts of some variety to collect waste materials. Hospital carts are frequently constructed in such a way that sanitizing them is impossible, thus providing surfaces where bacteria can multiply. The routing of carts into and through areas where freedom from contamination is critical increases the probability of contamination from wastes. In addition, individuals collecting wastes are repeatedly exposed to chemical and microbiological contamination and other hazards, but usually have minimal knowledge, skill, or equipment to protect themselves.

Gravity chutes are a simple and inexpensive means of transferring wastes vertically. However, the chutes are seldom constructed with mechanical exhausts, interlocking charging doors, or other systems for preventing the spread of microbiological contamination. In several instances, linen chutes are reserved for conveying solid wastes during certain times of the day, thus providing another potential way of spreading contamination. Chute usage has additional drawbacks: fire hazards, spilling of wastes during loading, blockages, difficulties in cleaning, and odors. Proper design and construction can help to prevent some of these, especially the fire hazard and cleaning problems. Others can be avoided by excluding certain wastes, especially grossly contaminated articles, and by exercising more care in the use of chutes.

Waste Processing and Disposal

Hospital wastes are disposed of in a number of ways, usually by the hospital's maintenance or engineering department. Eventually, almost two-thirds of the wastes leave the hospitals and go out into the community for disposal. Approximately 35 percent by weight, principally combustible rubbish and biological materials, are disposed of in hospital incinerators. Noncombustibles are usually separated and, along with the incinerator residue, leave the hospital to be disposed of on land.

Understandably, the presentation that follows addresses current hospital practices. Waste management methods include incineration, autoclaving, sanitary landfilling, sewer systems, chemical disinfection, thermal inactivation ionizing radiation, gas vapor sterilization, segregation, and bagging.[12] Table 5.2 lists typical treatment/disposal methods for each waste type. This table reveals that only 55 percent of hospitals that segregate infectious from noninfectious waste incinerate their infectious waste. Eighteen percent treat infectious waste by steam serialization and then incinerate or landfill the waste. Three percent of hospitals dispose of infectious waste in sanitary landfills without prior treatment.

Waste Management Programs

The large amounts of potentially contaminated wastes generated by hospitals raise the possibility that they are a concentrated source of environment health problems. Many hospital solid wastes are indeed contributing to occupational injuries, environmental pollution, and insect and rodent infestation. Some remedial steps that can be taken include the following:

1. Seal as many wastes as possible in disposable bags at the point of generation, or enclose them in such a way as to prevent or minimize contamination of the hospital environment.
2. Construct carts and other equipment used to handle waste so they are easy to keep in sanitary condition.
3. Construct and operate chutes in such a way as to prevent or minimize microbiological contamination of air, linen, and various areas of the hospital.

TABLE 5.2 **Approximate Percent of Hospitals Using Treatment/Disposal Methods for Each Waste Type**

Type of Waste	Incineration (%)	Sanitary Landfill (%)	Steam Sterilization (%)	Sewer (%)
Blood and blood products	58	12	25	23
Body fluids and wastes	58	32	11	6
Lab wastes	61	16	33	3
Pathological wastes	92	4	2	
Sharps	79	16	14	0
Animal wastes	81	2	2	0
Disposable materials	29	54	4	6

4. Reduce the danger to personnel handling wastes. Provide preventive health services such as immunization, as well as protective equipment such as gloves and uniforms. In the future, introduce equipment and systems that require less manpower.

5. Require higher qualifications for those handling wastes. Provide them with training on the hazards associated with hospital wastes and the means of protecting not only themselves but others in the hospital and the community.

6. Improve the operation of incinerators by training operators to keep loads within incinerator capacity and to maintain temperatures high enough for proper combustion.

7. Provide for safe management of hazardous wastes within the hospital so that they cannot pose a danger to the community.

Most hospitals have comprehensive and sound policies on solid waste management, including specific directives on segregation and special handling of hazardous materials. But, in practice, the policies break down. Employees fail to make the right judgments consistently, and stricter supervision is needed to ensure that employees maintain proper handling and disposal of pathological and sharps wastes, separate disposable wastes from reusable wastes such as dinnerware and linens, bag materials properly, and deposit chute materials promptly. In addition, storage, processing, and disposal areas should be supervised closely and security maintained so that unauthorized personnel cannot gain access.

Infectious Waste Management Programs

A waste management plan for an institution should be a comprehensive written plan that includes all aspects of management for different types of waste, including infectious, radioactive, chemical, and general wastes as well as wastes with multiple hazards (e.g., infectious and radioactive, infectious and toxic, infectious and radioactive and carcinogenic). In addition, it is appropriate for each laboratory or department to have specific, detailed, written instructions for the management of the types of waste that are generated in that unit. The waste management section would probably constitute one part of a general, more comprehensive document that also addresses other policies and procedures. Many such documents that include sections on the management of infectious waste have been prepared by various institutions and governmental agencies.[13]

An infectious waste management system should include the following elements:

1. Designation of infectious wastes
2. Handling of infectious wastes, including:
   a. Segregation
   b. Packaging
   c. Storage

    d. Transport and handling

    e. Treatment techniques

    f. Disposal of treated waste

3. Contingency planning

4. Staff training

Various options are available for the development of an infectious waste management system. Management options for an individual facility should be selected on the basis of what is most appropriate for the particular facility. Factors such as location, size, and budget should be taken into consideration. The selected options should be incorporated into a documented infectious waste management plan. An infectious waste management system cannot be effective unless it is fully implemented. Therefore, a specific individual at the generating facility should be responsible for implementation of the plan. This person should have the responsibility as well as the authority to make sure that the provisions of the management plan are being followed.

There are a number of areas in which alternative options are available in an infectious waste management system (e.g., treatment techniques for the various types of infectious waste, types of treatment equipment, treatment sites, and various waste handling practices). The selection of available options at a facility depends upon a number of factors, such as the nature of the infectious waste, the quantity of infectious waste generated, the availability of equipment for treatment onsite and offsite, regulatory constraints, and cost considerations. These factors are presented here in order to provide assistance in the development of an infectious waste management program.

Since treatment methods vary with waste type, the waste must be evaluated and categorized with regard to its potential to cause disease. Such characteristics as chemical content, density, water content, bulk, and so on are known to influence waste treatment decisions. For example, many facilities use a combination of treatment techniques for the different components of the infectious waste stream, for example, steam sterilization for laboratory cultures and incineration for pathological waste.

The quantity of each category of infectious waste generated at the facility may also influence the method of treatment. Decisions should be made on the basis of the major components of the infectious waste stream. Generally, it would be desirable and efficient to handle all infectious waste in the same manner. However, if a selected option is not suitable for treatment of all wastes, then other options must be included in the infectious waste management plan.

Another important factor in the selection of options for infectious waste management is the availability of onsite and offsite treatment. Onsite treatment of infectious waste provides the advantage of a single facility or generator maintaining control of the waste. For some facilities, however, offsite treatment may offer the most cost-effective option. Offsite treatment alternatives include such options as morticians (for pathological wastes), a shared treatment unit at another institution, and commercial or community treatment facilities. With offsite treatment, precautions should be taken in packaging and transporting to ensure containment of the infectious waste. In addition, generators should comply with all state and local regulations pertaining to the transport of regulated medical waste and ensure that the waste is being handled and treated properly at the offsite treatment facility.

It is also important to consider prevailing community attitudes in such matters as site selection for offsite treatment facilities. These include local laws, ordinances, and zoning restrictions as well as unofficial public attitudes that may result in changes in local laws.

Cost considerations are also important in the selection of infectious waste management options. Cost factors include personnel, equipment cost, hauling costs (for infectious waste and the residue from treatment), and, if applicable, service fees for the offsite treatment option.

As indicated earlier, the EPA recommends that each facility establish an infectious waste management plan. A responsible individual at the facility should prepare a comprehensive document that outlines policies and procedures for the management of infectious waste (including infectious wastes with multiple hazards). This recommendation is consistent with the standard of the Joint Commission on Accreditation of Hospitals (JCAH), which specifies a system "to safely manage hazardous materials and wastes."[12]

5.5 NUCLEAR WASTE MANAGEMENT[1]

As with many other types of waste disposal, radioactive waste disposal is no longer a function of technical feasibility but rather a question of social or political acceptability. The placement of facilities for the permanent disposal of municipal solid waste, hazardous chemical waste, and nuclear wastes alike has become an increasingly large part of waste management. Today a large percentage of the money required to build a radioactive-waste facility will be spent on the siting and licensing of the facility.

Nuclear or radioactive waste can be loosely defined as something that is no longer useful and that contains radioactive isotopes in varying concentrations and forms. Radioactive waste is then further broken down into categories that classify the waste by activity, by generation process, by molecular weight, and by volume.

Radioactive isotopes emit energy as they decay to more stable elements. The energy is emitted in the form of alpha particles, beta particles, neutrons, and gamma rays. The amount of energy that a particular radioactive isotope emits, the timeframe over which it emits that energy, and the type of contact with humans all help determine the hazard it poses to the environment. The major categories of radioactive waste that exist are high-level waste (HLW), low-level waste (LLW), transuranic waste (TRU), uranium mine and mill tailings, mixed wastes, and naturally occurring radioactive materials.

Current Status of Nuclear Waste Management

Nuclear or radioactive materials are used in many applications throughout today's society. Radioactive materials are used to generate power in nuclear power stations and are used to treat patients in hospitals. The generators of radioactive waste in today's society are primarily the federal government, electrical utilities, private industry, hospitals, and universities. Although each of these generators uses radioactive materials, the waste that is generated by each of them may be very different and must be handled accordingly. Any material that contains radioactive isotopes in measurable quantities is considered nuclear or radioactive waste. For the purposes of this chapter, the terms nuclear waste and radioactive waste will be considered synonymous.

Waste management is a field that involves the reduction, stabilization, and ultimate disposal of waste. Waste reduction is the practice of minimizing the amount of material that requires disposal. Some of the common ways in which waste reduction is accomplished are incineration, compaction, and dewatering. The object of waste disposal is to isolate the material from the biosphere, and in the case of radioactive waste allow it time to decay to sufficiently safe levels. Table 5.3 is a chronology of the laws that have affected radioactive waste management practices over the last 50 years.[14]

TABLE 5.3 Early Chronology of Major Events Affecting Nuclear Waste Management

Year	Event
1954	Atomic Energy Act is passed.
1963	First commercial disposal of LLW.
1967	DOE facilities begin to store TRU wastes retrievably.
1970	National Environmental Policy Act becomes effective; Environmental Protection Agency is formed.
1974	Atomic Energy Commission divides into the Nuclear Regulatory Commission (NRC) and the Energy Research and Development Administration (ERDA).
1975	Waste Isolation Pilot Plant (WIPP) proposed as unlicensed defense TRU disposal facility; West Valley, New York, low-level disposal facility closed.
1977	President Carter deferred reprocessing, pending the review of the proliferation implications of alternative fuel cycles.
1979	Three Mile Island, Unit 2 accident; report to the president of the Interagency Review Group on Radioactive Waste Management.
1980	Low-Level Waste Policy Act is passed; all commercial disposal of TRU wastes ends.
1982	Nuclear Waste Policy Act is passed; 10 CFR Part 61 issued as final regulation for LLW.
1985	Low-Level Radioactive Waste Policy Act Amendments.
1986	Reactor explosion at Chernobyl.
1987	Nuclear Waste Policy Act Amendments provide for the characterization of the proposed HLW repository at Yucca Mountain, Nevada.

The federal government has mandated that individual states or interstate compacts, which are formed and dissolved by Congress, be responsible for the disposal of the LLW generated within their boundaries. Originally, these states were to bring the disposal capacity online. Although access to the few remaining facilities is drawing to an end, none of the states or compacts have a facility available to accept waste. Some states are making progress, but none of the proposed facilities is currently in the construction phase.

Both the HLW and the TRU waste programs have sites defined for their respective facilities at Yucca Mountain and at the WIPP in Carlsbad, New Mexico. The WIPP facility is a Department of Energy (DOE) research and development facility that has been designed to accept 6 million ft^3 of contact-handled TRU waste, as well as 25,000 ft^3 of remote-handled TRU waste. The facility will accept defense-generated waste and place it into a retrievable geologic repository. A geologic repository is in this instance the salt formations located near Carlsbad. The facility has a design-based lifetime of 25 years.

Ramifications of Nuclear Accidents

The three largest radiological accidents of the last 20 years are the explosion at Chernobyl, the partial core meltdown at Three Mile Island Unit 2, and the mishandling of a radioactive source in Brazil. The least publicized, but perhaps the most appropriate of these accidents with respect to waste management, was the situation in Brazil.

An uncontrolled radiotherapy source was overlooked in an abandoned medical clinic and was eventually discarded as scrap. The stainless steel jacket and the platinum capsule surrounding the radioactive cesium were compromised by scavengers in a junkyard. The cesium was distributed among the people for use as "carnival glitter," because of its luminescent properties. The material was spread directly onto the individuals' skin and face, as well as their clothing. Severe illness was immediately evident in most of the exposed victims. Four people died from exposure by the spring of 1988, and it was estimated that an additional five persons would die over the next 5 years. Over 40 tons of material, including clothing, shoes, and housing materials, were contaminated from the release of less than 1 g of radioactive cesium.

Biological Effects of Radiation Although much still remains to be learned about the interaction between ionizing radiation and living matter, more is known about the mechanism of radiation damage on the molecular, cellular, and organ system level than most other environmental hazards. The radioactive materials warning sign is shown in Figure 5.1. A vast amount of quantitative dose–response data has been accumulated throughout years of studying the different applications of radionuclides. This information has allowed the nuclear technology industry to continue at risks that are no greater than any other technology. The following subsections will provide a brief description of the different types of ionizing radiation and the effects that may occur upon overexposure to radioactive materials.

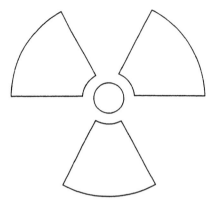

Figure 5.1 Radioactive materials warning sign.

Radioactive Transformations Radioactive transformations are accomplished by several different mechanisms, most importantly alpha particle, beta particle, and gamma ray emissions. These mechanisms are spontaneous nuclear transformations, whose result is the formation of different, but more stable elements. The kind of transformation that will take place for any given radioactive element is a function of the type of nuclear instability as well as the mass/energy relationship. The nuclear instability is dependent on the ratio of neutrons to protons; a different type of decay will occur to allow for a more stable daughter product. The mass/ energy relationship states that for any radioactive transformations the laws of conservation of mass and the conservation of energy must be followed.

An alpha particle is an energetic helium nucleus. The alpha particle is released from a radioactive element with a neutron to proton ratio that is too low. The helium nucleus consists of two protons and two neutrons. The alpha particle differs from a helium atom in that it is emitted without any electrons. The resulting daughter product from this type of transformation has an atomic number that is two less than its parent and an atomic mass number that is four less. Below is an example of alpha decay using polonium (Po); polonium has an atomic mass number (protons and neutrons) and atomic number of 210 and 84, respectively:

$$^{210}_{84}\text{Po} \longrightarrow {}^{4}_{2}\text{He} + {}^{206}_{82}\text{Pb} \tag{5.1}$$

The terms He and Pb represent helium and lead, respectively.

This is a useful example because the lead daughter product is stable and will not decay further. The neutron to proton ratio changed from 1.5 to 1.51, just enough to result in a stable element. Alpha particles are known as having a high LET, or linear energy transfer. The alphas will only travel a short distance while releasing energy. A piece of paper or the top layer of skin will stop an alpha particle. So, alpha particles are not external hazards, but can be extremely hazardous if inhaled or ingested.

Beta particle emission occurs when an ordinary electron is ejected from the nucleus of an atom. The electron (e), appears when a neutron (n) is transformed into a proton within the nucleus:

$$_0^1 n \longrightarrow \, _1^1 H + \, _{-1}^0 e \tag{5.2}$$

Note that the proton is shown as a hydrogen (H) nucleus. This transformation must conserve the overall charge of each of the resulting particles. Contrary to alpha emission, beta emission occurs in elements that contain a surplus of neutrons. The daughter product of a beta emitter remains at the same atomic mass number but is one atomic number higher than its parent. Many elements that decay by beta emission also release a gamma ray at the same instant. These elements are known as beta-gamma emitters. Strong beta radiation is an external hazard because of its ability to penetrate body tissue.

Similar to beta decay is positron emission, where the parent emits a positively charged electron. Positron emission is commonly called beta-positive decay. This decay scheme occurs when the neutron to proton ratio is too low and alpha emission is not energetically possible. The positively charged electron, or positron, will travel at high speeds until it interacts with an electron. Upon contact, each of the particles will disappear and two gamma rays will result. When two gamma rays are formed in this manner it is called annihilation radiation.

Unlike alpha and beta radiation, gamma radiation is an electromagnetic wave with a specified range of wavelengths. Gamma rays cannot be completely shielded against but can only be reduced in intensity with increased shielding. Gamma rays typically interact with matter through the photoelectric effect, Compton scattering, pair production, or direct interactions with the nucleus.

Dose–Response   The response of humans to varying doses of radiation is a field that has been widely studied. (See Chapter 7 for additional details on dose–response.) The observed radiation effects can be categorized as stochastic or nonstochastic effects, depending upon the dose received and the time period over which such dose was received. Contrary to most biological effects, effects from radiation usually fall under the category of stochastic effects. The nonstochastic effects can be noted as having three qualities: A minimum dose or threshold dose must be received before the particular effect is observed; the magnitude of the effect increases as the size of the dose increases; and a clear causal relationship can be determined between the dose and the subsequent effects. Cember[15] uses the analogy between drinking an alcoholic beverage and exposure to a noxious agent. For example, a person must exceed a certain amount of alcohol before he or she shows signs of drinking. After that, the effect of the alcohol will increase as the person continues to drink. Finally, if he or she exhibits drunken behavior, there is no doubt that this is a result of his or her drinking.

Stochastic effects, on the other hand, occur by chance. Stochastic effects will be present in a fraction of the exposed population as well as in a fraction of the unexposed population. Therefore, stochastic effects are not unequivocally related to a noxious agent as the above example implies. Stochastic effects have no threshold; any exposure will increase the risk of an effect but will not wholly determine if any effect will arise. Cancer and genetic effects are the two most common effects linked with exposure to radiation. Cancer can be caused by the damaging of a somatic cell, while genetic effects are caused when damage occurs to a germ cell that results in a pregnancy.

Sources of Nuclear Waste

Naturally Occurring Radioactive Materials[16]   Naturally occurring radioactive materials, or NORM, are present in the Earth's crust in varying concentrations. The major naturally occurring radionuclides of concern are radon, radium, and uranium. These radionuclides have been found to concentrate in water treatment plant sludges, petroleum scale, and phosphate fertilizers.

In the United States an estimated 40 billion gallons of water are distributed daily through public water supplies. Since water comes from different sources—streams, lakes, reservoirs, and aquifers—contains varying levels of naturally occurring radioactivity. Radioactivity is leached into ground or surface water while in contact with uranium- and thorium-bearing geologic materials. The predominant radionuclides found in water are radium, uranium, and radon, as well as their decay products.

For reasons of public health, water is generally treated to ensure its quality before consumption by the public. Water treatment includes passing the water through various filters and devices that rely on chemicals to remove any impurities and organisms. If water with elevated radioactivity is treated by one or more of these systems, there exists the possibility of generating waste sludges or brines with elevated levels of radioactive materials. These wastes may be generated even if it were not the original intention of the treatment process to remove radionuclides.

Mining of phosphate rock (phosphorite) is the fifth largest mining industry in the United States in terms of quantity of material mined. The southeastern United States is the center of the domestic phosphate rock industry, with Florida, North Carolina, and Tennessee having over 90 percent of the domestic rock production capacity.

Phosphate rock is processed to produce phosphoric acid and elemental phosphorus. These two products are then combined with other materials to produce phosphate fertilizers, detergents, animal feeds, other food products, and phosphorus-containing materials. The most important use of phosphate rock is the production of fertilizer, which accounts for 80 percent of the phosporite in the United States.

Uranium in phosphate ores found in the United States ranges from 20 to 300 parts per million (ppm), or about 7 to 100 pCi/g. Thorium occurs at a lower concentration between 1 and 5 ppm, or about 0.1 to 0.6 pCi/g. The unit picocuries per gram (pCi/g) represents a concentration of each radionuclide based on the activity of that radionuclide. The units of curies represent a fixed number of radioactive transformations in a second. Phosphogypsum is the principal waste by-product generated during the phosphoric acid production process. Phosphate slag is the principal waste by-product generated from the production of elemental phosphorous. Elevated levels of both uranium and thorium as well as their decay products are known to exist at elevated levels in these wastes. Since large quantities of phosphate industry wastes are produced, there is a concern that these materials may present a potential radiological risk to individuals that are exposed to these materials if distributed in the environment.

Fertilizers are spread over large areas of agricultural land. The major crops that are routinely treated with phosphate-based fertilizer include coarse grains, wheat, corn, soybeans, and cotton. Since large quantities of fertilizer are used in agricultural applications, phosphate fertilizers are included as a NORM material. The continued use of phosphate fertilizers could eventually lead to an increase in radioactivity in the environment and in the food chain.

Currently, there are no federal regulations pertaining directly to NORM-containing wastes. The volume of wastes produced is sufficiently large that disposal in a low-level waste facility is generally not feasible. A cost-effective solution must be implemented to both guard industry against large disposal costs and ensure the safety and health of the public.

Low-Level Radioactive Waste   Low-level radioactive waste (LLRW) is produced by a number of processes and is the broadest category of radioactive waste. Low-level waste is frequently defined for what it is not, rather than for what it is. According to the Low-Level Waste Policy Act of 1980, LLRW is defined as "radioactive waste not classified as highlevel radioactive waste, transuranic waste, spent nuclear fuel, or byproduct material as defined in Section 11 (e)(2) of the Atomic Energy Act of 1954."

This definition excludes high-level waste and spent nuclear fuel because of its extremely high activity. Transuranic wastes (those containing elements heavier than uranium) are excluded because of the amount of time needed for them to decay to acceptable levels. Finally, by-product material or mill tailings are excluded because of the very low concentrations of radioactivity in comparison to the extreme volume of waste that is present.

The generators of low-level waste include nuclear power plants, medical and academic institutions, industry, and the government. Low-level waste can be generated from any process in which radionuclides are used. A list of the different waste streams and the possible generators of each is presented in Table 5.4.[17]

Each of the aforementioned generators produces wastes that fall into the category of low-level waste. The waste streams identified in Table 5.4 are categorized by the generation process but may also, in some instances, be identified by the type of generating facility.

The disposal of low-level waste is accomplished through shallow land burial. This process usually involves the packaging of individual waste containers in large concrete overpacks. The overpack is designed to reduce the amount of water that may come into contact with the waste. Another function of the overpack is to guard against intruders coming into contact with the waste once institutional

TABLE 5.4   Typical Waste Streams by Generator Category

Waste Stream	Power Reactors	Medical & Academic	Industrial	Government
Compacted trash or solids	X	X	X	X
Dry active waste	X			
Dewatered ion exchange resins	X			
Contaminated bulk	X		X	X
Contaminated plant hardware	X		X	X
Liquid scintillation fluids		X	X	X
Biological wastes		X		
Absorbed liquids		X	X	X
Animal carcasses		X		
Depleted uranium MgF_2			X	

control of the facility is lost. When waste is delivered to the facility in drums, boxes, or in high-density polyethylene (HDPE) liners, they are placed in an overpack and sealed with cement before being buried in the landfill.

High-Level Radioactive Waste   High-level waste (HLW) consists of spent nuclear fuel, liquid wastes resulting from the reprocessing of irradiated reactor fuel, and solid waste that results from the solidification of liquid high-level waste. Spent reactor fuel is the fuel that has been used to generate power in a reactor. The spent fuel may be owned by a government reactor, a public utility reactor, or a commercial reactor. The wastes resulting from fuel reprocessing are either governmentally or commercially generated. Only a small fraction of the liquid HLW has been generated commercially. Approximately 600,000 gal of waste were produced in the nation's only commercial fuel reprocessing facility in West Valley, New York. The remainder of the HLW present in the United States today has been generated by the government in weapons facilities.

Spent nuclear fuel is removed from a reactor and stored in a pool of water on the site. The water in the spent fuel storage pools shields the workers and the environment from the fission products, as well as provides cooling to the fuel. The residual heat from a fuel assembly is quantified as approximately 6 percent of the operating power level of the reactor. Failure to provide additional cooling after the fission reaction has stopped was the reason for the fuel damage at Three Mile Island. Once a geologic repository is constructed, the spent fuel assemblies will be placed in a sealed canister and disposed of.

Most of the liquid high-level waste is stored in underground storage tanks. Many of these tanks are getting old and the availability of a geologic repository in the near future is doubtful. Many methods of solidifying the wastes for transport and ultimate disposal have been investigated. Plans are under way to store HLW in one central location in the United States. The chosen location is Yucca Mountain, Nevada.

Transuranic Waste Transuranic wastes are those wastes containing isotopes that are heavier than uranium, U. Generally, transuranic isotopes are not found in nature. These isotopes are manmade, produced by the irradiation of heavy elements, such as uranium and thorium. Transuranic wastes are

$$^{238}_{92}U + ^1_0n \longrightarrow ^{239}_{93}Np + ^0_{-1}e \longrightarrow ^{239}_{94}Pu + ^0_{-1}e \qquad (5.3)$$

where Np and Pu represent neptunium and plutonium, respectively. They are normally generated by the government, particularly from weapons testing. The transuranic waste is now being stored at a number of DOE facilities across the country, awaiting permanent disposal at WIPP in Carlsbad, New Mexico.

Radioactive Waste Treatment and Disposal

Many treatment processes can be employed to reduce the volume, or increase the stability, of waste that must ultimately be permanently disposed of. Landfill fees for radioactive waste is assessed largely on the volume of the waste to be disposed of. Current trends in the rising cost of waste disposal have led to the generators' implementing one or a number of waste minimization techniques. The physical form of the waste is a critical factor in determining the probability that the waste will remain isolated from the biosphere.

Compacting is a method of directly reducing the volume and increasing the specific weight of the resulting waste. Materials such as glass vials, protective clothing, and filter media can be compacted to reduce the volume. Compacting does not reduce the environmental hazard of the waste stream—its purpose is purely waste minimization.

Incineration of waste both reduces the volume and provides a more stable waste stream. Many biological wastes, including animal carcasses, are incinerated. The storage of animal carcasses in drums is generally not cost effective because of the gas generation of the materials as they decay biologically. A drum packed with animal carcasses must be filled with absorbent material so that the pressure inside the drum does not rise to unsafe levels. Incineration is a very cost-effective waste reduction technique for large generators of combustible materials.

Dewatering or evaporation is another waste minimization and stabilization technique that is practiced by waste generators. Evaporating sludges or slurries can greatly reduce the volume of the waste stream and stabilize the waste prior to disposal.

Future Trends

Current regulations call for each individual state or interstate compact to store and dispose of all of the LLW generated within its boundaries. An interstate compact consists of a group of states that have joined together to dispose of LLW. Interstate compacts can only be formed and dissolved by Congress. Many regulatory milestones have passed, leaving most states with restricted access to Barnwell,

South Carolina. Barnwell is the only remaining LLW disposal facility for such wastes. It is most certain that the Barnwell facility will close before most states have centralized storage capacity on line. Some states, like New York, have unsuccessfully attempted to sue the federal government, arguing that it is unconstitutional to mandate that states dispose of radioactive waste within their boundaries. Unless the individual states or compacts take immediate action to site and construct a permanent disposal facility or temporary centralized storage facility, waste generators will be forced to store either radioactive materials onsite or to stop generating radioactive wastes by ceasing all operations that utilize radioactive materials. While these two options may seem appropriate, neither of them will solve the problem of waste disposal for any extended period of time. Many radioactive waste generators, like hospitals, do not have the storage space allocated to handle onsite storage for periods exceeding 1 or 2 years. Much of the waste generated at hospitals is directly related to patient care, and it is unacceptable to assume that all processes, like chemotherapy, that produce radioactive waste will be stopped.

Both the HLW and TRU programs are limping along because of public concern for the areas surrounding the proposed facilities. The WIPP facility has performed some waste emplacement, but this has been accomplished only as a research and development activity. The HLW program has met drastic public opposition because of the amount of time that the waste will remain extremely hazardous. This time period is on the order of thousands of years. Opponents of this facility are arguing that the ability to label the disposal facility properly and guard against future intruders is lacking. Many symbols, such as thorns or unhappy faces, have been proposed.

The public at large will continue to oppose most activities involving nuclear waste until they are made aware of the unwanted characteristics of current more acceptable technologies as well as the extreme benefits that radioactive materials have made to society.

5.6 METALS[7]

Certain metals appear to be the major present environmental concern associated with solid wastes. Future nanoapplications may change that assessment. Until then, the review presented here for the four metals of greatest concern is warranted.

The dangers of human exposure to many metallic chemical elements have long been recognized. Metals such as lead, mercury, cadmium, and arsenic are toxic. Health effects range from retardation and brain damage, especially in children from lead poisoning, to the impairment of the central nervous system as a result of mercury exposure. Since these toxic metals are chemical elements, they cannot be broken down by any chemical or biological process. As a result of this, it is not uncommon to have metallic buildup or bioaccumulation.

Table 5.5 lists four common metals of environmental concern and the health effects associated with each. The following subsections will describe these concerns in more detail and explain how exposure to these substances can be reduced.

TABLE 5.5   Health Effects of Common Metals

Pollutant	Health Concerns
Lead	Retardation and brain damage, especially in children
Cadmium	Affects the respiratory system, kidneys, prostate, and blood
Mercury	Several areas of the brain as well as kidneys and bowels are affected
Arsenic	Causes cancer

Lead

Lead was once the most popular metal used in the gasoline and paint industries. In ancient times, it may have been the first metal to be smelted by humans. Lead pipes used by Romans are still in use today, and pottery glazed with lead oxide dates back to the bronze age.

Lead is a bluish-white metal with a bright luster. It is soft, malleable, and ductile. It is a poor conductor of electricity and is very resistant to corrosion. Lead ores commonly occur with zinc, copper, and pyrite ores. Galena (lead sulfide) accounts for more than 90 percent of primary lead production at present. There is also a large secondary lead industry that recycles lead from batteries and other lead scrap.

Lead in Drinking Water   The primary source of lead in drinking water is corrosion of plumbing materials, such as lead service lines and lead solder in water distribution systems and in houses and larger buildings. Virtually all public water systems serve households with lead solders of varying ages, and most faucets are made of materials that can contribute some lead to drinking water. One cannot see, smell, or taste lead, and boiling water will not get rid of lead. If one thinks the plumbing might have lead in it, the water should be tested for lead. The only certain way to know if there is lead in the water is to have it tested. The local health department or the water supplier can be called to see how to get it tested. Testing water is easy and cheap ($15 to $25). Household water will contain more lead if it has sat for a long time in the pipes or is hot or naturally acidic. If the water has not been tested or has high levels of lead, (a) do not drink, cook, or make baby formula with water from the hot water tap; (b) if cold water has not been used for more than 2 hours, run it for 30 to 60 s before drinking it or using it for cooking; and (c) consider buying a filter certified for lead removal. The EPA's Safe Drinking Water Hotline provides more information.[18]

The health effects related to the ingestion of too much lead are very serious and can lead to impaired blood formation, brain damage, increased blood pressure, premature birth, low birth weight, and nervous system disorders. Young children are especially at risk from high levels of lead in drinking water.[19]

Lead in Gasoline   Lead has long been used in gasoline to increase octane levels to avoid engine knocking. As described above, lead is a heavy metal that can cause serious physical and mental impairment; children are particularly vulnerable to the effects of high lead levels. Two efforts begun 25 years ago are responsible for a 95 percent reduction in the use of lead in gasoline.

Recognizing the health risks posed by lead, the EPA in the early 1970s required the lead content of all gasoline to be reduced over time. In addition to phasing down the lead in gasoline, the EPA's overall automotive emission control program required the use of unleaded gasoline in many cars beginning in 1975. Currently, all the gasoline sold in this country is unleaded.[18]

Lead in Paints Lead compounds were commonly used in the formulation of house paint. As the paint chipped, many young children ate the paint chips, which resulted in brain damage. Since the early 1940s, nontoxic pigments have gradually replaced lead in paints. In 1978, the Consumer Products Safety Commission (CPSC) banned the sale of lead for use in residential paints, and made it illegal to paint children's toys and household furniture with lead-based paints. Most states have banned the use of lead in house paints outright. If a home has been painted with lead-based paint, extreme caution must be taken when sanding or remodeling these surfaces. Lead dust levels are 10 to 100 times greater in homes where sanding or open-flame burning of lead-based paints has occurred.

If paint being removed is suspected of containing lead, have it tested. Lead-based paint should be left undisturbed, and not sanded or burned off. Lead-based paint should be covered with wallpaper or other building material. Moldings and other woodwork should be removed and chemically treated offsite.[20]

If a home was built before 1978, one should be concerned about lead-based paint hazards. The older one's house is, the more likely it is to contain lead-based paint. Even if the original paint has been covered with new paint or another covering, cracked or chipped painted surfaces can expose the older, lead-based paint layers, possibly creating a lead hazard.

If one is removing paint or breaking through painted surfaces, the individual should be concerned about lead-based paint hazards. If one's job involves removing paint, sanding, patching, scraping, or tearing down walls, the individual should be concerned about exposure to lead-based paint hazards. If one is doing other work, such as removing or replacing windows, baseboards, doors, plumbing fixtures, heating and ventilation duct work, or electrical systems, the individual should be concerned about lead-based paint hazards since you may be breaking through painted surfaces to do these jobs. Obtaining the right equipment and knowing how to use it are essential steps in protecting oneself during remodeling or renovating. A high-efficiency particulate air (HEPA) filter-equipped vacuum cleaner is a special type of vacuum cleaner that can remove very small lead particles from floors, window sills, and carpets and keep them inside the vacuum cleaner. Regular household or shop vacuum cleaners are not effective in removing lead dust since they blow the lead dust out through their exhausts and spread the dust throughout the home. HEPA vacuum cleaners are available through laboratory safety and supply catalogs and vendors. They can sometimes be rented at stores that carry remodeling tools. One will also need to use a properly fitted respirator with HEPA filters to filter lead dust particles out of the air that one breathes. Specific HEPA filters should be used—they are always purple. Dust filters and dust masks are not effective in preventing the breathing in lead particles. Protective clothes, such as coveralls, shoe covers, hats, goggles, and gloves should be used to help keep lead dust from being carried into areas outside of the worksite. Wet-sanding equipment, wet/dry abrasive paper, and wet-sanding sponges can be purchased at hardware stores; spray bottles for wetting surfaces to keep dust from spreading can also be purchased at general retail and garden supply stores.[21] Most importantly, all non-workers, especially children, pregnant women, and pets should be kept outside of the work areas while doing remodeling or renovation work until cleanup is completed. In some jurisdictions, even residential lead removal must be performed only by a licensed contractor.

Lead in Municipal Solid Waste Lead is widespread in the municipal solid waste (MSW) stream; it is in both the combustible and noncombustible portions of MSW. Discharges of lead in MSW are overwhelmingly greater than discards of cadmium.

Lead-acid batteries, primarily batteries for automobiles, rank first, by a wide margin, of the products containing lead that enter the waste stream. Trends in quantities of lead discarded in products in MSW, ranked by tonnage discarded in 1986, are shown in Table 5.6. The last two columns of the table indicate whether the total tonnage of lead in a product is generally increasing or decreasing, and whether the percentage of total lead contained in MSW contributed by a product is increasing or decreasing.

TABLE 5.6 Lead in Products Discarded in MSW,[a] 1970–2000

Products	1970	1986	2000[b]	Tonnage	Percentage
Lead-acid batteries	83,825	138,043	181,546	Increasing	Variable
Consumer electronics	12,233	58,536	85,032	Increasing	Increasing
Glass and ceramics	3,465	7,956	8,910	Increasing	Increasing; stable after 1986
Plastics	1,613	3,577	3,228	Increasing; decreasing after 1986	Fairly stable
Soldered cans	24,117	2,052	787	Decreasing	Decreasing
Pigments	27,020	1,131	682	Decreasing	Decreasing
All others	12,567	2,537	1,701	Decreasing	Decreasing
Totals	*164,840*	*213,652*	*181,887[b]*		

[a]Data are given in short tons.
[b]Estimated.

Lead discards in batteries are shown to be growing steadily, as are discards in consumer electronics. Discards of lead solder in cans and lead in pigments, however, virtually disappeared between 1970 and 1986. Lead discards in other products are known to be relatively small.

Findings about the individual products in MSW that contain lead are as follows:

1. *Lead-acid batteries* contributed 65 percent of the lead in MSW in 1986; this percentage has ranged between 50 and 85 percent during the 1970 to 1986 period studied. The tonnages in Table 5.6 represent discards after recycling but of all the products considered, only lead-acid batteries are recycled to a significant extent. Recycling rates, which have ranged from 52 to 80 percent, have a major effect on the tonnage of lead-acid batteries discarded.

2. *Consumer electronics* (television sets, radios, and video cassette recorders) accounted for 27 percent of lead discards in MSW in 1986. They contributed lead from soldered circuit boards, leaded glass in television sets, and plated steel chassis. Leaded glass accounts for most of the lead in these products.

3. *Glass and ceramics*, as reported here, include lead in products such as glass containers, tableware and cookware, and other items such as optical glass. These contributed 4 percent of lead discards in 1986. (Leaded glass in light-bulbs is included in the "All others" category in Table 5.6.)

4. *Plastics* use lead in two ways: as a heat stabilizer (primarily in polyvinyl chloride resins) and as a component of pigments in many resins. This category, which includes products such as nonfood packaging, clothing and foot-wear, housewares, records, furniture, appliances, and other miscellaneous products, accounted for about 2 percent of lead discards in 1986. Plastics in consumer electronics products are counted under that category.

5. *Soldered cans* have experienced a large decline in usage since 1970, when they contributed 14 percent of the lead in MSW. Leaded solder is currently used in steel food cans, general-purpose cans (like aerosols), and shipping containers.

6. *Pigments* containing lead compounds have declined greatly since 1970, drop-ping from 18 percent of total lead discards to less than 1 percent. This category includes pigments used in paints, printing inks, textile dyes, and so on. Pigments used in plastics, glass and ceramics, and rubber products are accounted for in those categories.

7. *All others* include brass and bronze products, lightbulbs (which contain lead in solder and in glass), rubber products, used oil, collapsible tubes, and lead foil wine bottle wrappers. Collapsible tubes contributed over 5 percent of total lead discards in 1970, but their use has declined dramatically since then. None of the other items has exceeded 1 percent of the total since 1970.[22]

Commonly Used Lead Compounds Lead is used in many compounds. Some of the more common sources are briefly described below.

Lead monoxide (PbO), commonly called *litharge*, is the most widely used inorganic lead chemical. It is used in storage battery plates, ceramics and glasses, paint, rubber, and other products. It is also used in the production of other lead chemicals such as lead orthoplumbate.

Lead dioxide (PbO$_2$) is used as an oxidizing agent in the manufacture of chemicals and dyes and as a curing agent for sulfide polymers. It is the active materials of the positive plate in electric storage batteries.

Lead orthoplumbate (Pb$_3$O$_4$) is commonly called *red lead*. It is a brilliant red pigment and is used as an inhibitor in surface coatings to prevent corrosion of metals. It is used in storage batteries, leaded glass, lubricants, and rubber. It is also used for making lead dioxide.

Lead sulfide (PbS), or *galena*, is a common lead mineral. Lead sulfide has semi-conducting properties. It is also used for mirror coatings and it is a component of blue lead pigments.

Lead metaborate (PbO·B$_2$O$_3$·H$_2$O) is used in glazes, enamels, and glasses.

Basic lead carbonate [2PbCO$_3$·Pb(OH)$_2$] is commonly called *white lead*. It is the most important basic salt of lead. It is a white pigment and is used in surface coatings, greases, and plastics stabilizers.

Lead silicates. The most common silicate of lead is lead metasilicate (PbSiO$_3$). The silicates are used in ceramics, glasses, paints, rubber, and as stabilizers in plastics.

Basic lead sulfates (XPbO · PbSO$_4$) are used as white or blue pigments in paints and as stabilizers for plastics. They are also used as a filler in rubbers and in inks.

Lead chromate (PbCrO$_4$) is an important pigment and is often formulated in combination with other lead compounds or with inorganic salts of other metals to make a range of colors, including chrome green, chrome yellow, and molybdate chrome orange. Chrome yellows contain lead sulfate; chrome greens contain iron cyanides; and molybdenum chrome oranges contain molybdate and often lead sulfate. These pigments are used in paints, coatings, inks, and leather goods.

Basic lead chromate (PbCrO$_4$·PbO) is used in pigments commonly called *chrome oranges.*

Lead chloride (PbCl$_2$) can be prepared by the reaction of lead monoxide or basic lead carbonate with hydrochloric acid. Most of its uses are industrial rather than in products that would commonly enter municipal solid waste. Lead chloride is used as a catalyst, as a cathode for seawater batteries, as a flame retardant in polycarbonates, as a flux for the galvanizing of steel, as a photochemical-sensitizing agent for metal patterns on printed circuit boards, and for other uses.

Lead salts are formed from lead and organic acids. Several lead salts of higher fatty acids (C$_{10}$ and over; commonly called lead soaps) have important uses as paint driers, stabilizers for plastics, additives to lubricating oil, and additives in rubber.[23–25]

Cadmium

Cadmium is a toxic metal most commonly known for its use in rechargeable nickel–cadmium batteries. It is a relatively rare metal that has some unique characteristics that make it useful in a variety of products. Cadmium is silvery-white in color and is soft, ductile, and easily worked. It has good electrical and thermal conductivity. When exposed to moist air, cadmium oxidizes slowly to form a thin coating of cadmium oxide, which protects the metal from further corrosion.

Cadmium usually occurs as the mineral greenockite (CdS). It is usually mined in association with zinc, but sometimes with lead and copper ores. It is almost never found alone in economical quantities. Secondary (recycled) cadmium production is of minor significance in the United States. Unlike lead, which has been used since ancient times, cadmium has been refined and utilized only relatively recently.

Consumption of cadmium by end use is reported by the Bureau of Mines.[24] The end-use categories are more limited than those reported for lead, however. The categories reported annually include coating and plating, batteries, pigments, plastics stabilizers, and other (including alloys).

While consumption of lead in the United States was over 1.2 million tons in 1986, consumption of cadmium was a relatively small 4800 tons. Overall, the domestic consumption of cadmium in the United States declined until 1983, but it has increased since then. In both percentage and tonnage, coating and plating and plastics stabilizers have declined since 1970. Use of cadmium in pigments grew in the early 1970s but has remained about stable since 1975. Domestic use of cadmium in nickel–cadmium batteries has been significant, although the tonnage has been fairly stable since 1976.

Nickel–Cadmium Batteries   Nickel–cadmium (Ni–Cd) batteries are a major consumer of cadmium in the United States. Ni–Cd batteries were invented in the early years of the twentieth century but were not used extensively until the mid-1940s when they came into use in the military and industrial sectors. Ni–Cd batteries are secondary batteries (rechargeable).

In the early 1960s Ni–Cd batteries were developed for consumer use, but they did not gain real popularity until the early 1970s. Ni–Cd batteries are now used in many products: pocket calculators; toys; microprocessors; hand tools such as portable drills, flashlights, screwdrivers, hedge trimmers, and soldering irons; and rechargeable appliances such as hand-held vacuums, mixers, can openers, VCRs, portable televisions, cameras, electric shavers, lawn mower engine starters, and alarm systems. Many consumer applications such as appliances have the Ni–Cd battery sealed in; the battery cannot be replaced by the owner.

New consumer, uses for Ni–Cd batteries, such as portable laptop computers and cellular telephones, are continually being developed. Ni–Cd batteries are also competing with mercury batteries in hearing aids and pocket calculators, and with carbon–zinc and other primary batteries, because their initial high cost can be offset as they are recharged. Military and industrial uses of nickel–cadmium batteries include railroad signaling, diesel locomotive starting, commercial and jet aircraft starting, satellites, missile guidance systems, naval signaling, television and camera lighting, portable hospital equipment, computer memories, pinball machines, and gasoline pumps.[26]

Use of Cadmium Pigments Yellow pigments composed of cadmium sulfide, barium sulfate, and zinc sulfate are used in textile printing. The quantities used are very small, however, because of high costs and poor tinting capabilities. The same is true for cadmium orange or red pigments (cadmium sulfoselenide or mercadmium compounds).

The plastics industry is the largest end user of cadmium pigments. These pigments disperse well in most polymers and give good color and high capacity and tinting strength. Cadmium pigments are also insoluble in organic solvents, and have good resistance to alkalis.

Cadmium reds and maroons, the most durable of the cadmium pigments, are used in automobile finishes. Cadmium reds are coprecipitated and cocalcined mixtures of cadmium sulfide and cadmium selenide. Mercury cadmium pigments are also used occasionally.[26]

Cadmium in Municipal Solid Waste Like lead, cadmium is widespread in products discarded into MSW, although it occurs in much smaller quantities overall. Since 1980, nickel–cadmium household batteries have been the number one contributor of cadmium in MSW.

Trends in quantities of cadmium discarded in products in MSW (ranked by tonnage discarded in 1986) are shown in Table 5.7. Discards of cadmium in household batteries were small in 1970, but then increased dramatically. Cadmium discards in plastics are relatively stable. Discards of cadmium in consumer electronics are shown to decrease over time, while the other categories are relatively small.

TABLE 5.7 **Cadmium in Products Discarded in MSW,[a] 1970 to 2000**

Products	1970	1986	2000[b]	Tonnage	Percentage
Household batteries	53	930	2,035	Increasing	Increasing
Plastics	342	520	380	Variable	Variable; decreasing after 1986
Consumer electronics	571	161	67	Decreasing	Decreasing
Appliances	107	88	57	Decreasing	Decreasing
Pigments	79	70	93	Variable	Variable
Glass and ceramics	32	29	37	Variable	Variable
All others	12	8	11	Variable	Variable
Totals	*1196*	*1788*	*2684[b]*		

[a]Data are given in short tons.
[b]Estimated.

Findings about cadmium discards in individual products in MSW are:

1. *Household batteries* (rechargeable nickel–cadmium batteries) have accounted for more than half of cadmium discards in the United States since 1980. This growth is projected to continue in this century, unless they are replaced by another type of battery.

2. *Plastics* continue to be an important source of cadmium in MSW, contributing 28 percent of discards in 1986. Cadmium is used in stabilizers in polyvinyl chloride (PVC) resins and in pigments in a wide variety of plastic resins. Cadmium is found in nonfood packaging, footwear, housewares, records, furniture, and other plastic products.

3. *Consumer electronics* (television sets and radios) formerly had cadmium-plated steel chassis in many cases. These chassis have been replaced by circuit boards, so cadmium discards in consumer electronics are declining as the older units are replaced. They contributed 9 percent of the total in 1986.

4. *Appliances* (dishwashers and washing machines) formerly had cadmium-plated parts to resist corrosion. This source of cadmium is declining as cadmium-plated parts are replaced by plastics, which are themselves another source of cadmium discards in appliances. Cadmium discards from appliances accounted for about 5 percent of total in 1986.

5. *Pigments* used in printing inks, textile dyes, and paints may contain cadmium compounds, although this is not a large source of cadmium in MSW (about 4 percent of the total).

6. *Glass and ceramics* may contain cadmium as a pigment, as a glaze, or as a phosphor. This is a relatively small source of cadmium in MSW.

7. *All other* sources of cadmium include rubber products, used oil, electric blankets, and heating pads. These contribute very small amounts of cadmium to MSW.

Commonly Used Cadmium Compounds   As with lead, cadmium is commonly used in many compounds. Some of the more common sources are briefly described next.

Cadmium chloride ($CdCl_2$) is used in the manufacture of nickel–cadmium batteries. It is also used as a pigment in dyeing and calico printing and in phosphors.

Cadmium oxide (CdO) has many uses, often in the preparation of cadmium products. It is used in processes in the manufacture of nickel–cadmium batteries, stabilizers for PVC, glass, phosphors, semiconductors, electroplating, and ceramic glass, among other uses.

Cadmium sulfide (CdS) is the most widely used cadmium compound. It is also called *cadmium yellow*, and is used in red and yellow pigments, in phosphors, as a photoconductor, and for other uses.

Cadmium hydroxide [$Cd(OH)_2$] is used mainly as the active material in the negative electrodes of nickel–cadmium batteries.

Cadmium nitrate [$Cd(NO_3)_2$] is used in the manufacture of nickel–cadmium batteries.

Cadmium sulfate ($CdSO_4$) is used in the manufacture of nickel–cadmium batteries.

Cadmium carboxcyclates are incorporated in stabilizers for PVC.

Cadmium acetate [$Cd(C_2H_3O_2)_2$] is used for iridescent effects on pottery and porcelain and in electroplating.

Cadmium fluoride (CdF_2) is used in the manufacture of phosphors and glass.

Cadmium selenide (CdSe) is used in photoconductors, semiconductors, and phosphors.

Cadmium telluride (CdTe) is used in phosphors and semiconductors.

Cadmium salts are used as light stabilizers in plastics; cadmium/barium salts are used as heat stabilizers in plastics.

Mercury

Mercury and most of its compounds are toxic substances. The dangers of human exposure to mercury have long been recognized. Nineteenth-century hat makers, for example, developed a characteristic shaking and slurring of speech from occupational exposure to large quantities of inorganic mercury during the manufacturing process—symptoms that gave rise to the phrase "mad as a hatter." Such problems result from impairment of the central nervous system. In addition, high levels of mercury can cause kidney damage, birth defects, and in extreme cases, death.

In the United States today, the groups most likely to be exposed to unacceptably high levels of mercury are Native American and recreational fishermen, who routinely eat large amounts of their catch. The U.S. Food and Drug Administration considers a mercury level of 1 ppm to be its action level—that is, the recommended upper limit for safe consumption. Several states have even stricter standards, which they use to issue health advisories.[27]

The most common source of the silvery liquid can be found in the bulbs of thermometers and thermostats. Large quantities of mercury can also be found in one type of barometer or another. Small quantities are also found in fluorescent lights, mercury switches, batteries, hearing aids, smoke detectors, watches, and cameras.

Mercury vapor in the atmosphere comes from a variety of natural and artificial sources: releases from the ocean and land surfaces, smelters, power plants, forest fires, mineral deposits, paint volatilization, disposal of batteries and fluorescent lamps, and the application of certain fungicides. In addition, other factors appear to be important in determining how readily mercury enters an aquatic food chain. It has been suggested, for example, that acidity may hasten the process, which has raised a further concern: that acid rain could lower the pH of vulnerable lakes and thus increase mercury contamination of the food chain. The potential importance of this effect, however, remains unclear. Neither the Everglades nor the pristine lakes of the upper Midwest have been substantially impacted by acid deposition.

Once deposited in a lake, mercury undergoes a complex series of physical and chemical changes. Some of it remains in solution, some volatizes back to the atmosphere, and some precipitates into sediments. While acidification has been suspected of accelerating uptake into the food chain, the exact nature of this role has not been clear. In particular, the effect could result either from pH changes or from the addition of sulfate ions in acid rain. The presence of dissolved organic carbon may either foster or inhibit uptake, depending on conditions, but again the reasons have remained unclear.

One of the most important issues is what environmental factors influence the formation of methyl mercury, the element's most toxic chemical compound and the one most easily incorporated into an aquatic food chain. More than 95 percent of the mercury in fish is in the form of methyl mercury. Most atmospheric mercury occurs simply in its elemental form, which must be oxidized before being carried to the earth in rain. In a lake, transformation of the oxidized form to methyl mercury, a process called methylation, takes place mainly through the action of bacteria. The amount of methyl mercury available in the water at any time depends on a balance between methylation and its opposite reaction, demethylation. Researchers are particularly interested in determining what factors can tip this balance one way or the other.

Fish can absorb the methyl mercury directly through their gills or ingest it by eating smaller organisms. Microscopic plants (phytoplankton) absorb considerable amounts of methyl mercury from the water, so the small fish that eat them receive an already concentrated dose. Large predatory fish then eat the smaller ones, and methyl mercury builds up in their bodies because it is difficult to excrete. Eventually, the magnitude of such bioconcentration becomes staggering. Fish at the top of a food chain may have levels of methyl mercury a million times the level in the surrounding water.[27]

Arsenic

Arsenic has long been known to cause cancer; however, most people associate arsenic with poisoning rather than with its toxic characteristics. The most common pesticide in farms and gardens was lead arsenate, but it has been almost entirely replaced by synthetic pesticides. Around the house, arsenic compounds can be found in medicine cabinets, in rat poison, and in plant killers. However, most of the products containing arsenic are being replaced with products with less toxic effects.

5.7 SUPERFUND

During the 1970s, people started to realize that the planet Earth and its environment had reached a critical point and pollution could cause potential health risks. If one examines the environmental laws passed during these times, one can see how the need for Superfund arose out of the public's concern for the environment.

The environmental problem became a major national issue of public concern because of a number of reasons. People were unable to fish or swim in some of the waterways and the air quality was poor. The health effects of smog and industrial air pollution alarmed the people and environmental concern initially moved toward clean air. In order to improve the nation's air quality, Congress passed the Clean Air Act in 1970, which was recently amended in 1990. As described in Chapter 1, this act reduced the pollutants being released into the air by forcing emission standards and regulations on individuals and private industry. Congress then responded in 1977 to the public's concern for clean water by passing the Clean Water Act. This act regulated safe drinking water by requiring secondary treatment on all wastewater facilities. The second provision of the Clean Water Act required previously polluted natural water bodies to become suitable for animal life. In 1976, Congress passed the Resources Conservation and Recovery Act (RCRA), which was one of the first laws to regulate solid and hazardous waste. Landfills had become a big problem since swampland, which was otherwise useless, was utilized for dumping waste. This led to groundwater contamination and other problems. RCRA addressed the issue of landfill sites and led to the issuance of permits for dumping hazardous waste.

During the late 1970s, the press and the American people's attention focused on hazardous waste sites like Love Canal and Times Beach. Love Canal is a hazardous waste site located in upstate New York, which at one time was merely an unfinished canal. It was never completed, and this large hole remained until a chemical company bought the property, used it as a landfill, and buried tons of hazardous waste chemicals. Once the dumping stopped, the land was covered with fill. The land was later used as a residential area, where houses and a school were built. The people living in the houses became ill, and they soon realized that their illness was caused by the dumping that had taken place 25 years earlier. The media began to publicize the story and stir up social concern as the public began to see the health risks of pollution. Because of this concern about hazardous waste sites, Congress passed the first law in 1980 to deal with the nation's hazardous waste sites. This law is called the Comprehensive Environmental Response, Compensation, and Liability Act (CERCLA), now commonly known as *Superfund*. In 1980, Superfund was given $1.6 billion by Congress to clean up the nation's highest risk hazardous waste sites. A method was sought to determine the worst sites. The act required every individual state to compile a list of the worst sites in its state and submit it to the Environmental Protection Agency. From this list, each site was evaluated and ranked on its risk to public health and to the environment. The sites were then placed in risk order from highest to lowest. This method of ranking sites was known as the National Priorities List (NPL). Congress initially had no idea how big the hazardous waste site problem was, and Superfund was not given enough money to clean up many of the sites now on the NPL. Congress extended Superfund in 1986 for 5 more years by passing the Superfund Amendments and Reauthorization Act (SARA). This gave Superfund $8.5 billion more in order to clean up hazardous waste sites. SARA also set up an infrastructure to run daily transactions and provided other means to obtain money for Superfund.

This section will examine the following topics:

1. Funding and legal considerations
2. The ranking systems

3. The cleanup process
4. The role of the private sector
5. The progress to date.

Funding of Superfund and Legal Considerations

When CERCLA was first passed in 1980, the law set up a trust fund of $1.6 billion, commonly known as *Superfund*. Congress initially obtained the money to fund the trust from taxes on crude oil and some commercial chemicals. Once this money ran out, Superfund was reauthorized by SARA in 1986, and Congress was again faced with the problem of how to fund the law. Congress decided that "these monies are to be made available to the Superfund directly from excise taxes on petroleum and feedstock chemicals, a tax on certain imported chemical derivatives, environmental tax on corporations, appropriations made by Congress from general tax revenues, and any monies recovered or collected from parties responsible for site contamination."[28]

CERCLA has three concepts that make it an unusual law once again. These are ex post facto, innocent landowner liability, and joint and several liability. These are once again discussed below (see Chapter 1 for more details).

Ex post facto means that, after the fact, a party can be liable for what was once legal but now is illegal. For example, a party could have legally disposed of waste at the time of disposal. However, this party could later be found liable under CERCLA for whatever that waste was and be legally responsible for its cleanup. Since it is necessary to obtain money for the cleanup of a sight, it is very important that the EPA find the parties responsible for the hazardous site. The Potentially Responsible Party (PRP) under Section 107(a) of CERCLA is defined as:

1. The current owner or operator of the site that contains hazardous substances
2. Any person who owned or operated the site at the time when hazardous substances were disposed of
3. Any person who arranged for the treatment, storage, or disposal of the hazardous substances at the site
4. Any generator who disposed of hazardous substances at the site
5. Any transporter who transported hazardous substances to the site

The persons listed above are liable for:

1. All costs of removal or remedial action incurred by the government
2. Any other necessary costs of response incurred by any other person consistent with the National Contingency Plan (NCP)
3. Damages for injury to, destruction of, or loss of natural resources, including the reasonable costs for assessing them
4. The costs of any health assessment or health effects study carried out under Section 104(i)

The second unique part of CERCLA is the innocent landowner liability. This states that anyone who buys property that is contaminated with a hazardous substance may be liable for the cost of cleanup even if he or she did not know the site was contaminated. The only way one might avoid liability is if he or she made an "all appropriate inquiry" before purchase and found nothing. The following factors are to be examined to see if an "all appropriate inquiry" was made:

1. Any specialized knowledge or experience on the part of the defendant
2. Commonly known or reasonably ascertainable information about the property
3. Relationship of the purchase price to the value of the property if uncontaminated
4. Obviousness of the presence or likely presence of contamination at the property
5. Ability to detect such contamination by appropriate inspection[29]

A third unique part of the law is the joint and several liability clause. This simply means that liability for a site can be shared between several PRPs or just one. "Each party could be liable for the same amount or one party may be liable for the entire amount even though the parties did not dispose of equal amounts."[29] Joint and several liability makes the enforcement side of CERCLA easier because the EPA can sue only one PRP and get all the money. In turn, that PRP can then sue the other contributors for their part. This saves the EPA money in legal fees. Under section 107(b) of CERCLA, there are only four legal defenses to avoid liability for hazardous site contamination. They are:

1. An act of God
2. An act of war
3. An act of omission of a third party
4. Any combination of the foregoing[29]

Enforcement and liability go hand in hand when cleaning up a Superfund site. Some of Superfund's goals are to encourage potentially responsible parties to finance and conduct the necessary response action and to recover the costs for response action(s) that were financed using the fund's money. The EPA has several enforcement options. Of the several options, the EPA usually seeks voluntary compliance. An enforcement agreement could be one of two options. The first is a judicial consent decree, which "is a legal document that specifies an entity's obligations when that entity enters into a settlement with the government."[29] The second is an administrative order that is a mutual agreement between the PRP and the EPA outside of court. If the PRP does not chose to reach an agreement with the EPA, then the EPA can issue a unilateral administrative order forcing the PRP to take charge. If the PRP still refuses, then the EPA can file a lawsuit. If the EPA wins, they may recover treble (triple) damages, which means that an uncooperative PRP could be charged three times what it cost the government to clean up the site. This is done to encourage the PRP to take responsibility early on in the cleanup process.

Ranking of Hazardous Waste Sites

When Superfund first began, every state was told to compile a list of its worst hazardous waste sites to be evaluated by the EPA for cleanup. It was soon realized that the number of sites were too large for federal action, so the government decided to rank the sites in order from the highest to the lowest risk to human health and the environment. This ranking system list is known as the *National Priorities List* (NPL). If a site makes the NPL, it is then eligible for federal money through the Superfund program. In order to be placed on the NPL, the site must meet at least one of the following three criteria:

1. Receive a health advisory from the Agency for Toxic Substances and Disease Registry (ATSDR) recommending that people be relocated away from the site.
2. Score 28.5 or higher in the Hazard Ranking System (HRS), which is the method that the EPA uses to assess the relative threat from a release, or potential release, of hazardous substances; HRS is the scoring system used to enhance the process for identifying the most hazardous and threatening sites for Superfund cleanup.
3. Be selected as the state top priority.[28]

Since risk is a very difficult quantity to measure, the HRS score was used to reflect the potential harm to human health and the environment from the migration of hazardous substances. To understand how waste can be ranked, it is necessary to look at the different types of hazardous wastes, and how they end up in the environment. The three media in which hazardous wastes can enter the environment are air, water, and soil. Hazardous wastes may leach, percolate, wash into ground or surface water, evaporate, explode, or they may get carried with the wind or rain into any media. Hazardous waste can bioaccumulate and end up in the food chain or water supply. The risk assessment looks at the waste quantity, where it is, who or what is near it, and relates these to its potential effects on the public health and the environment.

In 1990, the National Contingency Plan (NCP) was created to implement the response authorities and responsibilities created by Superfund and the Clean Water Act. The NCP outlines the steps that the federal government must follow in responding to situations in which hazardous substances are released or are likely to be released into the environment. The four basic components of the hazardous substance response provisions of the NCP are:

1. Methods for discovering sites at which hazardous substances have been disposed of
2. Methods for evaluating and remedying releases that pose substantial danger to public health and the environment
3. Methods and criteria for determining the appropriate extent of cleanup
4. Means of assuring that remedial action measures are cost effective[29]

To understand how Superfund works it is necessary to look at the structure of the Superfund program. The EPA was given responsibility as the designated manager of the trust fund by CERCLA. The policies Superfund follows comes from EPA headquarters. However, the EPA has 10 offices in different regional cities throughout the country. They have more control of the day-to-day program decisions and operations. This makes it easier to keep.a closer eye on what is going on at any particular site. Figure 5.2 shows the political structure of Superfund.

Cleanup process

The purpose of Superfund is to eliminate the short- and long-term effects of a hazardous waste. There are 10 basic steps in the cleanup process:

1. Site discovery
2. Preliminary assessment
3. Site inspection
4. Hazard ranking analysis
5. National Priorities List (NPL) determination
6. Remedial investigation and feasibility study (RI/FS)
7. Remedy selection/record of decision (ROD)
8. Remedial design (RD)
9. Remedial action (RA)
10. Project closeout

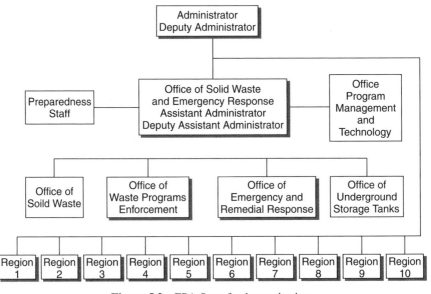

Figure 5.2 EPA Superfund organization.

The first five steps have been discussed previously in this section. Steps 6 through 10 will be discussed here. Once a site has been officially placed on the NPL, it is eligible for money from the fund. It first becomes necessary to determine which is the Lead agency. If there is a PRP, then fund money is not used, but the EPA follows up on the progress of the site. If a PRP cannot be found, or cannot pay, then the site becomes a State Lead, Federal Agency Lead, or a Fund Lead. Regardless of who the lead is, EPA's regional office will provide a remedial project manager (RPM) to coordinate between the EPA and the Lead agency. The RPM oversees the technical, enforcement, and financial aspects at the site until completion.

Once a site is placed on the NPL a remedial investigation and feasibility study (RI/FS) report is performed. A detailed study of the site is done in order to choose a remedy that would best protect the public health and the environment. The RI consists of sampling to determine a risk assessment. The risk assessment for a Superfund site has three different parts. The first is the Human Health and Environmental Evaluation, which examines baseline risks. This part measures levels of chemicals and helps in the evaluation of the site characteristics and the selection of possible response alternatives. The second part is the Health Assessment, which is conducted by The Agency for Toxic Substances and Disease Registry (ATSDR). ATSDR looks at risk in a qualitative way, and its effect on the neighboring people. The third part is the Endangerment Assessment, which is a legal determination of risk and the requirements to satisfy the RI/FS process.

The overall RI is made up of two phases: (1) site characterization, which involves field sampling and laboratory analyses, and (2) treatability investigations that examine how treatable the wastes are and the possible treatment technology alternatives.

The feasibility study (FS) works with the RI by taking the data from the RI and developing and screening different treatment alternatives for a given site. Once the FS is started, the RI continues as more sampling may be needed to make a determination. It then becomes necessary to look at the RI/FS and choose a remedy. This is done by a record of decision (ROD), which is a written report of the alternatives found in the RI/FS and reasons for the selection of a treatment technology. In the ROD, a remedy is proposed for the site and then it goes out for public evaluation. The comments are studied and then a final selection is made. In the ROD, there is a decision summary that explains the site characteristics and the determination of the method chosen for that site.

The next step in the treatment process is the Remedial Design (RD) phase. This is when the remedy selected in the ROD is engineered to meet the specifications and cleanup levels specified by law and the ROD. Detailed engineering plans are drawn up to implement the selected remedy and then the site enters the next phase, which is Remedial Action (RA). The RA phase is when the construction on the site begins and the treatment, removal, and other tasks are undertaken.

The last stage of cleanup is project closeout and deletion from the NPL.

Project closeout is divided into three phases: NPL deletion, operation and maintenance, and final project closeout. For a site to be eligible for deletion, at least one of the following three criteria must be met:

1. EPA, in consultation with the state, must have determined that responsible or other parties have implemented all appropriate response actions.
2. All appropriate fund-financed responses must have been implemented, and EPA, in consultation with the state, must have determined that no further response is appropriate.
3. Based on a remedial investigation, EPA, in consultation with the state, must have determined that no further response is appropriate.[29]

The EPA has recently created a new way to declare a site complete even if all the cleanup standards have not been met. It is called construction completion of a Superfund site, and it would occur after the RD/RA and before the actual deletion of the site from the NPL. A site may be declared construction complete if the entire RD/RA process is finished and the remedy has taken effect and has been proven to be working. The site may, however, need operations and maintenance (O&M) of the equipment, and an extended period of time to reach the cleanup levels. For example, a site that has contaminated groundwater could take years of pumping and treatment for actual cleanup to the standards specified in the ROD. All other procedures may be met and the treatment plant may be functioning, but the site is not eligible for deletion from the NPL until it reaches the specified cleanup levels. Declaring the site construction complete makes it possible for the site to be counted as complete as far as the public progress reports are concerned.

Role of the Private Sector

Superfund follows the policy that the public has a right to know what happens at a site. The EPA needs the public's help for many aspects of Superfund cleanups. Often, it is the private citizens of a community that report hazardous waste dumping. Superfund is mandated by law to involve the public in all aspects of a site except PRP legal activities. The public is given an information and comment period whenever a site makes it on to the NPL. Public concerns are analyzed and sometimes influence the decision toward an alternative remedy. Also, people in a town where a site is located are often helpful in locating a PRP. In this way, the public can help with liability and enforcement issues by assisting the EPA in finding a PRP.

To inform the public, the EPA creates a community relations plan that outlines the activities that will be used to inform the public of what is going on at a site. Once a plan is prepared, the EPA has a general informational meeting to explain to the public what will happen at the site. The EPA collects any comments and includes them in the ROD, as part of the considerations for selecting a remedy. The EPA then establishes an information repository, which contains the site updates, news releases, and phone numbers to call for questions or concerns about the site. Here are some of the things citizens can do:

1. Report hazardous waste dumping: Call the National Response Center at 1–(800)–424–8802.

2. Individual or organizations that suspect they are or may be affected by a hazardous waste release may petition the EPA to preform a Preliminary Assessment. Contact ATSDR at 1600 Clifton Road NE, Atlanta, GA 30333.
3. Find out when cleanup investigators will arrive and share information with them.
4. Get information from the EPA or state Superfund office.
5. Learn about the EPA's Community Involvement Programs.
6. Write the EPA for information on the status of any site.[28]

Progress to Date

Superfund often deals with dangerous contaminated sites that can have serious effects on the public health and the environment. There are nearly 1300 sites currently on the NPL and of those sites over 200 have made it through the entire cleanup process discussed earlier in this chapter. One of the common questions asked is what kind of progress is the EPA making on every site to deal with such a serious problem. First, the Superfund program is required to evaluate, stabilize, treat, or otherwise take action to make dangerous sites safe.[28] This is accomplished through emergency response action tailored to specific sites. After any emergency responses have been completed, a site goes through the 10 phases of cleanup discussed in the earlier sections of this chapter.

Regarding progress, the EPA has reported:

The net results of the work done at the NPL sites has been to reduce the potential risks from hazardous waste for an estimated 23.5 million people who live within 4 miles of these sites. Other results are to bring technology to bear by increased use of permanent treatment remedies at NPL sites, to remove contamination from the environment, and control the sources of contamination.[28]

The authors differ with the EPA's assessment since it is common knowledge that the bulk of the funding has been spent on assessment studies and lawyers' fees.

5.8 MONITORING METHODS[7]

This section focuses on methods used to obtain data regarding contamination of soil and/or land plus solid waste. Much of this information is based on the American Society for Testing and Materials (ASTM) standards and the EPA's Contract Laboratory Program, which was established under Superfund. A generic writeup on sampling is followed by a host of test procedures for soils. It should be noted that information on (new) sensors for nanoemissions in soils and/or solid waste was not available at the time this book was prepared. A number of additional references are provided for the reader.

Standard Practices for Sampling of Soils

There are two portions of the soil that are important to the environmental scientist. The surface layer (0 to 15 cm) reflects the deposition of airborne pollutants, especially those recently deposited pollutants. Pollutants that have been deposited by liquid spills or by long-term deposition of water-soluble materials may be found at depths ranging up to several meters. Plumes emanating from hazardous waste dumps or from leaking storage tanks may be found at considerable depths. The methods of sampling each of these are slightly different, but all make use of one of two basic techniques. Samples can be collected with some form of core sampling or auger device, or they may be collected by use of excavations or trenches. In the latter case, the samples are cut from the soil mass with spades or short punches. The ASTM has developed a number of methods that have direct application to soil sampling. Because the ASTM methods are designed primarily for engineering tests, they must often be modified slightly to meet the needs of the environmental scientist, who requires samples for chemical analyses. The techniques that are utilized should be closely coordinated with the analytical laboratory in order to meet the specific requirements of the analytical methods used.[30]

The methods outlined below are for the collection of soil samples alone. At times, it is desirable to collect samples of soil water. In these cases, use can be made of some form of suction collector. The statistical designs would be the same no matter which of the soil water collectors were used. In those cases where suction devices are used, the sampling medium is water, and not soil, even though the samples are a good reflection of soluble chemicals that may be moving through the soil matrix; those methods are not discussed below.

Surface Sampling   Surface soil sampling can be divided into two categories, namely, the upper 15 cm and the top meter. The very shallow pollution such as that found downwind from a new source or at sites of recent spills of relatively insoluble chemicals can be sampled by use of one of the methods described below. The deeper pollutants found in the top meter are the soluble, recent pollutants or those that were deposited on the surface a number of years ago. These have begun to move downward into the deeper soil layers. One of the methods discussed in the following subsection should be used in those cases.

A number of studies of surface soils have made use of a punch or thin-walled steel tube that is 15 to 20 cm long to extract short cores from the soil. The tube is driven into the soil with a wooden mallet, the core and the tube are extracted, and the soil is pushed out of the tube into a stainless steel mixing bowl and composited with other cores. The soil punch is fast and can be adapted to a number of analytical schemes, provided that precautions are taken to avoid contamination during shipping and in the laboratory. An example of how this method can be adapted would be to use the system to collect samples for volatile organic chemical analysis. The tubes could be sealed with a Teflon plug and coated with a vapor sealant such as paraffin, or better, some nonreactive sealant. These tubes could then be decontaminated on the outside and shipped to the laboratory for analyses.

Surface samples may also be collected using a seamless steel ring, approximately 15 to 30 cm in diameter. The ring is driven into the soil to a depth of 15 to 20 cm. The ring is extracted as a soil-ring unit, and the soil is removed for analysis. These large cores should be used where the results are going to be expressed on a per-unit area basis. This allows a constant area of soil to be collected each time. Removal of these cores is often difficult in very loose sandy soil and in very tight clayey coils. Loose soil will not stay in the ring. Clayey soil is often difficult to break free from the underlying soil layers, and therefore the ring must be removed with a shovel.

Perhaps the most undesirable sample collection device is the shovel or scoop. This technique is often used in agriculture, but where samples are being taken for chemical pollutants, the inconsistencies are too great. Samples can be collected using a shovel or trowel if area and/or volume are not critical. Usually the shovel is used to mark out a boundary of soil to be sampled. The soil scientist attempts to take a constant depth of soil, but the reproducibility of sample sizes is poor; thus, the variation is often considerably greater than with one of the methods listed above.

Shallow Subsurface Sampling   Precipitation may move surface pollutants into the lower soil horizons or move them away from the point of deposition by surface runoff. Sampling pollutants that have moved into the lower soil horizons requires the use of a device that will extract a longer core than can be obtained with the short probes or punches. Three basic methods are used for sampling these deeper soils:

1. Soil probes or soil augers
2. Power-driven corers
3. Trenching

The soil probe collects 30 to 45 cm of soil in intact, relatively undisturbed soil cores, whereas the auger collects a "disturbed" sample in approximately the same increments as the probe. Power augers can extract cores up to 60 cm long. With special attachments, Longer cores can be obtained with the power auger using special attachments, if this is necessary. The requirements for detail often desired in research studies or in cases where the movement of the pollutants is suspected to be through very narrow layers cannot be met effectively with the augers. In those cases, some form of core sampling or trenching should be used.

Two standard tools used in soil sampling are the soil probe (often called a King tube) and the soil auger. These tools are designated to acquire samples from the upper 2 m of the soil profile. The soil probe is nothing more than a stainless steel or brass tube that is sharpened on one end and fitted with a long, T-shaped handle. These tubes usually have an approximately 2.5-cm inside diameter, although larger tubes can be obtained. The cores collected by the tube sampler or soil probe are considered to be "undisturbed" samples, although in reality this is probably not the case. The tube is pushed into the soil in approximately 20- to 30-cm increments. The soil core is then removed from the probe and placed either in the sample container or in a mixing bowl for compositing.

The auger is approximately 3 cm in diameter and is used to take samples when the soil probe will not work. The samples are disturbed; therefore, this method should not be used when it is necessary to have a core to examine or when very fine detail is of interest. The auger is twisted or screwed into the soil and then extracted. Because of the length of the auger and the force required to pull the soil free, only about a 20- to 30-cm maximum length can be extracted at one time. In very tight clays, it may be necessary to limit the length of each pull to about 10 cm. Consecutive samples are taken from the same hole, and thus cross-contamination is a real possibility. The soil is compacted into the threads of the auger and must be extracted with a stainless steel spatula. Larger-diameter augers such as the bucket auger, the Fenn auger, and the blade augers can also be used if larger samples are needed. These range in size from 8 to 20 cm in diameter.

If distribution of pollutant with depth is of interest, the augers and the probes are not recommended because they tend to contaminate the lower samples with material from the surface. Also, many workers have sustained back injuries trying to extract a hand auger or soil probe that has been inserted too far into the soil. A foot jack is a necessary accessory if these tools are to be used. The foot jack allows the tube to be removed from the soil without use of the back muscles.

Trenching is used for careful removal of sections of soil during studies where a detailed examination of pollutant migration patterns and detailed soil structure are required. It is perhaps the least cost-effective sampling method because of the relatively high cost of excavating the trench from which the samples are collected. It should therefore be used only in those cases where detailed information is desired.

A trench approximately 1 m wide is dug to a depth approximately 1 ft below the desired sampling depth. The maximum effective depth for this method is about 2 m, unless done in some stepwise fashion. Where a number of trenches are to be dug, a backhoe can greatly facilitate sampling. The samples are taken from the side of the pit using the soil punch or a trowel.

The sampler takes the surface 15-cm sample using the soil punch or by carefully excavating a 10-cm slice of soil that is 10 cm^2 on the surface. The soil can be treated as an individual sample or can be composited with other samples collected from each face of the pit. After this initial sample is taken, the first layer is completely cut back, thereby exposing clean soil at the top of the second layer to be sampled. Care must be exercised to ensure that the sampling area is clear of all material from the layers above. The punch or trowel is then used to take samples from the shelf created by the excavation from the side of the trench. This process is repeated until all samples are taken. The resulting hole appears as a set of steps cut into the side of the trench.

An alternate procedure that is also effective results from using the punch to remove soil cores from the side of the trench at each depth to be sampled. Care must be taken to guard against soil sloughing down the side of the hole. A shovel should be used to clean the soil sampling area carefully prior to driving the punch into the side of the trench.

Sampling for Underground Plumes   Sampling for underground plumes is perhaps the most difficult of all of the soil sampling methods. Often it is conducted along with groundwater and hydrological sampling. The equipment required usually consists of large, vehicle-mounted augers and coring devices, although there are some small tripod-mounted coring units available that can be carried by several people using backpacks.

The procedure listed here closely follows ASTM method D1586-67 in many respects. The object of the sampling is to take a series of 45.7-cm (18-in.) or 61-cm (24-in.) undisturbed cores with a split spoon sampler. (Longer cores can be obtained by combining several of the shorter tubes into one long split spoon). A 15.2-cm (6-in.) auger is used to drill down to the desired depth for sampling. The split spoon is then driven to its sampling depth through the bottom of the augered hole and the core extracted.

The ASTM manual calls for the use of a 63.5-kg (140-lb) hammer to drive the split spoon. The hammer is allowed to free fall 76 cm (30 in.) for each blow to the spoon. The number of blows required to drive the spoon 15.2 cm (6 in.) is counted and recorded. The blow counts are a direct reflection of the density of the soil, and they can be used to obtain some information on the soil structure below the surface. Unless this density information is needed for interpretive purposes, it may not be necessary to record the blow counts. In soft soils, the split spoon can often be forced into the ground by the hydraulic drawdown on the drill rig. This is faster than the hammer method and does not require the record-keeping necessary to record the blow counts.

Samples should be collected at least every 1.5 m (5 ft) or in each distinct stratum. Additional samples should be collected where sand lenses or thin silt and sand layers appear in the profile. This sampling is particularly important when information on pollution migration is critical. Soluble chemicals are likely to move through permeable layers such as sand lenses. This appears to be especially important in tight clay layers where the main avenue of water movement is through the porous sandy layers.

Detailed core logs should be prepared by the technical staff present at the site during the sampling operation. These logs should note the depth of sample, the length of the core, and the depth of any features of the soil such as changes in physical properties, color changes, the presence of roots, rodent channels, and so on. If chemical odors are noted or unusual color patterns are detected, these should be noted also. Blow counts from the hammer should be recorded on the log along with the data mentioned above.

The procedure using samples collected every 1.5 m (5 ft) is most effective in relatively homogeneous soils. A variation in the method is to collect samples of every distinct layer in the soil profile. Large layers may be sampled at several points if they are unusually thick. A disadvantage of this approach is the cost for the analyses of the additional samples acquired at a more frequent interval. The soil horizons, or strata, are the avenues through which chemical pollutants are likely to migrate. Some are more permeable than others and thus are more likely to contain traces of the chemicals if they are moving through the soil. Generally speaking, the sands and gravels are more prone to contamination than are the clays because of increased permeability. This is especially true out on the leading edges of the plume and shortly after a pollutant begins to move. Low levels found in these layers can often serve as a warning of a potential problem at a later date.

A disadvantage of this type of sampling is the impact of the vehicle on yards and croplands. Special care must be taken to protect yards, shrubs, fences, and crops. The yards must be repaired, all holes backfilled, and all waste removed. Plastic sheeting should be used under all soil handling operations such as subsampling, compositing, and mixing.

Miscellaneous Tools   Hand tools such as shovels, trowels, spatulas, scoops, and pry bars are helpful for handling a number of sampling situations. Many of these can be obtained in stainless steel for use in sampling hazardous pollutants. A set of tools should be available for each sampling site where cross-contamination is a potential problem. These tool sets can be decontaminated on some type of schedule to avoid having to purchase an excessive number of these items.

A hammer, screwdriver, and wire brushes are helpful when working with the split spoon samplers. The threads on the connectors often get jammed because of soil in them. This soil can be removed with the wire brush. Pipe wrenches are also a necessity, as is a pipe vise or a plumber's vise.

A series of inexpensive and dependable tests can be run on soil that may be treated thermally. Some are thermal related, while others are classic civil/soil property tests. These tests can be used quantitatively; for example, the heating value can be used to judge the amount of auxiliary fuel required to treat the soil, or the amount of dilution required if concentrations are too high. They can be used to define the amount of drying agent required to reduce the moisture content below the plastic limit to make material handling easier. The results of these tests can also be used to make qualitative judgments as to the appropriate material handling, thermal processing, and air pollution control equipment. McGowan[31] has summarized many of the key tests that are presently available. Details are provided below.

ASTM Combustion Tests

Listed below are standard ASTM tests that should be run on organic contaminated soils. These provide the engineering design data for cost estimating thermal treatment projects and choosing appropriate equipment.

ASTM D5142 Proximate, for moisture content (mc), volatile matter (vm), fixed carbon (fc), ash, at 950°C

ASTM D3176 Ultimate Analysis modified, for C, H, N, Cl, S, O by difference, ash, mc

Higher Heating Value—ASTM D1989 via automatic calorimeter

Ash Fusion Temperature—ASTM D1857 (oxidizing and reducing)

Moisture Content—Gravimetric @105°C

Elemental Analysis—ASTM D2795 (optional)

Major and Minor Elements on Ash—D3862 (optional)

Trace Metals Analysis—ASTM D3863 (optional)

Screen Fractionation (particle size)—ASTM D410

Bulk Density—ASTM D292 (optional)

Ash Content—ASTM D5142 (or as part of proximate)

TGA/DSC (thermo gravimetric analysis, differential scanning calorimetry)

EPA 160.4 TVS/LOI for water samples/sewage sludge.

On soils of low heating value, the calorimeter test must be performed using a combustion aid, usually benzoic acid or mineral oil. Without the combustion aid, soils of high moisture and low (<2000 Btu/lb) heating value will not burn and give false low readings on heat content. The calorimeter test will yield reliable data on total fuel value in the soil if done properly.

EPA Test Protocols For Organic Contaminants

EPA Test Protocols

Total organic chlorine. Optional if chlorine does not exist in samples. Accomplished by total chlorine via ASTM D-808, oxygen bomb for sample prep, followed by D-4327, ion chromatography. D-4327 is repeated on raw sample for inorganic chlorine. Organic chlorine is by difference

Total organic sulfur (requires inorganic and total to find organic)

Total recoverable petroleum hydrocarbons—method 418.1

Oil and grease, similar to total recoverable petroleum hydrocarbons (optional)

TOC, total organic carbon in soil, EPA 9060

A qualitative assessment of soil contaminates is required based on history of the site, types of chemicals used, and concentration variation (e.g., uniform or isolated hot spots). Large samples (e.g., 5 gal) give a better picture of soil properties.

Civil Engineering/Soil Tests

The Atterberg limits and related civil/soils engineering tests are useful in determining soil properties for thermal treatment projects. Proctor density may be required for refilling and building on the site. The amount of clay and silt determines the fines content and aids in predicting the particulate loading to the air pollution control system. The liquid, plastic, and sticky limits give a qualitative feel for the stickiness of the soil and its propensity for jamming and bridging on conveyors and feeders.

The following are definitions of the soil test terms, ranked in increasing order of moisture content from the cohesion limit through the liquid limit:

Cohesion limit   Soil particles just begin to stick together.

Sticky limit   Soil sticks to a steel plate.

Shrinkage limit	No further volume reduction occurs as moisture content is decreased.
Plastic limit	Soil begins to exhibit plasticity.
Liquid limit	A rise in moisture content renders the soil a viscous liquid rather than a solid.
Proctor density	Allows prediction of compaction/moisture content relationships for landfilling of soil or ash.

Most of these tests are standard, such as ASTM 423-66 for the liquid limit; the plastic limit test, ASTM D424-59; and the Proctor moisture density test, ASTM D698. These tests can be run by qualified soil laboratories.

The moisture content of the treated soil must not exceed the soil's Atterberg plastic limit (measured by method ASTM D4318) where auger conveyors are used because cohesion and packing of the soil in the auger can result. If backfilling and compaction are required, Proctor density curve (ASTM D698) must be known. Typical specs are that enough water must be added (approaching from the dry side of the maximum density point on the curve) to achieve 90 or 95 percent maximum Proctor density.

Another soil test that may be of use is the USCS (unified soil classification system) soil classification (ASTM D2487). This defines the soil in terms of particle size, clay, silt, sand, and gradation.

5.9 SUMMARY

1. The predominating compounds in waste usually determine the treatment processes that will be required for that particular waste. An important factor for waste management planning in the future is the rise in the costs of disposal. Costs vary depending on the type of waste, the quantity (bulk), its containment, and its method of disposal.

2. Solid wastes encompass all of the wastes arising from human and animal activities that are normally solid, and that are discarded as unwanted or useless. The term solid waste is all-inclusive of the heterogeneous mass of throwaways from urban areas, as well as the more homogenous accumulation of agricultural, industrial, and mineral wastes. The hierarchy of solid waste management, in order of decreasing importance, is source reduction, recyling, waste transformation, and ultimate disposal.

3. In 1990, Americans generated over 195 million tons of municipal solid waste, with ever increasing amounts expected in years to come. There is significant public opposition to the siting of any type of municipal solid waste management facility. In the future, the public needs to be educated and informed so that these facilities can be properly located.

4. Medical wastes are not only generated by hospitals but also by laboratories, animal research facilities, and by other institutional sources. Hospital wastes are

disposed of in a number of ways, usually by the hospital's maintenance or engineering department. Eventually, almost two-thirds of the wastes leave the hospitals and go out into the community for disposal.

5. If an individual state or an interstate compact is denied access to the current nationwide disposal facilities for nuclear wastes, that state's generators will be forced to store LLW on site. The only other alternative is to stop generating waste until such time as the state or compact develops and constructs and appropriate disposal or storage facility. Neither of these options constitutes an appropriate choice for generators such as hospitals that offer nuclear medical services.

6. The dangers of human exposure to many metallic chemical elements have long been recognized. Metals such as lead, mercury, cadmium, and arsenic are toxic. Health effects range from retardation and brain damage, especially in children from lead poisoning, to the impairment of the central nervous system as a result of mercury exposure.

7. Because of concern about hazardous waste sites, Congress passed a law in 1980 to deal with the nation's hazardous waste sites. This law is called the *Comprehensive Environmental Response, Compensation, and Liability Act* (CERCLA), now commonly known as *Superfund*. Two ideas to improve Superfund are to base remedies for cleaning toxic sites on "probable future use" and to reduce wasteful transaction costs. However, based on past history, the EPA continues to move toward a legally based rather than a technology-driven agency.

8. Two portions of the soil that are important in sampling are: (a) the surface layer, which reflects the deposition of airborne pollutants, and (b) the soil at various depths, which represents soluble pollutants or traveling plumes of pollutants.

REFERENCES

1. M.K. Theodore and L. Theodore, *Major Environmental Issues Facing the 21st Century.* Theodore Tutorials, East Williston, NY, 1996.

2. R. Perry and D. Green, "Perry's Chemical Engineers' Handbook", 7th ed., section 25—Waste Management (L. Theodore, editor) McGraw-Hill, New York, 1997.

3. J. Santoleri, J. Reynolds, and L. Theodore, *Introduction to Hazardous Waste Incineration*, 2nd ed., Wiley, Hoboken, NJ, 2000.

4. G. Tchobanoglous, H. Theisen, and S. Vigil, *Integrated Solid Waste Management*, McGraw-Hill, New York, 1993.

5. N. Nemerow, *Liquid Waste of Industry: Theories and Treatment*, Syracuse University, New York, 1971.

6. U.S. EPA. "Safer Disposal for Solid Waste", Document EPA/530SW91092, March, 1993.

7. G. Burke, B. Singh, and L. Theodore, *Handbook of Environmental Management and Technology*, 2nd ed., Wiley, Hoboken, NJ, 2000.

8. R. Dupont, L. Theodore, and R. Ganesan, *Pollution Prevention*, CRC/Lewis Publishers, Boca Raton, FL, 2000.

9. L. Theodore and E. Moy, *Hazardous Waste incineration*, a Theodore Tutorial, Theodore Tutorials, East Williston, NY, 1994.

10. G. Geiger, "Incineration of Municipal and Hazardous Waste." *Natl. Environ. J.*, **1**(2), Nov/Dec (1991).

11. L. Theodore, *Air Pollution Control for Hospitals and Other Medical Facilities*, Theodore Tutorials, East Williston, NY, 1988.

12. L. Doucet, "Update of Alternative and Emerging Medical Waste Treatment Technologies," AHA Technical Document Series, 1991.

13. U.S. Department of Health and Human Resources, Army Medical Research Institute of Infectious Diseases, 1978.

14. R. Berlin and C. Stanton, *Radioactive Waste Management*, Wiley, Hoboken, NJ, 1989.

15. H. Cember, *Introduction to Health Physics*, McGraw-Hill, New York, 1992.

16. U.S. EPA, *Draft: Diffuse NORM Wastes Waste Characterization and Risk Assessment*, Washington, DC, May 1991.

17. EG&G Idaho, Inc. *The State by State Assessment of Low Level Radioactive Wastes Shipped to Commercial Disposal Sites*, DOE/LL W50T, December 1985.

18. U.S. EPA, *Lead Poisoning and Your Children*, EPA–800–B–92-0002, September 1992.

19. U.S. EPA, "Environmental Progress and Challenges", *EPA's Update*, EPA–230–07–88–033, August 1988.

20. U.S. EPA, *The Inside Story, A Guide to Indoor Air Quality*, EPA/400/1–88/004, September 1988.

21. U.S. EPA, *Reducing Lead Hazards When Remodeling Your Home*, EPA–747–R–94–002, April 1994.

22. F. Smith, Jr., *A Solid Waste Estimation Procedure: Materials Flows Approach*, U.S. EPA Office of Solid Waste (SW-147), May 1975.

23. U.S. EPA, *Second Report to Congress: Resource Recovery and Source Reduction*, Office of Solid Waste Management Programs (SW-161), 1975.

24. U.S. EPA, *Fourth Report to Congress: Resource Recovery and Source Reduction*, Office of Solid Waste (SW-600), 1977.

25. U.S. EPA, *Characterization of Products Containing Lead and Cadmium in Municipal Solid Waste in the United States, 1970 to 2000, Final Report*, EPA/530–SW–89–015C, January 1989.

26. U.S. EPA, *Characterization of Products Containing Lead and Cadmium in Municipal Solid Waste in the United States, 1970 to 2000, Final Report*, EPA/530–SW–89–015A, January 1989.

27. U.S. EPA, "Mercury in the Environment," *EPA Journal*, 16(8), December 1991.

28. U.S. EPA. "Focusing on the Nation at Large." *EPA's Update*, EPA/540/8–91/016, September 1991.

29. T. Wagner, *The Complete Guide to the Hazardous Waste Regulation*, 2nd ed., Van Nostrand Reinhold, New York, 1991.

30. U.S. EPA "Preparation of Soil Sampling Protocol," U.S. EPA 600–483–020, August 1983.

31. T. McGowan, Personal note (with permission), 2004.

CHAPTER 6

MULTIMEDIA ANALYSIS

6.1 INTRODUCTION

It is important that those individuals responsible for addressing present-day and future environmental problems arising from nanoapplications realize the need and importance of multimedia analysis of processes/systems. Both industry and the EPA did not properly address this concern during the 1970s. These misunderstandings, to some extent, still exist today. The need for applying this multimedia approach to nanoapplications is an absolute necessity.

It is now increasingly clear that some treatment technologies (as introduced in the previous three chapters) while solving one pollution problem can create others. Most contaminants present problems in more than one medium. Since nature does not recognize neat jurisdictional compartments, these same contaminants are often transferred across media. Air pollution control devices or industrial wastewater treatment plants prevent wastes from going into the air and water, respectively, but the toxic ash and sludge that these systems produce can themselves become

---

A significant amount of material in this chapter has been drawn (with permission) from a paper entitled "Educational Aspects of Multimedia Pollution Prevention" by Dr. Thomas Shen that was presented at the International Pollution Prevention Conference in Washington, D.C., in 1990, personal notes of L. Theodore; M.K. Theodore and L. Theodore, *Major Environmental Issues Facing the 21st Century*, Theodore Tutorials, East Williston, New York, 1996; and G. Burke, B. Singh, and L. Theodore, *Handbook of Environmental Management and Technology*, 2nd ed., Wiley, Hoboken, NJ, 2000.

Nanotechnology: Environmental Implications and Solutions, L. Theodore and R. G. Kunz.
ISBN 0-471-69976-4 Copyright © 2005 John Wiley & Sons, Inc.

hazardous waste problems. For example, removing trace metals from a flue gas usually transfers the products to a liquid or solid phase. Does this fact in effect exchange what had been an air quality problem for a liquid or solid waste management problem? Waste disposed of on the land or in deep wells can convert solid or liquid waste into air pollution problems.[1] This thought process should be extended to nanotechnology. However, the problem is compounded in that (as described in the last three chapters) emissions from future nanoapplications are not clearly defined.

Control of cross-media pollutants cycling in the environment is therefore an important step in the management of environmental quality. Pollutants that do not remain where they are released or where they are deposited move from a source to receptors by many routes, including air, water, and land. Unless information is available on how pollutants are transported, transformed, and accumulated after they enter the environment, they cannot be controlled effectively. A better understanding of the cross-media nature of pollutants and their major environmental processes—physical, chemical, and biological—is required. Thus, any strategy of an environmental management nature, as applied to nanotechnology systems, has to consider multimedia concerns and approaches.

The need for a multimedia approach in many pollution prevention programs should also be obvious. The advantage of a multimedia pollution control approach is its ability to manage the transfer of pollutants so they will not continue to cause pollution problems. Among the possible steps in the multimedia approach are understanding the cross-media nature of pollutants, modifying pollution control methods so as not to shift pollutants from one medium to another, applying available waste reduction technologies, and training environmental professionals in a total environmental concept. This chapter introduces the multimedia analysis concept and applies it to three separate situations: (1) a chemical plant, (2) products and services, and (3) a hazardous waste incineration facility.

6.2 HISTORICAL PERSPECTIVE

The EPA's own single-media offices, often created sequentially as individual environmental problems were identified and responded to through legislation, have played a role in impeding development of cross-media multimedia prevention strategies. In the past, innovative cross-media agreements involving or promoting pollution prevention, as well as involving or promoting voluntary arrangements for overall reductions in the releases of pollutants, have as a result not been encouraged. However, newer initiatives have been characterized by the use of a wider range of tools, including market incentives, public education and information, small business grants, technical assistance, and research and technology applications, as well as the more traditional regulations and enforcement.

In the past, the responsibility for pollution prevention and/or waste management was delegated to engineers functioning as the equivalent of an environmental control department. The personnel involved were thus skilled in engineering treatment techniques but had almost no responsibility for and thus no control over what went into the plant that generated the waste they were supposed to manage. In addition, most engineers are trained to make a product work, not to minimize or prevent pollution. There is still little emphasis on pollution prevention in the educational curricula of engineers. Business school students, the future business managers, also have not had the pollution prevention ethic instilled in them.

The reader should also note that the federal government, through its military arm, is responsible for some major environmental problems. It has further compounded these problems by failing to apply a multimedia or multiagency approach. The following are excerpts from a front-page article by Keith Schnedier in the August 5, 1991, edition of the *New York Times*:

> A new strategic goal for the military is aimed at restoring the environment and reducing pollution at thousands of military and other government military–industrial installations in the United States and abroad ... the result of environment contamination on a scale almost unimaginable. The environmental projects are spread through four federal agencies and three military services, and are directed primarily by deputy assistant sec-retaries. Many of the military–industry officials interviewed for this article said the scat-tered environmental offices were not sharing information well, were suffering at times from duplicated efforts, and might not be supervising research or contractors closely enough. Environmental groups, state agencies, and the Environmental Protection Agency began to raise concerns about the rampant military–industrial contamination in the 1970s, but were largely ignored. The Pentagon, the Energy Department, the National Aeronautics and Space Administration (NASA) and the Coast Guard considered pollution on their property to be a confidential matter. Leaders feared not only the embar-rassment from public disclosure, but also that solving the problems would divert money from projects they considered more worthwhile. Spending on military–environmental projects is causing private companies, some of them among the largest contractors for the military industry, to establish new divisions to compete for government contracts, many of them worth $100 million to $1 billion.

This lack of communication and/or unwillingness to cooperate among the various agencies within the federal government has created a multimedia problem that has just begun to surface. The years of indifference and neglect have allowed pollutants/wastes to contaminate the environment significantly beyond what would have occurred had the responsible parties acted sooner.

6.3 MULTIMEDIA APPLICATION: A CHEMICAL PLANT

A multimedia approach can be applied to any type of system, product, service, or process. In the discussion in this section, a relatively simple example involving a chemical plant is discussed. This is followed in the next section by a more detailed analysis of the manufacture, use, and ultimate disposal of a product. The chapter then concludes with a review of potential emissions to the environment from a hazardous waste incineration (HWI) facility.

Regarding the application to a chemical plant, the reader is referred to Figure 6.1. This relatively simple method of analysis, as applied to a chemical plant alone, enables one to understand and quantify how various chemical inputs to the plant—chemicals, fuel, additives, and so forth—are chemically transformed within the facility and partitioned through plant components into various chemical outputs—the gaseous, aqueous, and solid discharge streams. The principles of the law of conservation of mass clearly show that if a substance is removed from one discharge stream without chemical change (as opposed to one that undergoes a combustion reaction or is subjected to an incineration process), the source and fate of the remaining chemicals in the process streams of a chemical plant may accordingly be determined and quantified. Chemical changes can be accounted for through chemical kinetic considerations.[2]

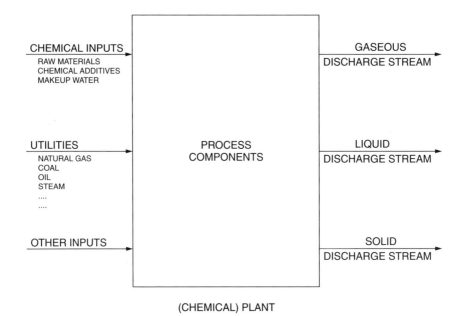

Figure 6.1   Multimedia approach.

A multimedia approach to the chemical plant normally starts with a detailed analysis of the discharge and/or emissions from the plant. These potentially harmful streams need to be examined not only in terms of the types of control that may be required (and their corresponding discharges) but also in terms of the impact that the changes in feed/inlet streams and design/operating conditions can have. Katin[3] has suggested that the following pollution prevention procedures be employed:

1. Inventory management
2. Raw material substitution
3. Process design and operation

4. Volume reduction

5. Recycling

6. Chemical alteration

A short description of each of these procedures is provided below.

1. Inventory all raw materials and purchase only the minimum amount required. As new plants, e.g., those involving nanotechnology applications, are built and as processes change, the storeroom is often the last to find out. The original raw materials needed and the quantity required may change over time. There is no need to purchased chemicals that are no longer needed in the process. Likewise, if more is purchased than is needed, the shelf life may expire before the material is completely consumed. By ordering only the amount required, the disposal of unused raw materials as hazardous waste can be minimized. Improve material receiving, storage, and handling procedures. Develop a "first-in, first-out" (FIFO) program, placing newly purchased raw materials at the back of the shelf and bringing the older stock forward. This reduces the probability of having to dispose of a container of raw material before it is ever opened because its shelf life has expired.

2. Focus on the one or two streams that have the largest volume or that account for the greatest degree of hazard. Waste streams with the highest costs for treatment and disposal should receive priority. The largest single waste stream is often the most economical to modify. Substituting raw materials can be a very expensive endeavor—a significant amount of labor will be expended, process design changes may be required, and equipment may need to be modified or replaced. Therefore, it is imperative that waste streams be assessed and prioritized. Also, attempt to purchase fewer materials that are hazardous; instead, buy either nonhazardous or less hazardous materials.

3. Install equipment that produces minimal or no waste, and modify equipment to enhance recovery or recycling operations. Redesign equipment or production lines to produce less waste. The generation of paint waste can be minimized by using high-volume, low-pressure paint guns or electrostatic paint equipment. Using proportional paint-mixing equipment to prepare two- or three-component paints can also reduce the generation of hazardous paint waste. Also, maintain a strict preventive maintenance program.

4. Segregate wastes by type for recovery. Users of large quantities of solvents should segregate, not mix, solvents so that they can be recycled.

5. Common industrial solvents can be recycled by distillation. One DuPont plant used 30,000 gal/year of 1.1.1-trichloroethane to degrease parts after forming and plating. A distillation system was installed that saved $148,000 in its first year. The still bottoms were incinerated, reducing fuel requirements for the process furnace.

6. Hazardous waste can be encapsulated so that it is sealed with a material that makes the exterior of the waste nonhazardous. An example of this is the encapsulation of nuclear waste in borosilicate glass. Adding lime to a spent catalyst that contains clay and phosphoric acid allows the material to be sold as concrete for parking

lots and sidewalks. Waste oil can be chemically treated and turned into a saleable aromatic distillate. At a 1-billion-lb/year ethylene plant, 10 million lb/year of waste oil was so treated for a cost savings of $2 million/year.

6.4 MULTIMEDIA APPLICATION: PRODUCTS AND SERVICES

Perhaps a more meaningful understanding of the multimedia approach can be obtained by examining the production and ultimate disposal of a product or service. A flow diagram representing this situation is depicted in Figure 6.2. Note that each of the 10 steps in the overall "process" has potential inputs of mass and energy, and may produce an environmental pollutant and/or a substance or form of energy that may be used in a subsequent or later step. Traditional partitioned approaches to control can provide some environmental relief, but a total systems approach is required if optimum improvements in terms of pollution/waste reduction are to be achieved.

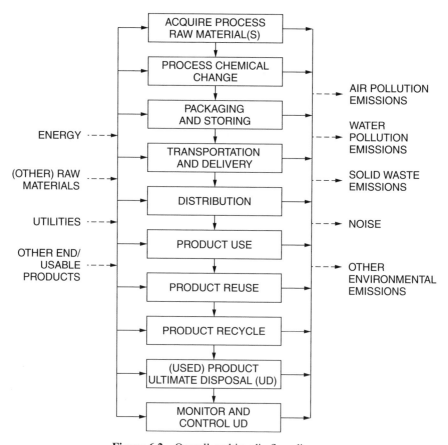

Figure 6.2 Overall multimedia flow diagram.

One should note that a product and/or service is usually conceived to meet a specific market need with little thought given to manufacturing parameters. At this stage of consideration, it may be possible to avoid some significant waste generation problems in future operations by answering a few simple questions:

1. What are the raw materials used to manufacture the product?
2. Are there any toxic or hazardous chemicals likely to be generated during manufacturing?
3. What performance regulatory specifications must the new product(s) and/or service(s) meet? Is extreme purity required?
4. How reliable will the delivery manufacturing/distribution process be? Are all steps commercially proven? Does the company have experience with the operations required?
5. What type of wastes are likely to be generated? What is their physical and chemical form? Are they hazardous? Does the company currently manage these wastes onsite or offsite?

6.5 MULTIMEDIA APPLICATION: A HAZARDOUS WASTE INCINERATION FACILITY

One may also examine an HWI facility using a comprehensive, multimedia approach. This system is presented in Figure 6.3. The reader is referred to the literature[4] for more details on incinerators. In addition to the various items of equipment contained in this plant facility, the major environmental emission discharge points are also indicated. The reader is left the exercise of determining other potential emission locations.

Note that emissions can appear from a variety of sources and locations, not only from the incinerator. The major emission points are the fan (noise), stack (gaseous), quench (liquid), water treatment (liquid and solid), and land disposal (solid). A hazard operability study, often referred to as a HAZOP, can be performed in order to locate (potential) emissions and hazardous problems.[5]

Looking at a facility such as this from a comprehensive, multimedia perspective allows one to see the forest rather than the trees. It permits the development of cost-effective strategies for managing chemical discharges to minimize risks to public health and the environment while avoiding unnecessarily expensive control requirements.

6.6 EDUCATION AND TRAINING

The role of environmental professionals in waste management and pollution control has been changing significantly in recent years. Many talented, dedicated environmental professionals in academia, government, industry, research institutions, and

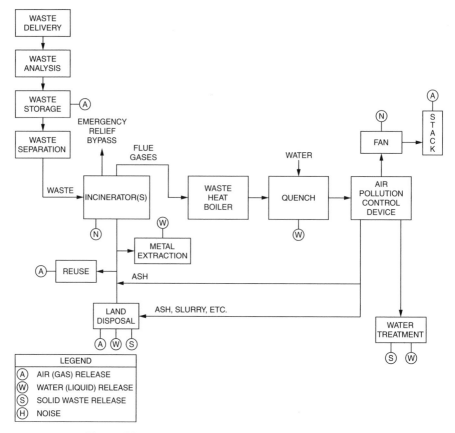

Figure 6.3 Emission release locations for an HWI facility.

private practice need to cope with this change and extend their knowledge and experience from media-specific, "end-of-pipe," treatment-and-disposal strategies to multimedia pollution prevention management. The importance of this extension and reorientation in education, however, is such that the effort cannot be further delayed. Many air pollution, water pollution, and solid waste supervisors in government agencies spend their entire careers in just one function because environmental quality supervisors usually work in only one of the media functions. Some may be reluctant to accept such activities. This is understandable given the fact that such a reorientation requires time and energy to learn new concepts and that time is a premium for them. Nevertheless they must support such education and training in order to have well-trained young professionals.

Successful implementation of multimedia programs will require well-trained environmental professionals who are fully prepared in the principles and practices of such programs. These programs need to develop a deep appreciation of the necessity for multimedia pollution prevention in all levels of society, which will require a high priority for educational and training efforts. New instructional materials and tools are needed for incorporating new concepts in the existing curricula of elementary and secondary education, colleges and universities, and training institutions. The use of computerized automation offers much hope.

Governmental agencies need to conduct a variety to activities to achieve three main educational objectives:

1. Ensure an adequate number of high-quality environmental professionals.
2. Encourage groups to undertake careers in environmental fields and to stimulate all institutions to participate more fully in developing environmental professionals.
3. Generate databases that can improve environmental literacy of the general public and especially the media.

The above objectives are related to, and reinforce, one another. For example, improving general environmental literacy should help to expand the pool of environmental professionals by increasing awareness of the nature of technical careers. Conversely, steps taken to increase the number of environmental professionals should also help improve the activities of general groups and institutions. Developing an adequate human resource base should be the first priority in education. The training environmental professionals receive should be top quality.

There is significant need to provide graduate students with training and experience in more than one discipline. The most important and interesting environmental scientific/technological questions increasingly require interdisciplinary and/or multidisciplinary approaches. Environmental graduate programs must address this aspect. Most practicing environmental professionals face various types of environmental problems that they have not been taught in the universities. Therefore, continuing education opportunities and cross-disciplinary training must be available for them to understand the importance of multimedia pollution prevention principles and strategies, as well as to carry out such principles and strategies.

The education and training plan of multimedia pollution prevention may be divided into technical and nontechnical areas. Technical areas include:

1. Products—Lifecycle analysis methods, trends-in-use patterns, new products, product life-span data, product substitution, and product applicability. (A product's lifecycle includes its design, manufacture, use, maintenance and repair, and final disposal.)
2. Processes—Feedstock substitution, waste minimization, assessment procedures, basic unit process data, unit process waste generation assessment methods, materials handling, cleaning, maintenance, and repair.
3. Recycling and Reuse—Market availability, infrastructure capabilities, new processes and product technologies, automated equipment and processes, distribution and marketing, management strategies, automation, waste stream segregation, onsite and offsite reuse opportunities, closed-loop methods, waste recapture, and reuse.

Nontechnical areas include:

1. Educational programs and dissemination of information
2. Incentives and disincentives
3. Economic costs and benefits
4. Sociological human behavioral trends
5. Management strategies including coordination with various concerned organizations.

6.7 SUMMARY

1. It is now increasingly clear that some treatment technologies, while solving one pollution problem, have created others. For this reason, there is a need for a multimedia approach in many pollution prevention programs.

2. The EPA's single-media offices, often created sequentially as individual environmental problems were identified and responded to in legislation, have played a role in impeding the development of cost-effective multimedia strategies. In addition, most engineers are challenged to make a product work, not to prevent pollution.

3. A multimedia approach can be applied to any type of system, product, or process. A multimedia analysis of a chemical plant enables one to understand and quantify how various chemical inputs to the plant are chemically transformed and positioned through plant components into various chemical outputs.

4. A multimedia approach can also be applied to any product or service. Traditional partitioned approaches to environmental control can provide some environmental relief, but a total systems approach is required if optimum improvements are to be achieved.

5. Emissions can appear from a variety of sources and locations. A hazard operability study (HAZOP) can be performed to locate emissions and hazardous problems.

6. The role of the environmental professional in waste management and pollution control has been changing significantly in recent years. Many talented, dedicated environmental professionals in academia, government, industry, research institutions, and private practice need to cope with the change and extend their knowledge and experience from media-specific, "end-of-pipe," treatment-and-disposal strategies to multimedia pollution prevention management.

REFERENCES

1. R. Dupont, K. Ganesan, and L. Theodore, *Pollution Prevention: The Waste Management Option for the 21st Century*, CRC, Boca Raton, FL, 2000.

2. L. Theodore, *Chemical Reaction Kinetics*, A Theodore Tutorial, Theodore Tutorials, East Williston, NY, 1998.

3. R. A. Katin, *Pollution Prevention at Operating Chemical Plants*, CEP, July 1991.

4. J. Santoleri, J. Reynolds, and L. Theodore, *Introduction to Hazardous Waste Incineration*, 2nd ed., Wiley-Interscience, Hoboken, NJ, 2000.

5. L. Theodore, J. Reynolds, and F. Taylor, *Accident and Emergency Management*, Wiley-Interscience, Hoboken, NJ, 1989.

CHAPTER 7

HEALTH RISK ASSESSMENT

7.1 INTRODUCTION

This chapter is concerned with health effects in general and, to a certain degree, the health effects associated with nanotechnology. (Some additional details are provided in Section 2.7.) At the time of the preparation of this book, the risks of nanotechnology were not known, and it appears that they will not be known for some time. However, it should also be noted that health benefits, if any, are also not known. Furthermore, there are no specific nano-health-related regulations or rules at EPA, OSHA, or other organizations, and it may be years before any definite nanoregulations are promulgated.

It is fair to say that, nanotechnology has the potential to change one's comprehension of nature and life, develop unprecedented manufacturing tools and medical procedures, and even change societal and international relations. And, since nanoapplications involve "exact" manufacturing, the consumption of raw materials, energy, water, and so forth is conserved while minimizing waste/pollution. Enter Murphy's law. The dark clouds in this case are the environmental health impacts associated with these new and unknown operations, and the reality is that there is a serious lack of information on these impacts. Risk assessment studies in the future will be the path to both understanding and minimizing these effects.

As described earlier, perhaps the greatest danger from nanomaterials may be their escape and persistence in the environment, the food chain, and human and animal tissues. Like the unintended consequences of bioengineered, genetically modified

Nanotechnology: Environmental Implications and Solutions, L. Theodore and R. G. Kunz.
ISBN 0-471-69976-4 Copyright © 2005 John Wiley & Sons, Inc.

crops (dubbed *Frankenfoods* by opponents),[1,2] the potential exists for results contrary to one's intuition and experience gained from handling materials on a macroscale. The blessing that nanomaterials offer of being able to penetrate cell barriers in a targeted drug delivery system, for example, quickly becomes a curse when these same tiny particles find their way into unwitting subjects.

Experimental evidence has begun to surface showing toxic effects in various organs of laboratory animals deliberately exposed to nanomaterials[3] (lungs,[4,5] livers,[6] brains[7,8]). Nonetheless, the data for risk assessment are far from conclusive at this time.[9] The alarm has already been sounded in the mainstream media[10] to alert the public to at least a perceived risk from nanotechnology efforts currently underway. The Etc. Group, a Canadian environmentalist organization,[11] has issued a call for an immediate moratorium on commercial production of new nanomaterials and an evaluation of their socioeconomic, health, and environmental implications.[6] Meanwhile, development continues at a rapid pace, with at least some U.S. government-sponsored research into environmental impacts being undertaken or contemplated.[1,12–15]

There is, however, a middle ground between the extremes of "full speed ahead" and "Let's shut the whole thing down—now!" It behooves commercial interests with a monetary stake in nanotechnology to embrace a proactive approach regarding environmental management coupled with adequate safeguards lest public outcry and/or governmental action decide the issue, as in the case of asbestos, DDT, and chlorofluorocarbon refrigerants.[9,13,16]

Although the potential pollutants and the tools for dealing with them may be different, the methodology and protocols developed for conventional materials will be the same, bearing in mind that some instrumentation that is required may not yet be available.[17] Thus, environmental risk assessment remains *environmental risk assessment*, using the same techniques regardless of the size of the alleged causative agent, as discussed in this chapter.

How is it possible to make decisions dealing with environmental risks from a new application, for example, nanotechnology, in a complex society with competing interests and viewpoints, limited financial resources, and a lay public that is deeply concerned about the risks of cancer and other illness? Risk assessment and risk management, taken together, constitute a decision-making approach that can help the different parties involved and thus enable the larger society to work out its environmental problems rationally and with good results.

Risk assessment and risk management also provides a framework for setting regulatory priorities and for making decisions that cut across different environmental areas. This kind of framework has become increasingly important in recent years for several reasons, one of which is the considerable progress made in environmental control. Thirty-plus years ago, it was not difficult to figure out where the first priorities should be. The worst pollution problems were all too obvious.

As a practical matter, it often comes down to the question of whether the final increment of a control program is cost-effective, given the resources available, or whether those same resources would be better spent on other, more pressing environmental problems. For example, it is known that the last 5 percent of pollution control is usually the most difficult and the most costly on a percentage basis. Is it worth it? Risk assessment and risk management help answer such pragmatic questions, and also enable evaluation of regulatory efforts to ensure that the environment is being made safer and that, as discussed in the previous chapter, pollution is not just moving from one place to another.

Environmental risk assessment may be broadly defined as a scientific enterprise in which facts and assumptions are used to estimate the potential for adverse effects on human health or the environment that may result from exposures to specific pollutants or other toxic agents. Risk management, as the term is used by the EPA and other regulatory agencies, refers to a decision-making process that involves such considerations as risk assessment, technological feasibility, statutory requirements, public concerns, and other factors. Risk communication is the exchange of information about risk.

Risk assessment may be also defined as the characterization of potential adverse effects to humans or to an ecosystem resulting from exposure to environmental hazards. Risk assessment supports risk management, the set of choices centering on whether and how much to control future exposure to the suspected hazards. Risk managers face the necessity of making difficult decisions involving uncertain science, potentially grave consequences to health or the environment, and large economic effects on industry and consumers. What risk assessment provides is an orderly, explicit, and consistent way to deal with scientific issues in evaluating whether a hazard exists and what the magnitude of the hazard may be. This evaluation typically involves large uncertainties because the available scientific data are limited, and the mechanisms for adverse impacts or environmental damage are only imperfectly understood.

When one looks at risk, how does one decide how safe is safe, or how clean is clean? To begin with, the engineer/scientist must look at both sides of the risk equation—that is, both the toxicity of a pollutant and the extent of public exposure. Both *current* and *potential* exposure should be estimated/calculated, considering *all* possible exposure pathways. In addition to human health risks, the potential *ecological* or other *environmental* effects should be analyzed. Even in conducting the most comprehensive risk assessment, there are always uncertainties, and one must make assumptions.

From a risk management standpoint, whether dealing with a site-specific situation or national standard, the deciding question is ultimately, "What degree of risk is acceptable?" In general, this does not mean a "zero-risk" standard, but rather a concept of *negligible* risk: At what point is there really no significant health or environmental risk, and at what point is there an adequate safety margin to protect public health and the environment? In addition, some environmental statutes require consideration of benefits together with risks in making risk management decisions.

It is possible to promote the goal of zero risk by emphasizing preventive policies so that pollution does not occur in the first place. For example, regulators can strive to ensure that pesticides do not have the potential to leach into groundwater. However, it is simply not possible to develop zero-risk environmental programs across the board. Public health and environmental risks can be minimized, but it is not likely that all such risks can be eliminated. This is not a risk-free society.

The reader should note the differences described above between health risk assessment and hazard risk assessment. As defined by Theodore,[18] hazard risk assessment is used for any hazard, particularly accidents. There are hazards practically everywhere, including chemical plants, factories, and homes. Health risk assessment (HRA) deals specifically with the health effects that can result from chronic or continuous (usually) human exposure to chemicals, whereas hazard risk assessment (HZRA) deals with acute (short-term) exposures, and/or incidents. Both risk assessment approaches are very valuable because they thoroughly examine a potential hazard. They explore all possible situations and their probabilities and consequences. Also, they provide a risk characterization value that can be used to compare different hazards. This is important in relating a hazard to the risk that it represents.[18]

This chapter explores the theory and practice of heath risk assessment/risk management and attempts to put this problem-solving approach in the context of today's environmental challenges. It describes the challenges to environmental decision making today, including a national tradition of focusing narrowly on separate environmental problems, the danger of stalemate on crucial environmental issues, and a divergence of public and scientific views on what are the most risky health environmental problems. The risk assessment/risk management decision-making approach is explained in terms of its relevance to today's environmental needs, how it works—generally and in specific situations—and how it can be applied to nanotechnology.

Topics common to both this and the next chapter are treated in this chapter. This includes sections on the public's perception of risk and on risk communication, a term widely used to refer to public discussion about environmental risks. The risk communication section provides an explanation of why this tool is important and how it works.

The next chapter addresses primarily "acute" exposures, while this chapter examines "chronic" exposures. Since both classes of exposure ultimately lead to the subject of risk and risk assessment, the overlap between the two exposures can create problems in their presentation in textual form. For purposes of this text, the chronic and acute subjects are described as HRA (health) and HZRA (hazard), respectively. Much of the material presented has been drawn from several of the references cited in this chapter.

7.2 HEALTH RISK ASSESSMENT EVALUATION PROCESS[18,19]

Health risk assessments provide an orderly, explicit way to deal with scientific issues in evaluating whether a hazard exists and what the magnitude of the hazard may be. Typically, this evaluation involves large uncertainties because the available scientific data are limited, and the mechanisms for adverse health impacts or environmental damage are only imperfectly understood. As stated earlier, when examining risk, how does one decide how safe is safe, or how clean is clean? To begin with, one has to look at both sides of the risk equation, that is, both the toxicity of a pollutant and the extent of public exposure. Information is required at both the current and the potential exposure, considering all possible exposure pathways. In addition to human health risks, one needs to look at potential ecological or other environmental effects. In conducting a comprehensive risk assessment, it should be remembered that there are always uncertainties, and these assumptions must be included in the analysis.[20]

In recent years, several guidelines and handbooks have been produced to help explain approaches for doing health risk assessments. As discussed by a special National Academy of Sciences committee convened in 1983, most human or environmental health hazards can be evaluated by dissecting the analysis into four parts: hazard identification, dose–response assessment or hazard assessment, exposure assessment, and risk characterization (see Fig. 7.1). For some perceived hazards, the risk assessment might stop with the first step, hazard identification, if no adverse effect is identified or if an agency elects to take regulatory action without further analysis.[21] Regarding hazard identification, a hazard is defined as a toxic agent or a set of conditions that has the potential to cause adverse effects to human health or the environment. Hazard identification involves an evaluation of various forms of information in order to identify the different hazards. Dose–response or toxicity assessment is required in an overall assessment; responses/effects can vary widely since all chemicals and contaminants vary in their capacity to cause adverse effects. This step frequently requires that assumptions be made to relate experimental data from animals and humans. Exposure assessment is the determination of the magnitude, frequency, duration, and routes of exposure of human populations and ecosystems. Finally, in risk characterizations, toxicology

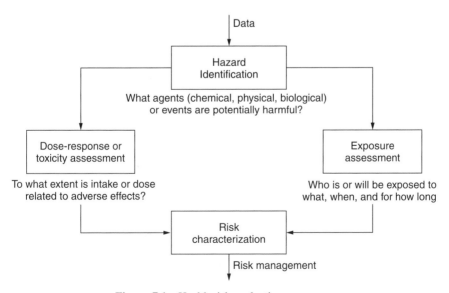

Figure 7.1 Health risk evaluation process.

and exposure data/information are combined to obtain a qualitative or quantitative expression of risk.

Risk assessment also involves the integration of the information and analysis associated with the above four steps to provide a complete characterization of the nature and magnitude of the risk and the degree of confidence associated with this characterization. A critical component of the assessment is a full elucidation of the uncertainties associated with each of the major steps. All of the essential problems of toxicology under this broad concept of risk assessment are encompassed. Risk assessment takes into account all of the available dose–response data. It should treat uncertainty not by the application of arbitrary safety factors but also by stating them in quantitatively and qualitatively explicit terms, so that they are not hidden from decision makers. Risk assessment, defined in this broad way, forces an assessor to confront all the scientific uncertainties and to set forth in explicit terms the means used in specific cases to deal with these uncertainties.[22]

An expanded presentation on each of the four health assessment steps is provided below.

Hazard Identification

Hazard identification is the most easily recognized of the actions taken by regulatory agencies. It is defined as the process of determining whether human exposure to an agent could cause an increase in the incidence of a health condition (cancer, birth defect, etc.) or whether exposure by a nonhuman receptor, for example, fish, birds, or other wildlife, might adversely be affected. It involves characterizing the nature and strength of the evidence of causation. Although the question of whether a substance causes cancer or other adverse health effects in humans is theoretically a yes–no question, there are few chemicals or physical agents for which the human data are definitive. Therefore, the question is often restated in terms of effects in laboratory animals or other test systems: "Does the agent induce cancer in test animals?" Positive answers to such questions are typically taken as evidence that an agent may pose a cancer risk for any exposed human. Information for short-term in vitro tests and structural similarity to known chemical hazards may, in certain circumstances, also be considered as adequate information for identifying a hazard.[21]

A hazards identification for a chemical plant or industrial applications can include information about:

1. Chemical identities
2. The location of facilities that use, produce, process, or store hazardous materials
3. The type and design of chemical containers or vessels
4. The quantity of material that could be involved in airborne release
5. The nature of the hazard (e.g., airborne toxic vapors or mist, fire, explosion, large quantities stored or processed, most likely to accompany hazardous materials spills or releases)[23]

An important aspect of hazards identifications is a description of the pervasiveness of the hazard. For example, most environmental assessments require knowledge of the concentration of the material in the environment, weighted in some way to account for the geographical magnitude of the site affected; that is, a l-acre or 300-acre site, a 1000 gal/min or 1,000,000 gal/min stream. All too often environmental incidents regarding chemical emission have been described by statements like "concentrations as high as 150 ppm" of a chemical were measured at a 1000-acre waste site. However, following closer examination, one may find that only 1 of 200 samples collected on a 20-acre portion of a 1000-acre site showed this concentration, and that 2 ppm was the geometric mean level of contamination in the 200 samples.

An appropriate sampling program is critical in the conduct of a health risk assessment. This topic could arguably be part of the exposure assessment, but it has been placed within hazard identification because if the degree of contamination is small, no further work may be necessary. Not only is it important that samples be collected in a random or representative manner. but the number of samples must be sufficient to conduct a statistically valid analysis. The number needed to ensure statistical validity will be dictated by the variability between the results. The larger the variance, the greater the number of samples needed to define the problem.[21]

The means of identifying hazards is complex. Different methods are used to collect and evaluate toxic properties (those properties that indicate the potential to cause biologic injury, disease or death under certain exposure conditions). One method is the use of epidemiological studies that deal with the incidence of disease among groups of people. Epidemiological studies attempt to correlate the incidence of cancer from an emission by an evaluation of people with a particular disease and people without the disease. Long-term animal bioassays are the most common method of hazard determination. (A bioassay as referred to here is an evaluation of disease in a laboratory animal.) Increased tumor incidence in laboratory animals is the primary health effect considered in animal bioassays. Exposure testing for a major portion of an animal's lifetime (2 to 3 years for rats and nice) provides information on disease and susceptibility, primarily for carcinogenicity (the development of cancer).

The understanding of how a substance is handled in the body, transported, changed and excreted, and of the response of both animals and humans, has advanced remarkably. There are many questions concerning these animal tests as to what information they provide, which kinds of studies are the best, and how the animal data compare with human data. In an attempt to answer these questions, epidemiological studies and animal bioassays are compared to each other to determine if a particular chemical is likely to pose a health hazard to humans. Many assumptions are made in hazard assessments. Fox example, it is assumed that the chemical administered in a bioassay is in a form similar to that present in the environment. Another assumption is that animal carcinogens are also human carcinogens. An example is that there is a similarity between animal and human metabolism, and so on. With these and other assumptions, and by analyzing hazard identification procedures, lists of hazardous chemicals have been developed.

Dose–Response

Toxicology and (later) epidemiology are certain to play an important role in assessing environmental effects of present and future nanoapplications. A short discussion of both these sciences follows.

Toxicology is the science of poisons, that is, the study of chemical or physical agents that produce adverse responses in biological systems. Together with other scientific disciplines (such as epidemiology, the study of the cause and distribution of disease in human populations, and risk assessment), toxicology can be used to determine the relationship between an agent of interest and a group of people or a community. Of the many different types of toxicology (see Table 7.1),[24] all types, or different applications of the science, start from a common nomenclature and set of cardinal principles.[25]

Of interest to the engineer and scientist are the regulatory and environmental applications of the discipline. The former is of use in interpreting the setting of standards for allowable exposure levels of a given contaminant or agent in an ambient or occupational environment; the latter is of use in estimating the persistence and movement of an agent in a given environment. Both applications can be of direct use to risk assessment activities, and both regulatory toxicology and environmental toxicology closely involve other branches of the discipline. The relationship is particularly close for the regulatory toxicologist, who depends largely on the products of descriptive toxicology when making decisions on the risk posed by a specific agent.[25]

As discussed above, epidemiology is a discipline within the health sciences that deals with the study of the occurrence of disease in human populations. The term is derived from the Greek words *Epi* (upon) and *Demos* (people) or diseases upon people. Whereas physicians are generally concerned with the single patient, epidemiologists are generally concerned with groups of people who share certain

TABLE 7.1   Types of Toxicology

Type	Purpose
Clinical toxicology	To determine the effects of chemical poisoning and the treatment of poisoned people
Descriptive toxicology	To test the toxicity of chemicals
Environmental toxicology	To determine the environmental fate of chemicals and their ecological and health effects
Forensic toxicology	To answer medicolegal questions about health effects
Industrial toxicology	To determine health effects of occupational exposures
Mechanistic toxicology	To describe biochemical mechanisms that cause health effects
Regulatory toxicology	To assess the risk involved in marketing chemicals and products and establish their subsequent regulation by governmental agencies

characteristics. A good example would be the interest epidemiologists show in characteristics associated with adverse health effects, for example, smoking and lung cancer, asbestos exposures and asbestosis, or noise and hearing loss.[25]

Epidemiology operates within the context of public health with a strong emphasis on the prevention of disease through the reduction of factors that may increase the likelihood that an individual or group will suffer a given disease. Implicit in the practice of epidemiology is the need for different disciplines in studying the influence of occupation on human health. Epidemiologic data come from many different sources. Acquiring reliable, accurate, and complete data describing occupational health problems is a key concern of the epidemiologist. A primary and continuing problem is the ascertainment of occupational disease. Ascertainment is the identification of diseases that are, in this case, of occupational origin.[25]

Dose–response assessment is the key element of any study involving toxicology. It is defined as the process of characerizing the relation between the dose of an agent administered or received and the incidence of an adverse health effect in exposed populations, and estimating the incidence of the effect as a function of exposure to the agent. This process considers such important factors as intensity of exposure, age pattern of exposure, and possibly other variables that might affect response, such as sex, lifestyle, and other modifying factors. A dose–response assessment usually requires extrapolation from high to low doses and extrapolation from animals to humans, or from one laboratory animal species to a wildlife species. A dose–response assessment should describe and justify the methods of extrapolation used to predict incidence, and it should characterize the statistical and biological uncertainties in these methods. When possible, the uncertainties should be described numerically rather than qualitatively.

Toxicologists tend to focus their attention primarily on extrapolations from cancer bioassays. However, there is also a need to evaluate the risks of lower doses to see how they affect the various organs and systems in the body. This approach needs to be applied to nanofiltrations and emissions. Many scientific papers focus on the use of a safety factor or uncertainty factor approach, since all adverse effects other than cancer and mutation-based developmental effects are believed to have a threshold—a dose below which no adverse effect should occur. Several researchers have discussed various approaches to setting acceptable daily intakes or exposure limits for developmental and reproductive toxicants. It is thought that an acceptable limit of exposure could be determined using cancer models, but today they are considered inappropriate because of thresholds.[21]

For a variety of reasons, precise evaluation of toxic responses caused by acute exposures to hazardous materials is difficult. First, humans experience a wide range of acute adverse health effects, including irritation, narcosis, asphyxiation, sensitization, blindness, organ system damage, and death. In addition, the severity of many of these effects varies with intensity and duration of exposure. Secondly, there is a high degree of variation in response among individuals in a typical population. Thirdly, for the overwhelming majority of substances encountered in industry, there are not enough data on toxic responses of humans to permit an accurate or precise assessment of the substance's hazard potential. Fourthly, many releases involve multiple components. There are presently no rules on how these types of releases should be evaluated. Lastly, there are no toxicology testing protocols that exist for studying episodic releases on animals. In general, this has been a neglected area of toxicology research. There are many useful measures available to employ as benchmarks for predicting the likelihood that a release event will result in serious injury or death. Several works in the literature[26,27] review various toxic effects and discuss the use of various established toxicological criteria.

Dangers are not necessarily defined by the presence of a particular chemical, but rather by the amount of that substance one is exposed to, also known as the dose. A dose is usually expressed in milligrams of chemical received per day per kilogram of body weight. For toxic substances other than carcinogens, a threshold dose must be exceeded before a health effect will occur, and for many substances, there is a dosage below which there is no harm, that is, no health effect occurring or at least not detected at the threshold. For carcinogens, it is assumed that there is no threshold, and, therefore, any substance that produces cancer is assumed to produce cancer at any concentration. It is vital to establish the link to cancer and to determine if that risk is acceptable. Analyses of cancer risks are much more complex than those for noncancer risks.

Not all contaminants or chemicals are equal in their capacity to cause adverse effects. Thus, cleanup standards or action levels are based in part on the compounds' toxicological properties. Toxicity data employed are derived largely from animal experiments in which the animals (primarily mice and rats) are exposed to increasingly higher concentrations or doses. As described above, responses or effects can vary widely from no observable effect to temporary and reversible effects, to permanent injury to organs, to chronic functional impairment, and, ultimately, to death.

Exposure Assessment

Exposure assessment is the process of measuring or estimating the intensity, frequency, and duration of human or animal exposure to an agent currently present in the environment or of estimating hypothetical exposures that might arise from the release of new chemicals into the environment. In its most complete form, an exposure assessment should describe the magnitude, duration, schedule, and route of exposure, size, nature, and classes of human, animal, aquatic, or wildlife populations exposed, and the uncertainties in all estimates. The exposure assessment can often be used to identify feasible prospective control options and to predict the effects of available control technologies for controlling or limiting exposure.[21]

Much of the attention focused on exposure assessment has come recently. Many of the risk assessments performed in the past used too many conservative assumptions, which caused an overestimation of the actual exposure. Obviously, without exposure(s) there are no risks. To experience adverse effects, one must first come into contact with the toxic agent(s). Exposures to chemicals can occur via inhalation of air (breathing), ingestion of water and food (eating and drinking), or absorption through the skin. These are all pathways to the human body.

Generally, the main pathways of exposure considered in this step are atmospheric transport, surface and groundwater transport, ingestion of toxic materials that have passed through the aquatic and terrestrial food chain, and dermal absorption. Once an exposure assessment determines the quantity of a chemical with which human populations may come into contact, the information can be combined with toxicity data (from the hazard identification process) to estimate potential health risks.[20] Thus, the primary purpose of an exposure assessment is to determine the concentration levels over time and space in each environmental medium where human and other environmental receptors may come into contact with the chemicals of concern. In addition, there are four components of an exposure assessment: potential sources, significant exposure pathways, populations potentially at risk, and exposure estimates.[21]

The two primary methods of determining the concentration of a pollutant to which target population are exposed are direct measurement and computer analysis, also known as the computer dispersion modeling. Measurement of the pollutant concentration in the environment is used for determining the risk associated with an exiting discharge source. Receptors are placed at regular intervals from the source, and the concentration of the pollutant is measured over a certain period of time (usually several months or a year). The results are then related to the size of the local population. This kind of monitoring, however, is expensive and time-consuming. Many measurements must be taken because exposure levels can vary under different atmospheric conditions or at different times of the year.

Computer dispersion modeling predicts environmental concentrations of pollutants. In the prediction of exposure, computer dispersion modeling focuses on the discharge of a pollutant and the dispersion of that discharge by the time it reaches the receptor. This method is used primarily for assessing risk from a proposed facility or discharge. Sophisticated techniques are employed to relate reported or measured emissions to atmospheric, climatological, demographic, geographic, and other data in order to predict a population's potential exposure to a given chemical.

Exploring the various exposure routes of nanoemissions will unquestionably become a major area of research in the future. It is necessary to understand the relative importance of these pathways and how they work. However, at this time, it is difficult, if not impossible, to ascertain what negative (or positive) effects will occur at different exposure levels from these different pathways. In addition, little is known about the fate, transport, and physical/chemical changes that occur once these emissions enter the environment. This information is absolutely necessary if one is to obtain qualitative estimates of exposure.

Risk Characterization

Risk characterization is the process of estimating the incidence of a health effect under the various conditions of human or animal exposure described in the exposure assessment. It is performed by combining the exposure and dose–response assessments. The summary effects of the uncertainties in the preceding steps should also be described in this step.

The quantitative estimate of the risk is the principal interest to the regulatory agency or risk manager making the decision. The risk manager must consider the results of the risk characterization when evaluating the economics, societal aspects, and various benefits of the assessment. Factors such as societal pressure, technical uncertainties, and severity of the potential hazard influence how the decision makers respond to the risk management. There is room for improvement in this step of the risk assessment.[21]

As indicated above, a risk estimate indicates the likelihood of occurrence of the different types of health or environmental effects in exposed populations. Risk assessment should include both human health and environmental evaluations (i.e., impacts on ecosystems). Ecological impacts include actual or potential effects on plants and animals (other than domesticated species). The number produced from the risk characterization, representing the probability of adverse health effects being caused, must be evaluated. This is performed because certain agencies will look only at risks of specific numbers and act on them.

There are two major types of risk: maximum individual risk and population risk. Maximum individual risk is defined exactly as it implies, that is, the maximum risk to an individual person. This person is considered to have a 70-year lifetime of exposure to a process or a chemical. Population risk is basically the risk to a population. It is expressed as a certain number of death per thousand or per million people. These risks are often based on very conservative assumptions, which may yield too high a risk.

7.3 WHY USE RISK-BASED DECISION MAKING?

The use of the risk-based decision-making process allows for efficient allocation of limited resources, such as time, money, regulatory oversight, and qualified professionals. Advantages of using this process include:

1. Decisions are based on reducing the risk of adverse human or environmental impacts.
2. Site assessment activities are focused on collecting only that information that is necessary to make risk risk-based corrective action decisions.
3. Limited resources are focused on those sites that pose the greatest risk to human health and the environment at any time.
4. Compliance can be evaluated relative to site-specific standards applied at site-specific point(s) of compliance.
5. Higher quality, and in some cases faster, cleanups may be achieved than are currently possible.
6. Documentation is developed that can demonstrate that the remedial action is protective of human health, safety, and the environment.

By using risk-based decision making, decisions are made in a consistent manner. Protection of both human health and the environment is accounted for.

A variety of EPA programs involved in the protection of groundwater and cleanup of environmental contamination utilize the risk-based decision-making approach. Under the EPA's regulations dealing with cleanup of underground storage tank (UST) sites, regulators are expected to establish goals for cleanup of UST releases based on consideration of factors that could influence human and environmental exposure to contamination. Where UST releases affect groundwater being used as public or private drinking water sources, EPA generally recommends that cleanup goals be based on health-based drinking water standards: even in such cases, however, risk-based decision making can be employed to focus corrective action.

In the Superfund program, risk-based decision making plays an integral role in determining whether a hazardous waste site belongs on the National Priorities List. Once a site is listed, qualitative and quantitative risk assessments are used as the basis for establishing the need for action and determining remedial alternatives. To simplify and accelerate baseline risk assessments at Superfund sites, EPA has developed generic soil screening guidance that can be used to help distinguish between contamination levels that generally present no health concerns and those that generally require further evaluation. The Resource Conservation and Recovery Act (RCRA) corrective action program also uses risk-based decision making to set priorities for cleanup so that high-risk sites receive attention as quickly as possible, to assist in the determination of cleanup standards, and to prescribe management requirements for remediation of wastes.

7.4 RISK-BASED CORRECTIVE ACTION APPROACH[18,19,28,29]

The risk-based corrective action (RBCA) process is implemented in a tiered approach, with each level involving increasingly sophisticated methods of data collection and analysis. As the analysis progresses, the assumptions of earlier tiers are replaced with site-specific data and information. Upon evaluation of each tier, the results and recommendations are reviewed, and it is determined whether more site-specific analysis is required. Generally, as the tier level increases, so do the costs of continuing the analysis. This approach will probably be employed for nanoapplications.

The first step is the site assessment, which is the identification of the sources of the chemical(s) of concern, any obvious environmental impacts, any potentially impacted human and environmental receptors (e.g., workers, residents, lakes, streams, etc.), and potentially significant chemical transport pathways (e.g., groundwater flow, atmospheric dispersion, etc.) The site assessment also includes information collected from historical records and a visual inspection of the site. An example of criteria used for a site classification in outline form follows.

Example of Site Classification—Criteria and Prescribed Scenarios[28]

1. Immediate threat to human health, safety, or sensitive environmental receptors.
    a. Explosive levels, or concentrations, of vapors that could cause acute health effects are present in a residence or other building.

b. Explosive levels of vapors are present in subsurface utility system(s), but no building or residences are impacted.

c. Free-product is present in significant quantities at ground surface, on surface water bodies, in utilities other than water supply lines, or in surface water runoff.

d. An active public water supply well, public water supply line, or public surface water intake is impacted or immediately threatened.

e. Ambient vapor/particulate concentrations exceed concentrations of concern from an acute exposure or safety viewpoint.

f. A sensitive habitat or sensitive resources (sport fish, economically important species, threatened and endangered species, etc.) are impacted and affected.

2. Short-term (0 to 2 years) threat to human health, safety, or sensitive environmental receptors.

a. There is potential for explosive levels, or concentrations of vapors that could cause acute effects, to accumulate in a residence or other building.

b. Shallow contaminated surface soils are open to public access, and dwellings, parks, playgrounds, daycare centers, schools, or similar use facilities are within 500 ft (152 m) of those soils.

c. A nonpotable water supply well is impacted or immediately threatened.

d. Groundwater is impacted, and a public or domestic water supply well producing from the impacted aquifer is located within 2 years' projected groundwater travel distance down-gradient of the known extent of chemical(s) of concern.

e. Groundwater is impacted, and a public or domestic water supply well producing from a different interval is located within the known extent of chemicals of concern.

f. Impacted surface water, storm water, or groundwater discharges within 500 ft (152 m) of a sensitive habitat or surface water body used for human drinking water or contact recreation.

3. Long-term (>2 years) threat to human health, safety, or sensitive environmental receptors.

a. Subsurface soils [>3 ft (0.9 m) below ground surface] are significantly impacted, and the depth between impacted soils and the first potable aquifer is less than 50 ft (15 m).

b. Groundwater is impacted, and potable water supply wells producing from the impacted interval are located >2 years' groundwater travel time from the dissolved plume.

c. Groundwater is impacted, and nonpotable water supply wells producing from the impacted interval are located >2 years' groundwater travel time from the dissolved plume.

d. Groundwater is impacted, and nonpotable water supply wells that do not produce from the impacted interval are located within the known extent of chemical(s) of concern.

e. Impacted surfae water, storm water, or groundwater discharges within 1500 ft (457 m) of a sensitive habitat or surface-water body used for human drinking water or contact recreation.

f. Shallow contaminated surface soils are open to public access, and dwellings, parks, playgrounds, daycare centers, schools, or similar use facilities are more than 500 ft (152 m) from those soils.

4. No demonstrable long-term threat to human health or safety or sensitive environmental receptors. Priority 4 scenarios encompass all other conditions not described in priorities 1, 2, and 3 and that are consistent with the priority description given above. Some examples are as follows:

a. Nonpotable aquifer with no existing local use impacted.

b. Impacted soils located more than 3 ft (0.9 m) below ground surface and greater than 50 ft (15 m) above nearest aquifer.

c. Groundwater is impacted, and nonpotable wells are located down-gradient and outside the known extent of the chemical(s) of concern, and they produce from a nonimpacted zone.

Once the applicable criteria are met, the site is then classified according to the urgency of need for initial response action, based on information collected during the site assessment. Associated with site classifications are initial response actions that are to be implemented simultaneously with the RBCA process. Sites should be reclassified as actions are taken to resolve concerns or as better information becomes available.

A tier 1 evaluation is then conducted using a "look-up table." The look-up table contains screening level concentrations for the various chemicals of concern. The *look-up table* is defined as a tabulation for potential exposure pathways (e.g., inhalation, ingestion, etc.), media (e.g., soil, water, and air), a range of incremental carcinogenic risk levels that are used as target levels for determining remediation requirements, and potential exposure scenarios (e.g., residential, commercial, industrial, and agricultural). If a look-up table is not provided by the regulatory agency or available from a previous evaluation, the person conducting the RBCA analysis must develop the look-up table. If a look-up table is available, the user is responsible for determining that the risk-based screening levels (RBSLs) in the table are based on currently acceptable methodologies and parameters.

The RBSLs are determined using typical, non-site-specific values for exposure parameters and physical parameters for media. The RBSLs are calculated according to methodology suggested by the EPA.[30,31] The value of creating a look-up table is that users do not have to repeat the exposure calculations for each site encountered. The look-up table is altered only when reasonable maximum exposure parameters, toxicological information, or recommended methodologies are updated. Some states have compiled such tables that, for the most part, contain identical values (since they are based on the same assumptions). The look-up table is used to determine whether site conditions satisfy the criteria for a quick regulatory closure or warrant a more site-specific evaluation.

If further evaluation is required, a tier 2 evaluation provides the user with an option to determine site-specific target levels (SSTLs) and point(s) of compliance.

It is important to note that both tier 1 RBSL and tier 2 SSTLs are based on achieving similar levels of protection of human health and the environment. However, in tier 2 the non-site-specific assumptions and point(s) of exposure used in tier 1 are replaced with site-specific data and information. Additional site assessment data may be needed. For example, the tier 2 SSTL can be derived from the same equations used to calculate the tier 1 RBSL, except that site-specific parameters are used in the calculations. The additional site-specific data may support alternate fate and transport analysis. At other sites, the tier 2 analysis may involve applying tier 1 RBSLs at more probable point(s) of exposure.

At the end of tier 2, if it is determined that more detailed evaluation is again warranted, a tier 3 evaluation is then conducted. A tier 3 evaluation provides the user with an option to determine SSTLs for both direct and indirect pathways using site-specific parameters and point(s) of exposure and compliance when it is judged that tier 2 SSTLs should not be used as target levels. Tier 3, in general, can be a substantial incremental effort relative to tiers 1 and 2, since the evaluation is much more complex and may include additional site assessment, probabilistic evaluations, and sophisticated chemical fate/transport models.

With the RBCA process, the user compares the target levels (RBSLs or SSTLs) to the concentrations of the chemical(s) of concern at the point(s) of compliance at the conclusion of each tier evaluation. If the concentrations of the chemical(s) of concern exceed the target levels at the point(s) of compliance, then either remedial action, interim remedial action, or further tier evaluation should be conducted. When it is judged that no further assessment is necessary or practicable, a remedial alternatives evaluation should be conducted to confirm the most cost-effective option for achieving the final remedial action target levels (RBSLs or SSTLs as appropriate).

Detailed design specifications may then be developed for installation and operation of the selected measures. The selected measures may include some combination of source removal, treatment, and containment technologies, as well as engineering and institutional controls. Examples of theses include the following: soil venting, bioventing, air sparging, "pump-and-treat," and natural attenuation/ passive remediation. The remedial action must continue until such time as monitoring indicates that concentrations of the chemical(s) of concern are not above the RSBL or SSTL, as appropriate, at the points of compliance or source area(s), or both. When concentrations of chemical(s) of concern no longer exceed the target levels at the point of compliance, then the user may elect to deem the RBCA process complete. If achieving the desired risk reduction is impracticable because of technology or resource limitations, an interim remedial action, such as removal or treatment of "hot spots," may be conducted to address the most significant concerns, change the site classification, and facilitate reassessment of the tier evaluation.

After completion of the RBCA activities, most regulatory agencies require the submission of an RBCA report. The RBCA report typically contains a variety of site characterization items, including a site description; summaries of the site ownership and use, past releases or potential source areas, and the current and completed site activities; a description of regional hydrogeologic conditions and site-specific hydrogeologic conditions; and a summary of beneficial use. A site map of the location should be provided, which includes designations for local land use and groundwater supply wells, as well as the location of structures, above-ground storage tanks, buried utilities and conduits, and suspected/confirmed sources. Site photos should also be provided. The report also typically provides a discussion of the RBCA process, including a summary and discussion of the risk assessment (hazard identification, dose–response assessment, exposure assessment, and risk characterization), and the methods and assumptions used to calculate the RBSL and/or SSTL; a summary of the tier evaluation; a summary of the analytical data and the appropriate RBSL or SSTL used; and a summary of the ecological assessment. Additional data needed for the analysis, such as (a) groundwater elevation map(s), geologic cross section(s), and dissolved plume map(s) of the chemical(s) of concern, should also be included in the report.

In conclusion, risk-based decisions and assessments will be important to ascertain cost–benefit information on existing and future nanoapplications. Unknown health and hazard effects can lead to disastrous results at later times if not properly evaluated and studied early on. There have been repeated calls for funding in this area, some of which may be justified. Considering that this is an evolving and futuristic industry, present-day approaches that have been routinely applied should suffice at this time. However, initial and/or early analyses of risks can prove economically beneficial in the long run.

7.5 STATUTORY REQUIREMENTS INVOLVING ENVIRONMENTAL COMMUNICATIONS

The last four sections of this chapter are concerned with the general subject of communicating environmental risks. This section on regulatory requirements is followed by a section dealing with the public perception of risk. The chapter concludes with sections concerned with risk communication and the seven cardinal rules of risk communication. These four topics, common to both this and the next chapter, will unquestionably find application in nanotechnology.

There are several major statutes and regulations that mandate communication activities with the general public. Since EPA's founding in December 1970, it has been given the task of implementing a wide range of environmental statutes. Many of these statutes owe their very existence to years of citizen lobbying, so it is no surprise that some should specifically mandate "public participation." The Federal Water Pollution Control Act of 1972 (later known as the Clean Water Act), the Safe Drinking Water Act of 1974, and the Resource Conservation and Recovery Act of 1976 all contain language requiring EPA to involve the public in their implementation.

The expressed philosophy behind public participation can be found in Section 101(e) of the Federal Water Pollution Control Act of 1972: "A high degree of informed public participation in the control process is essential to the accomplishment of the objectives we seek—a restored and protected natural environment."[32] In what way essential? Legislators—inspired by the dedication of clean air activists in the so-called *Breathers' Lobby*—envisioned the public as the conscience of EPA.

Their hope was that concerned citizens, both individually and in groups, would monitor EPA and ensure that the agency actually did its job.[33]

The Comprehensive Environmental Response, Compensation, and Liability Act and the Superfund Amendments and Reauthorization Act of 1986 encourage community residents to participate in the process of determining the best way to clean up an abandoned site. To ensure effective and substantive two-way communications from the outset at each remedial response site, a community relations program is tailored to local circumstances. Often, EPA or the state will interview residents, local officials, and civic leaders to learn everything possible about the site and about the community's concerns. These interviews are conducted before and during field work on the remedial investigation. The new Superfund thus formalized existing EPA community relations policy and public participation requirements outlined in the National Contingency Plan. It also required EPA to

1. Publish a notice and brief analysis of the proposed remedial action plan.
2. Provide an opportunity for the public to comment on that plan.
3. Provide an opportunity for a public meeting to allow for two-way communication on the remedial action plan.
4. Make a copy of the transcript of the public meeting available to the public.
5. Prepare a response to each significant comment made on the proposed remedial action plan.

Community relations activities are somewhat different during a removal action, where human health and the environment must be protected from an immediate threat. During the initial phase of these response actions, the EPA's primary responsibility is to inform the community about actions being taken and the possible effect on the community.

The new Superfund also required the EPA to develop a grant program to make funding for technical assistance available to those who may be affected by a release. The purpose of these grants was to help concerned citizens understand and interpret technical information on the nature of the hazard and recommended alternatives for cleanup. Grants are limited by law to one grant of no more than $50,000 per NPL site. In addition, the grant recipient must contribute at least 20 percent of the total cost of the grant.[34] The Emergency Planning and Community Right-to-Know Act created a new relationship among government at all levels, business and community leaders, environmental and other public-interest organizations, and individual citizens. For the first time, the law made citizens full partners in preparing for emergencies and managing chemical risks. Each of these groups and individuals has an important role in making the program work.[35]

1. *Local communities and states* have the basic responsibility at the local level for understanding risks posed by chemicals, for managing those risks, for reducing those risks, and for dealing with emergencies. By developing emergency planning and chemical risk management at the levels of government

closest to the community, the law helps to ensure the broadest possible public representation in the decision-making process.

2. *Citizens, health professionals, public-interest and labor organizations, the media, and others* are working with government and industry to use the information for planning and response at the community level. The law gives everyone involved access to more of the facts needed to determine what chemicals mean to the public health and safety.

3. *Industry* is responsible for (a) operating as safely as possible using the most appropriate techniques; (b) gathering information on the chemicals it uses, stores, and releases into the environment and providing it to governmental agencies and local communities; and (c) helping set up procedures to handle chemical emergencies. Beyond meeting the letter of the law, some industry groups and individual companies are reaching out to their communities by explaining the health hazards involved in using chemicals, by opening communications channels with community groups, and by considering changes in their practices to reduce any potential risks to human health or the environment.

4. *The federal government* is responsible for providing national leadership and assistance to states and communities so they will have the tools and expertise they need to (a) receive, assimilate, and analyze all Title III data; and (b) take appropriate measures in accidental risk and emission reduction at the local level. The EPA is also working to ensure that (a) industry complies with the law's requirements; (b) the public has access to information on annual toxic chemical releases; and (c) the information is used in various EPA programs to protect the nation's air, water, and soil from pollution. The EPA is also working with industry to encourage voluntary reductions in the use and release of hazardous chemicals wherever possible.[35]

The Emergency Planning and Community Right-to-Know Act has forged a closer, more equal relationship among citizens, health professionals, industry, public-interest organizations, and the local, state, and federal governmental agencies responsible for emergency planning and response, public health, and environmental protection.

In the past, most of the responsibility for these activities fell upon experts in government and industry. To the extent that citizens or their representatives participated, it was generally "from the outside looking in," as they did what they could to influence decisions that were, for the most part, out of their hands, But under the provisions of the Emergency Planning and Community Right-to-Know Act, all of these groups, organizations, and individuals have vital roles to play in making the law work for the benefit of everyone. The law requires facilities to provide information on the hazardous chemicals in communities directly to the people who are most affected, both in terms of exposure to potential risks and the effects of those risks on public health and safety, the environment, jobs, the local economy, property values, and other factors.

These "stakeholders" are also the people who are best able to do something about assessing and managing risks—through inspections, enforcement of local codes, reviews of facility performance, and, when appropriate, political and economic pressures.

This relationship between the Title III data and community action can best occur at the local level, through the work of the local emergency planning committee (LEPC). For example, if a local firm has reported the presence of extremely hazardous substances at its facility, several accidents, substantial quantities of chemicals, and continuing releases of toxic chemicals, a community has the data it needs to seek appropriate corrective action. In short, the law opens the door to community-based decision making on chemical hazards for citizens and communities throughout the nation.

The LEPCs are crucial to the success of the Emergency Planning and Community Right-to-Know Act. Appointed by State Emergency Response Commissions (SERCs), local planning committees must consist of representatives of all of the following groups and organizations: elected state and local officials; law enforcement, civil defense, firefighting, first aid, health, local environmental and transportation agencies; hospitals; broadcast and print media; community groups; and representatives of facilities subject to the emergency planning and community right-to-know requirements.[35]

It was clear from the outset that the public could not put persistent and informed pressure on EPA without a steady flow of information and guidance from the EPA. Meeting that need has been the purpose of EPA's public participation programs. Their mission is threefold:

1. To keep the public informed of important developments in EPA's program areas.
2. To provide technical information and, if necessary, translate that information into plain English.
3. To ensure that the agency takes community viewpoints into account in implementing these programs.

7.6 PUBLIC PERCEPTION OF RISK[18]

Public concern about risk stems from earthquakes, fires, and hurricanes to asbestos, chemical and radon emissions, ozone depletion, toxins in food and water, and so on. Many of the public's worries are out of proportion, with the fear being overestimated or at times underestimated. The risks given the most publicity and attention receive the greatest concern, while the ones that are more familiar and accepted are given less thought.

A large part of what the public knows about risk comes from the media. Whether it is newspapers, magazines, radio, or television, the media provides information about the nature and extent of specific risks. They also help shape the perception of the danger involved within a given risk.

Laypeople and experts disagree on risk estimates for many environmental problems. This creates a problem since the public generally does not trust the experts. This section concentrates on how the public views risk and lays the groundwork for what the future of public risk perception may be for nanotechnology.

The public often worries about the largely publicized risks and thinks little about those that they face regularly. A study was recently performed and compared the responses of two groups, 15 national risk assessment experts and 40 members of the League of Women Voters.[36] This search produced striking discrepancies, as presented in Table 7.2. The League members rated nuclear power as the number 1 risk, while experts numbered it at 20, and the League ranked X-rays at 22 while the experts gave it a rank of 7.

There are various reasons for the differences in risk perception. Governmental regulators and industry officials look at different aspects in assessing a given risk than would members of the community. The "experts" will look at the mortality rates to assess risk, while the "laypeople" worry about their children and the potential long-term health risks. Another reason for the difference is that people take reports of bad news more to heart than they would a report that might increase their trust.

Problems exist with risk estimates because the substance or process in question may be calculated to present too high a risk. To understand the significance of risk analyses, a list of everyday risks derived from actual statistics and reasonable estimates is presented in Table 7.3. A lifetime risk of 70×10^{-6} means that 70 out of one million people will die from that specific risk.

Risk managers in government and industry have started turning to risk communication to bridge the gap between the public and the "experts." Table 7.3 enables the public to see that certain everyday risks are higher than some dreaded environmental risks. It shows that eating peanut butter possesses a greater risk than toxins in the air or water.

In 1987, the EPA released a report titled "Unfinished Business: A Comparative Assessment of Environmental Problems" in order to apply the concepts of risk assessment to a wide array of pressing environmental problems. It is difficult to make direct comparisons of different environmental problems because most of the data are usually insufficient to quantify risks. Also, risks associated with some problems are incomparable with risks of others. The study was based on a list of 31 environmental problems. Each was analyzed in terms of four different types of risks: cancer risks, noncancer health risks, ecological effects, and welfare effects (visibility impairment, materials damage, etc.)

The ranking of cancer was probably the most straightforward part of the study, since the EPA had already established risk assessment procedures and there are considerable data already available from which to work. Two problems were considered at the top of the list: the first was worker exposure to chemicals, which does not involve a large number of individuals but does result in high individual risks to those exposed; the other problem was radon exposure, which is causing considerable risk to a large number of people. The results from the cancer report are provided in Table 7.4.

The other working groups had greater difficulty when ranking the 31 environmental problem issues because there are no accepted guidelines for quantitatively

TABLE 7.2   **Ranking Risks: Reality and Perception**

League of Women Voters	Activity or Technology	Experts
1	Nuclear power	20
2	Motor vehicles	1
3	Handguns	4
4	Smoking	2
5	Motorcycles	6
6	Alcoholic beverages	3
7	General (private) aviation	12
8	Police work	17
9	Pesticides	8
10	Surgery	5
11	Fire fighting	18
12	Large construction	13
13	Hunting	23
14	Spray cans	26
15	Mountain climbing	29
16	Bicycles	15
17	Commercial aviation	16
18	Electric power (nonnuclear)	9
19	Swimming	10
20	Contraceptives	11
21	Skiing	30
22	X-rays	7
23	High school and college football	27
24	Railroads	19
25	Food preservatives	14
26	Food coloring	21
27	Power motors	28
28	Prescription antibiotics	24
29	Home appliances	22
30	Vaccinations	25

Note: The rankings are of perceived risks for 30 activities and technologies, based on averages in a survey of a group of experts and a group of informed laypeople, members of the League of Women Voters. A ranking of 1 denotes the highest level of perceived risk.

assessing relative risks. As noted in EPA's study, the following general results were produced for each of the four types of risks described earlier in this section.[38]

1. No problems rank high in all four types of risk, or relatively low in all four.
2. Problems that rank relatively high in three of the four types, or at least medium in all four, include criteria air pollutants, stratospheric ozone depletion,

TABLE 7.3   Lifetime Risks to Life Commonly Faced by Individuals[37]

Cause of Risk	Lifetime (70-year) Risk per Million
Cigarette smoking	252,000
All cancers	196,000
Construction	42,700
Agriculture	42,000
Police killed in the line of duty	15,400
Air pollution (eastern United States)	14,000
Motor vehicle accidents (traveling)	14,000
Home accidents	7,700
Frequent airplane traveler	3,500
Pedestrain hit by motor vehicle	2,940
Alcohol, light drinker	1,400
Background radiation at sea level	1,400
Peanut butter, four tablespoons per day	560
Electrocution	370
Tornado	42
Drinking water containing chloroform at allowable EPA limit	42
Lightning	35
Living 70 years in a zone of maximum impact from a modern municipal incinerator	1
Smoking 1.4 cigarettes per day	1
Drinking 0.5 liters of wine per day	1
Traveling 10 miles by bicycle	1
Traveling 30 miles by car	1
Traveling 1000 miles by jet plane (air crash)	1
Traveling 6000 miles by jet plane (cosmic rays)	1
Drinking water containing trichloroethylene at maximum allowable EPA limit	0.1

pesticide residues on food, and other pesticide risks (runoff and air deposition of pesticides).

3. Problems that rank relatively high in cancer and noncancer health risks but low in ecological and welfare risks include hazardous air pollutants, indoor radon, indoor air pollution other than radon, pesticide application, exposure to consumer products, and worker exposures to chemicals.

4. Problems than rank relatively high in ecological and welfare risk but low in both health risks include global warming, point and non-point sources of surface water pollution, physical alteration of aquatic habitats (including estuaries and wetlands), and mining wastes.

5. Areas related to groundwater consistently rank medium or low.

TABLE 7.4 Consensus Ranking of Environmental Problem Areas on the Basis of Population Cancer Risk

Rank	Problem Area	Selected Comments
1 (tied)	Worker exposure to chemicals	About 250 cancer cases per year estimated based on exposure to 4 chemicals; but workers face potential exposures to over 20,000 substances. Very high individual risk possible.
1 (tied)	Indoor radon	Estimated 5000–20,000 lung cancers annually from exposure in homes.
3	Pesticide residues on foods	Estimated 6000 cancers annually, based on exposure to 200 potential carcinogens.
4 (tied)	Indoor air pollutants (nonradon)	Estimated 3500–6500 cancers annually, mostly due to tobacco smoke.
4 (tied)	Consumer exposure to chemicals	Risk from 4 chemicals investigated is about 100–135 cancers annually; an estimated 10,000 chemicals in consumer products. Cleaning fluids, pesticides, particleboard, and asbestos-containing products especially noted.
6	Hazardous/toxic air pollutants	Estimated 2000 cancers annually based on an assessment of 20 substances.
7	Depletion of stratospheric ozone	Ozone depletion projected to result in 10,000 additional annual deaths in the year 2100; not ranked higher because of the uncertainties in future risk.
8	Hazardous waste sites, inactive	Cancer incidence of 1000 annually for 6 chemicals assessed. Considerable uncertainty since risk based on extrapolation from 35 sites to about 25,000 sites.
9	Drinking water	Estimated 400–1000 annual cancers, mostly from radon and trihalomethanes.
10	Application of pesticides	Approximately 100 cancers annually; small population exposed but high individual risks.
11	Radiation other than radon	Estimated 360 cancers per year. Mostly from building materials. Medical exposure and natural background levels not included.
12	Other pesticide risks	Consumer and professional exterminator uses; estimated cancers of 150 annually. Poor data.
13	Hazardous waste sites, active	Probably fewer than 100 cancers annually; estimates sensitive to assumptions regarding proximity of future wells to waste sites.
14	Nonhazardous waste sites, industrial	No real analysis done; ranking based on consensus of professional opinion.
15	New toxic chemicals	Difficult to assess; done by consensus.

(*continued*)

TABLE 7.4 *Continued*

Rank	Problem Area	Selected Comments
16	Nonhazardous waste sites, municipal	Estimated 40 cancers annually not including municipal surface impoundments.
17	Contaminated sludge	Preliminary results estimate 40 cancers annually, mostly from incineration and landfilling.
18	Mining waste	Estimated 10–20 cancers annually, largely due to arsenic. Remote locations and small population exposure reduce overall risk though individual risk may be high.
19	Releases from storage tanks	Preliminary analysis, based on benzene, indicates low cancer incidence (<1).
20	Non-point-source discharges to surface waters	No quantitative analysis available; judgment.
21	Other groundwater contamination	Lack of information; lifetime individual risks considered less than 10^{-6}, with rough estimate of total population risk at <1.
22	Criteria air pollutants	Excluding carcinogenic particles and VOCs (included under hazardous/toxic air pollutants); ranked low because remaining criteria pollutants have not been shown to be carcinogens.
23	Direct point-source discharges to surface water	No quantitative assessment available. Only ingestion of contaminated seafood was considered.
24	Indirect point-source discharges to surface water	Same as above.
25	Accidental releases, toxics	Short duration exposure yields low cancer risk; noncancer health effects of much greater concern.
26	Accidental releases, oil spills	See above. Greater concern for welfare and ecological effects.

Not ranked: Biotechnology; global warming; other air pollutants; discharges to estuaries, coastal waters, and oceans; and discharges to wetlands.

Although there were great uncertainties involved in making these assessments, the divergence between the EPA effort and relative risks in noteworthy. From this study, areas of relatively high risk but low EPA effort/concern include indoor radon, indoor air pollution, stratospheric ozone depletion, global warming, nonpoint sources, discharges to estuaries, coastal waters and oceans, other pesticide risks, accidental releases of toxics, consumer products, and worker exposures. The EPA gives high effort but relatively medium or low risks to RCRA sites, Superfund sites, underground storage tanks, and municipal nonhazardous waste sites.

The perception of a given risk is amplified by what are known as "outrage" factors. These factors can make people feel that even small risks are unacceptable. More than 20 outrage factors have been identified; a few of the main ones are available in the literature.[18,19] These include, for example, the perceived voluntariness, controllability, fairness, morality, and familiarity of the risk.

Regarding nanotechnology, the American Association for the Advancement of Science offers the following comments.[39]

> Perhaps most important, hard data on the environmental effects of nanomaterials would also go a long way in building the public's trust. Recent news coverage of genetically modified foods, for example, points to a great dichotomy of the human race: we are both fearful and curious about the unknown. Science is the embodiment of our curiosity, and the nanoscience community has a responsibility to ask the hard questions about nanomaterials before the public forces it to answer. Not only are nanoscientists best positioned to clearly answer the questions surrounding the potential environmental and health effects of nanomaterials, we also have the most to lose from unfounded fear of the unknown.

7.7 RISK COMMUNICATION[18]

Environmental risk communication is one of the more important problems that this country faces. Since 1987, public concerns about the environment have grown faster than concern about virtually any other national problem.[19] Some people are suffering (and in some cases dying) because they do not know when to worry and when to calm down. They do not know when to demand action to reduce risk or when to relax because risks are trivial or even nonexistent. The key, of course, is to pick the right worries and right actions. Unfortunately, when it comes to health and the environment, society does not do that well. The government and media together have failed to communicate clearly what is a risk and what is not a risk.

There are two major categories of risk: nonfixable and fixable. Nonfixable risks can never be substantially reduced, such as cancer-causing sunlight or cosmic radiation. Fixable risks can be reduced, and include those risks that are both large and small. There are so many of these fixable risks that all of them can never be successfully attacked, so choices must be made. When it comes to risk reduction, the outcome should be to obtain the most reduction possible, taking into account that people fear some risks more than others. This means essentially that the technical community should concentrate on the big fixable targets, and leave the smaller ones to later.

Risk communication comes into play because citizens ultimately determine which risks government agencies attack. On the surface, it appears practical to remedy the most severe risks first, leaving the others until later or perhaps, if the risks are small enough, never remedying the others at all. However, the behavior of individuals in everyday life often does not conform with this view. Consider now two environmental issues: gasoline that contains lead, and ocean incineration.[40]

According to the EPA, lead in gasoline poses very large risks: risks of learning disabilities, mental retardation, and worse to hundreds of thousands of children. The EPA's decision to reduce lead in gasoline is the most significant protective action the agency has undertaken in a long time. The only difference encountered on this issue was that the public acted with virtual indifference.

On the other hand, citizens threatened to lie down bodily in front of trucks and blockade harbors to stop the EPA's proposal to allow final testing of ocean incineration. The public reacted irrationally here. Every indication showed that the risk involved was small, and that the technology would be replacing more risky alternatives now in use.[40]

Why is there such an imbalance on the perception of risk? Ironically, part of the reason lies in the fact that the people responsible for communicating this information did their job too well. They achieved their objective of getting the information out to the public. Unfortunately, their objectives did not include effective communication of risk.

The professionals at the EPA are quite precise in the statements they deliver concerning risks and their apparent hazards. Their job is to present a scientifically defensible product, so they add qualifiers and use scientific terms. The problem with this is that the public often receives a misunderstanding of the actual risk.[40]

The challenge of risk communication is to provide the information in ways that can be incorporated in the views of people who have little time or patience for arcane scientific discourse. Success in risk communication is not to be measured by whether the public chooses to set the outcomes that minimize risk as estimated by the experts; it is achieved instead when those outcomes are knowingly chosen by a well-informed public.[19]

When citizens understand a risk, and the cost of reducing it, they can determine for themselves if control actions are too lax, too stringent, or just right. The two previous cases that were used as an example demonstrate that the risk message is not getting through to people who need to know when to demand action and when to calm down. The answer is not to communicate more information but more pertinent and understandable information. All the public needs to know is the following three pieces of information: How big is the risk? What is being done about it? What will it cost?[40]

7.8 SEVEN CARDINAL RULES OF RISK COMMUNICATION[41–43]

There are no easy prescriptions for successful risk communication. However, those who have studied and participated in recent debates about risk generally agree on seven cardinal rules. These rules apply equally well to the public and private sectors. Although may of these rules may seem obvious, they are continually and consistently violated in practice. Thus, a useful way to read these rules is to focus on why they are frequently not followed.[41]

1. Accept and involve the public as a legitimate partner. A basic tenet of risk communication in democracy is that people and communities have a right to participate in decisions that affect their lives, their property, and the things they value.

Guidelines: Demonstrate your respect for the public and underscore the sincerity of your effort by involving the community early, before important decisions are made. Involve all parties that have an interest or stake in the issue under consideration. If you are a government employee, remember that you work for the public. If you do not work for the government, the public still holds you accountable.

Point to Consider: The goal in risk communication in a democracy should be to produce an informed public that is involved, interested, reasonable, thoughtful, solution-oriented, and collaborative; it should not be to diffuse public concerns or replace action.

2. Plan carefully and evaluate your efforts. Risk communication will be successful only if carefully planned.

Guidelines: Begin with clear, explicit risk communication objectives, such as providing information to the public, motivating individuals to act, stimulating response to emergencies, and contributing to the resolution of conflict. Evaluate the information you have about the risks and know its strengths and weaknesses. Classify and segment the various groups in your audience. Aim your communications at specific subgroups in your audience. Recruit spokespeople who are good at presentation and interaction. Train your staff, including technical staff, in communication skills; reward outstanding performance. Whenever possible, pretest your messages. Carefully evaluate your efforts and learn from your mistakes.

Points to Consider: There is no such entity as "the public"; instead, there are many publics, each with its own interests, needs, concerns, priorities, preferences, and organizations. Different risk communication goals, audiences, and media require different risk communication strategies.

3. Listen to the public's specific concerns. If you do not listen to the people, you cannot expect them to listen to you. Communication is a two-way activity.

Guidelines: Do not make assumptions about what people know, think, or want done about risks. Take the time to find out what people are thinking: Use techniques such as interviews, focus groups, and surveys. Let all parties know that have an interest or stake in the issue be heard. Identify with your audience and try to put yourself in their place. Recognize people's emotions. Let people know that you understand what they said, addressing their concerns as well as yours. Recognize the "hidden agendas," symbolic meanings, and broader economic or political considerations that often underlie and complicate the task of risk communication.

Point to Consider: People in the community are often more concerned about such issues as trust, credibility, competence, control, voluntariness, fairness, caring, and compassion than about mortality, statistics, and the details of quantitative risk assessment.

4. Be honest, frank, and open. In communicating risk information, trust and credibility are your most precious assets.

Guidelines: State your credentials; but do not ask or expect to be trusted by the public. If you do not know an answer or are uncertain, say so. Get back to people with answers. Admit mistakes. Disclose risk information as soon as possible (emphasizing any reservations about reliability). Do not minimize or exaggerate the level of risk. Speculate only with great caution. If in doubt, lean toward sharing more information, not less, or people may think you are hiding something. Discuss data uncertainties, strengths and weaknesses, including the ones identified by other credible sources. Identify worst-case estimates as such, and cite ranges of risk estimates when appropriate.

Point to Consider: Trust and credibility are difficult to obtain. Once lost, they are almost impossible to regain completely.

5. *Coordinate and collaborate with other credible sources.* Allies can be effective in helping you communicate risk information.

Guidelines: Take time to coordinate all interorganizational and intraorganizational communications. Devote effort and resources to the slow, hard work of building bridges with other organizations. Use credible and authoritative intermediates. Consult with others to determine who is best able to answer questions about risk. Try to issue communications jointly with other trustworthy sources (e.g., credible university scientists and/or professors, physicians, or trusted local officials).

Point to Consider: Few things make risk communication more difficult than conflicts or public disagreements with other credible sources.

6. *Meet the needs of the media.* The media are prime transmitters of information on risks; they play a critical role in setting agendas and in determining outcomes.

Guidelines: Be open and accessible to reporters. Respect their deadlines. Provide risk information tailored to the needs of each type of media (e.g., graphics and other visual aids for television). Prepare in advance and provide background material on complex risk issues. Do not hesitate to follow up on stories with praise or criticism, as warranted. Try to establish long-term relationships of trust with specific editors and reporters.

Point to Consider: The media are frequently more interested in politics than in risk, more interested in simplicity than in complexity, and more interested in danger than in safety.

7. *Speak clearly and with compassion.* Technical language and jargon are useful as professional shorthand, but they are barriers to successful communication with the public.

Guidelines: Use simple, nontechnical language. Be sensitive to local norms, such as speech and dress. Use vivid, concrete images that communicate on a personal level. Use examples and anecdotes that make technical risk data come alive. Avoid distant, abstract, unfeeling language about deaths, injuries, and illnesses. Acknowledge and respond (both in words and with action) to emotions that people express—anxiety, fear, anger, outrage, helplessness. Acknowledge and respond to the distinctions that the public views as important in evaluating risks, for example, voluntariness, controllability, familiarity, dread, origin (natural or man-made), benefits, fairness, and catastrophic potential. Use risk comparisons to help put risks in perspective, but avoid comparisons that ignore distinctions that people consider important. Always try to include a discussion of actions that are under way or can be taken. Tell people what you cannot do. Promise only what you can do, and be sure to do what you promise.

Points to Consider: Regardless of how well you communicate risk information, some people will not be satisfied. Never let your efforts to inform people about risks prevent you from acknowledging and saying that any illness, injury, or death is a tragedy. Any finally, if people are sufficiently motivated, they are quite capable of understanding complex risk information, even if they may not agree with you.

The preceding seven cardinal rules of risk communication seem only logical. It is when they are violated that the proper and necessary communication will fail. Because it is the public that determines which risks will be remedied first, it is important to work with them, getting them involved in the decision-making process before it is too late. When one has the cooperation of the public, carefully state the objectives. Work with these objectives to provide necessary information, and motivate the involved individuals to act. Be a listener as well as a talker. Find out what the people want to know and let their voices be heard. Be honest with the issues at hand. State the truth and do not tell people what they may want to hear; they usually only want to know the truth. Work with, not against or in competition with, other credible sources. Get the message across in all possible ways, whether it be pamphlets, radio, or television. Most importantly, speak clearly and in terms that can be understood by everyone. Take into account the concerned people, and work on a personal level. All this is necessary for communicating and being heard. Successful communication will surely result if the seven rules are followed.

7.9 SUMMARY

1. Environmental risk assessment may be broadly defined as a scientific enterprise in which facts and assumptions are used to estimate the potential for adverse effects on human health or the environment that may result from exposures to specific pollutants or other toxic agents.

2. The health risk evaluation process consists of four steps: hazard identification, dose–response assessment or hazard assessment, exposure assessment, and risk characterization. In hazard identification, a hazard is a toxic agent or a set of conditions that has the potential to cause adverse effects to human health or the environment. In dose–response assessment, effects are evaluated and these effects vary widely because of differing capacities to cause adverse effects. Exposure assessment is the determination of the magnitude, frequency, duration, and routes of exposure to human populations and ecosystems. In risk characterization, the toxicology and exposure data are combined to obtain a quantitative or qualitative expression of risk.

3. Risk is the probability that individuals or the environment will suffer adverse consequences as a result of an exposure to a substance. Risk assessment is the procedure used to attempt to quantify or estimate risk.

4. The use of the risk-based decision-making process allows for efficient allocation of limited resources, such as time, money, regulatory oversight, and qualified professionals.

5. Risk communication is the exchange of information about risk.

6. People often overestimate the frequency and seriousness of dramatic, sensa-tional, dreaded, well-publicized causes of death and underestimate the risks from more familiar, accepted causes that regularly claim lives.

7. The seven rules of risk communication are as follows: Accept and involve the public as a legitimate partner; plan carefully and evaluate your efforts; listen to the public's specific concerns; be honest, frank, and open; coordinate and collaborate with other credible sources; meet the needs of the media; and speak clearly and with compassion.

REFERENCES

1. V. Colvin, "Responsible Nanotechnology: Looking Beyond the Good News," *EurekAlert! Nanotechnology in Context*, American Association for the Advancement of Science, Washington, DC, available on-line at *www.eurekalert.org* (Nov. 2002).

2. J. A. Moore, "New Technologies: The Public Is Listening But Are Scientists Talking?" USG Interagency Meeting on Nanotechnology & the Environment: Applications & Implications National Science Foundation, Washington, DC (Sept. 15, 2003).

3. R. F. Service, "Nanomaterials Show Signs of Toxicity," *Science*, **300**(5617), 243 (April 11, 2003).

4. D. B. Warheit, B. R. Laurence, K. L. Reed, D. H. Roach, G. A. M. Reynolds, and T. R. Webb, "Comparative Pulmonary Toxicity Assessment of Single-Wall Carbon Nanotubes in Rats," *Toxicological Sciences*, **77**, 117–125 (2004).

5. C. Lam, J. T. James, R. McClusky, and R. L. Hunter, "Pulmonary Toxicity of Single-Wall Carbon Nanotubes in Mice 7 and 90 Days after Intratracheal Instillation," *Toxicological Sciences*, **77**, 126–134 (2004).

6. The Etc. Group, "No Small Matter! Nanotech Particles Penetrate Living Cells and Accumulate in Animal Organs," *Communiqué*, **76**; available on line at *www.etcgroup.org* (May/June 2002).

7. J. Hempel, "Nanotech: Is It Nasty or Nice?" *Business Week*, 16 (May 3, 2004).

8. B. Halford, "From the ACS Meeting: Buckyballs Damage Bass Brains," *Chemical & Engineering News (C&EN)*, **82**(14), 14 (April 5, 2004).

9. R. Dagani, "Nanomaterials: Safe or Unsafe?" *Chemical & Engineering News (C&EN)*, **81**(17), 30–33 (April 26, 2003).

10. See, for example, B. J. Feder, "Research Shows Hazards in Tiny Particles," *New York Times* (April 14, 2003).

11. Rice University Center for Biological and Environmental Nanotechnology, (CBEN), "Environmental and Health Effects of Nanomaterials," available on-line at *www.ruf. rice.edu/~cben*.

12. T. Masciangioli and W. Zhang, "Environmental Technologies at the Nanoscale," *Environmental Science & Technology*, **37**(5), 102A–108A (Mar. 1, 2003).

13. V. L. Colvin, "Testimony before the U.S. House of Representatives Committee on Science in Regard to 'Nanotechnology Research and Development Act of 2003'" (April 9, 2003).

14. House Committee on Science, "More Research on Societal and Ethical Impacts of Nano-technology Is Needed to Avoid Backlash; Experts Say H.R. 766 Is 'Central to Goal,'" *Latest News and Information* (April 19, 2003).

15. U.S. Environmental Protection Agency, National Center for Environmental Research, "Research Opportunities: Impacts of Manufactured Nanomaterials on Human Health and the Environment," EPA, Washington, DC, closing date Dec. 11, 2003.

16. W. G. Sculz, "Nanotechnology under the Scope," *Chemical & Engineering News (C&EN)*, **80**(49), 23–24 (Dec. 9, 2002).

17. S. K. Friedlander, Workshop Chair, "Emerging Issues in Nanoparticle Aerosol Science and Technology (NAST)," NSF Workshop Report, University of California, Los Angeles, June 27–28, 2003, p. 15.

18. Adapted from M. K. Theodore and L. Theodore, *Major Environmental Issues Facing the 21st Century*, Theodore Tutorials, East Williston, NY, 1996.

19. Adapted from G. Burke, B. Singh, and L. Theodore, *Handbook of Environmental Management and Technology*, 2nd ed., Wiley, Hoboken, NJ, 2000.

20. L. Theodore and J. Reynolds, *Health, Safety and Accident Prevention: Industrial Applications*, A Theodore Tutorial, East Williston, NY, 1996.

21. D. Paustenbach, *The Risk Assessment of Environmental and Human Health Hazards: A Textbook of Case Studies*. Wiley, Hoboken, NJ, 1989.

22. J. Rodricks and R. Tardiff, *Assessment and Management of Chemical Risks*, American Chemical Society, Washington, DC, 1984.

23. EPA, *Technical Guidance for Hazards Analysis*, EPA, Washington, DC, December 1987.

24. D. Gute and N. Hanes, *An Applied Approach to Epidemiology and Toxicology for Engineers*, NIOSH, Cincinnati, OH, 1993.

25. L. Hadden and S. Hadden, "Handling Hazardous Materials: The Regulatory Picture," *Health and Environment Digest*, **2**(12), 1–3 (1989).

26. D. B. Clayson, D. Krewski, and I. Munro, *Toxicological Risk Assessment*, CRC, Boca Raton, FL, 1985.

27. V. Foa, E. A. Emmett, M. Maron, and A. Colombi, *Occupational and Environmental Chemical Hazards*, Ellis Horwood, Chichester, England, 1987.

28. *Standard Guide for Risk-Based Corrective Action Applied to Petroleum Release Sites*, ASTM E1729-95, American Society for Testing and Materials, Philadelphia, PA, 1995.

29. *Ecological Assessment of Hazardous Waste Sites: A Field and Laboratory Reference Document*, EPA/600/3-89/013, NTIS No. PB-89205967, Environmental Protection Agency, Washington, DC, March 1989.

30. *Integrated Risk Information System (IRIS)*, Environmental Protection Agency, Washington, DC, October 1993.

31. *Health Effects, Assessment Summary Tables (HEAST)* OSWER OS-230, Environmental Protection Agency, Washington, DC, March 1992.

32. *Federal Water Pollution Control Act*, Sect. 101(e), 1972.

33. EPA, Office of Public Affairs, *EPA Journal*, **11**(10), December 1985.

34. EPA. *The New Superfund. What It Is. How It Works*, August 1987.

35. EPA, *Chemicals in Your Community, A Guide to Emergency Planning and Community Right-to-Know Act*, September 1988.

36. D. Goleman, "Assessing Risk: Why Fear May Outweigh Harm," *New York Times*, Feb. 1, 1994.

37. Charles T. Main, Inc., *Health Risk Assessment for Air Emissions of Metals and Organic Compounds from the PERC Municipal Waste to Energy Facility*, Prepared for Penobscot Energy Recovery Company (PERC), Boston, MA, December 1985.

38. G. Masters, *Introduction to Environmental Engineering and Science*, Prentice Hall, Englewood Cliffs, NJ, 1991.

39. Information provided by the American Association for the Advancement of Science, 2002.

40. M. Russell, "Communicating Risk to a Concerned Public," *EPA Journal*, November, 1989.

41. EPA. "Seven Cardinal Rules of Risk Communication," EPA OPA/8700, April 1988, including periodic updates.

42. V. T. Covello, D. von Winterfeldt, and P. Slovic, "Risk Communication: A Review of the Literature,"*Risk Abstracts*, **3**(4), 171–182, 1986.

43. C. J. Davies, V. T. Covello, and F. W. Allen, *Risk Communication, Proceedings of the National Conference on Risk Communication, January 29–31, 1986*, Conservation Foundation, Washington, DC, 1987.

CHAPTER 8

HAZARD RISK ASSESSMENT[1,2]

8.1 INTRODUCTION

As indicated in the previous chapter, many practitioners and researchers have confused health risk with hazard risk, and vice versa. Although both employ a four-step method of analysis, the procedures are quite different, with each providing different results, information, and conclusions. The reader is referred to Sections 7.1 and 7.2 for details differentiating health risk from hazard risk. Introductory information is provided in Section 2.7.

As with health risk, there is a serious lack of information on the hazards and associated implications of these hazards with nanoapplications. The unknowns in this risk area are both larger in number and greater in potential consequences. It is the authors' judgment that hazard risk has unfortunately received something less than the attention it deserves. However, hazard risk analysis details are available and the traditional approaches successfully applied in the past can be found in this section. Future work will almost definitely be based on this methodology.

Much has been written about Michael Crichton's powerful science-thriller novel titled *Prey* (The book was not only a best seller, but the movie rights were sold for $5 million.) In it, Crichton provides a frightening scenario in which swarms of nanorobots, equipped with special power generators and unique software, prey on living creatures. To compound the problem, the robots continue to reproduce without any known constraints. This scenario is an example of an accident and represents only one of a near infinite number of potential hazards that can arise in any nanotechnology application. Although the probability of the horror scene portrayed by

Nanotechnology: Environmental Implications and Solutions, L. Theodore and R. G. Kunz.
ISBN 0-471-69976-4   Copyright © 2005 John Wiley & Sons, Inc.

Crichton, as well as other similar events, is extremely low, steps and procedures need to be put into place to reduce, control, and, it is hoped, eliminate these events from actually happening. This chapter attempts to provide some of that information.

The previous chapter (see Section 7.1) defined both "chronic" and "acute" problems. As indicated, when the two terms are applied to emissions, the former usually refers to ordinary, round-the-clock, everyday emissions while the latter term deals with short, out-of-the-norm, accidental emissions. Thus, acute problems normally refer to accidents and/or hazards. The Crichton scenario discussed above is an example of an acute problem, and one whose solution would be addressed/ treated by a hazardous risk assessment, as described in Section 8.7, rather than the health risk approach provided in the previous chapter.

Accidents are a fact of life, whether they are a careless mishap at home, an unavoidable collision on the freeway, or a miscalculation at a chemical plant. Even in prehistoric times, long before the advent of technology, a club-wielding caveman might have swung at his prey and inadvertently toppled his friend in what can only be classified as an "accident." As Man progressed, so did the severity of his misfortunes. The "Modern Era" has brought about assembly lines, chemical manufacturers, nuclear power plants, and so on, all carrying the capability of disaster. To keep pace with the changing times, safety precautions must constantly be upgraded. It is no longer sufficient, as with the caveman, to shout the warning, "Watch out with that thing!" Today's problems require more elaborate systems of warnings and controls to minimize the chances of serious accidents.[1]

Industrial accidents occur in many ways—a chemical spill, an explosion, a nuclear plant out of control, and so on. There are often problems in transport, with trucks overturning, trains derailing, or ships capsizing. There are "acts of God," such as earthquakes and storms. The one common thread through all of these situations is that they are rarely expected and frequently mismanaged.[2]

Much of the material presented has been drawn from several references cited in this chapter.

Early Accidents[3]

Accidents have occurred since the birth of civilization and were just as damaging then as they are today. Anyone who crosses a street, rides in a car, or swims in a pool runs the risk of injury through carelessness, poor judgment, ignorance, or other circumstances. This has not changed throughout history. In the following pages, a number of accidents and disasters that took place before the advances of modern technology will be examined.

Catastrophic explosions have been reported as early as 1769, when one-sixth of the city of Frescia, Italy, was destroyed by the explosion of 100 tons of gunpowder stored in the state arsenal. More than 3000 people were killed in this, the second deadliest explosion in history.

The worst explosion in history occurred in Rhodes, Greece, in 1856. A church on this island, which had gunpowder stored in its vaults, was struck by lightning. The resulting blast killed an estimated 4000 people. This remains the highest death toll for a single explosion.[4]

One of the most legendary disasters occurred in Chicago, Illinois, in October 1871. The "Great Chicago Fire," as it now known, is alleged to have started in a barn owned by Patrick O'Leary, when one of his cows overturned a lantern. Whether or not this is true, the O'Leary house escaped unharmed because it was upwind of the blaze, but it left the barn destroyed along with 2124 acres of Chicago real estate.

Catastrophic accidents have occurred on the sea as well as on land. In 1947 an unusual incident involved both. The French freighter *Grandcamp* arrived at Texas City, Texas, to be loaded with 1400 tons of ammonium nitrate fertilizer. Sometime that night, a fire broke out in the hold of the *Grandcamp*. The ship's crew made only limited attempts to fight the flames, apparently fearing water would damage the rest of the cargo. Because the *Grandcamp* was docked only 700 ft from a Monsanto chemical plant that produced styrene, a highly combustible ingredient in synthetic rubber, the *Grandcamp* was ordered towed from the harbor.

As tugboats prepared to hook up their lines, the *Grandcamp* exploded in a flash of fire and steel fragments. The blast rattled windows 150 miles away, registered on a seismograph in Denver, and killed many people standing on the dock. The Monsanto chemical plant exploded minutes later, killing many survivors of the first blast, shattering most of the Texas City business district, and setting fires throughout the rest of the city. As the fires burned out of control, the freighter *High Flyer*, also loaded with nitrates, exploded in the harbor. This third explosion proved too much for the people of Texas City, who had responded so efficiently to the initial two blasts. Hundreds of people were forced to leave the city, letting the fire burn itself out. The series of explosions had killed 468 people and seriously injured 1000 others. The final death toll may have been as high as 1000 because the dock area contained a large population of migrant workers without permanent address or known relatives. This disaster was probably caused by careless smoking aboard the *Grandcamp*.[4,5]

Recent Major Accidents[3]

The advances of modern technology have brought about new problems. Perhaps the most serious of these is the threat of a nuclear power plant accident, known as a *meltdown*. In this section, several of this era's most infamous (nonnuclear) accidents will be examined.

An explosion at the Nypro Ltd. caprolactum factory at Flixborough, England, on June 1, 1947, was one of the most serious in the history of the chemical industry and the most serious that has occurred in the history of the United Kingdom. Of those working on the site at the time, 28 were killed and 36 others injured. Outside the plant, 53 people were reported injured, while 1821 houses and 167 shops suffered damage. The estimated cost of the damage was well over $100 million.

The worst disaster in the recent history of the chemical industry occurred in Bhopal, in central India, on December 3, 1984. A leak of methyl isocyanate from a chemical plant, where it was used as an intermediate in the manufacture of a pesticide, spread into the adjacent city and caused the poisoning death of over 2500 people and injuries to approximately 20,000 others. The owner of the plant, Union Carbide Corporation, reported that the accident was "the result of a unique combination of unusual events," although there has also been talk of possible sabotage. No data have been presented to support this latter claim. The cause of the accident was attributed to faults in the design of the plant's safety system—which was the responsibility of Union Carbide.[6]

A later accident occurred that, although it did not involve the loss of human life, can still be considered a disaster. On January 2, 1988, a 48-ft-high fuel tank at the Ashland Oil terminal in Pennsylvania ruptured. Nearly 3.9 million gallons of No. 2 distilled fuel oil poured out. The force caused the tank to jump backward 100 ft and sent a wave 35 ft high crashing into another tank, 100 ft away. Much of the spilled fuel was trapped by a containment dike; however, 600,000 gal escaped into the Monongahela River at Floreffe, about 25 miles upstream from Pittsburgh. Soon after the spill, a rumor began circulating that there was a possible gasoline leak as well. This raised concerns about a fire, leading to the evacuation of 250 homes. The fuel that had spilled into the frigid water began to emulsify and sink. The extremely cold weather caused ice to form on the river. It is nearly impossible to recover oil that sinks below the skirts of the recovery booms or that becomes trapped in the ice. Various methods were used to remove the oil from the river. Chemists developed a method that mixed the contaminated water with powdered carbon and bentonite, which gives the slurry higher absorbency. The mixture was then pumped to a treatment plant where other chemicals were added to balance acidity and make the oil coagulate in a settling tank. This treatment is not new, but the chemists had to come up with the right combinations of chemicals to handle the oil. At one point, the EPA allowed the use of a substance called Elastol for the first time. Elastol congeals spilled hydrocarbons into a mass that can be easily recovered.

Accidents have also occurred in the "air," an area that would be classified as a transportation accident. An example of this is the Trans World Airlines (TWA) Flight 800 from New York to Paris. On July 17, 1996, a Boeing 747 sat on the runway at JFK airport for an extended period of time. The heat generated from the idle plane was sufficient to vaporize residual fuel in what was considered an "empty" fuel tank. This created a very dangerous, flammable mixture of fuel vapor and air. Approximately 10 min after departure from JFK airport, TWA Flight 800 exploded in midair. The most likely source of ignition for the flammable mixture of fuel vapor and air was an electrical wiring problem. All 230 people aboard perished when the jet crashed off the coast of Long Island, New York. However, this tragedy could have been avoided by preventing the flammable mixture in the fuel tank from developing.

In response to the TWA Flight 800 crash, the Federal Aviation Administration (FAA) has moved toward requiring airlines to pump nitrogen gas into empty fuel tanks to keep the pressure inside the tanks above the vapor pressure of the fuel and, hence, keep the fuel from vaporizing. This prevents a flammable mixture from developing inside the tanks. Unfortunately, it took the deaths of the 230 people aboard TWA Flight 800 to bring prior FAA philosophies that allowed aircraft to fly with flammable fuel tanks under scrutiny.[7,8]

An example of the effects of poor planning is described by what has become known as the Y2K (Year 2000) computer bug. Computers work with numbers and a very important and widely used number in both computer hardware and software is the date, particularly the year. This information, like any other data used by a computer, is stored in the computer's memory. When computers first started being used and programs were being written for them, computer storage—random access memory (RAM) and disk memory—was very expensive. In an effort to reduce the amount of memory required to store the year, only the last two digits of the year were used. This system worked fine until the year 2000. On December 31, 1999, as the digits were to change from 99 to 00, the year was actually going to change from 1999 to 1900 (instead of 2000). This would affect every business that was linked to a computer, and the commotion and hysteria that surrounded the Y2K bug were due to the present age's dependence on computers. For example, it was feared that the government's computer system would print social security checks with amounts seen in the year 1900 and that banks would calculate dividends based on interest rates from the year 1900. The problems could have been catastrophic. Fortunately, the response to the "potential" Y2K disaster was effective in that most corrections or "patches" were made before the year 2000. However, these corrections were costly and required large amounts of money to be spent.[7,9]

Advances in Safety Features[6]

Today's sophisticated equipment and technologies require equally sophisticated means of accident prevention. Unfortunately, the existing methods of detection and prevention are often assumed to be adequate until proven otherwise. This latter means of determining a technology's effectiveness can sometimes be costly and can often lead to loss of life. Chemical manufacturers and power plants are businesses and thus are not as likely to update their present controls "unnecessarily."

Prior to the advent of technology, there was still a need for safety features and warnings; yet these did not exist. Many accidents occurred because of a lack of knowledge of the system, process, or substance being dealt with. Many of the pioneers of modern science were sent to an early grave by their experiments. Karl Wilhelm Scheele, the Swedish chemist who discovered many chemical elements and compounds, often sniffed or tasted his finds. He eventually died of mercury poisoning. Likewise, Madame Curie died from leukemia contracted from overexposure to radioactive elements. Had either of these brilliant scientists any idea of the danger of their work, their methods would certainly have changed significantly. In those days, a safety precaution was often devised by trial and error; if inhaling a certain gas was found to make someone sick, the prescribed precaution was not to smell it. Today, the physical properties of most known compounds are readily found in handbooks, so that proper care can be exercised when working with these chemicals. Labs are equipped with exhaust hoods and fans to minimize a buildup of gases; in addition, safety glasses and eyewash stations are required, as are gloves and smocks.

8.2 SUPERFUND AMENDMENTS AND REAUTHORIZATION ACT OF 1986[10]

The Superfund Amendments and Reauthorization Act (SARA) of 1986 renewed the national commitment to correcting problems arising from previous mismanagement of hazardous wastes. (Note that this Act is being revisited, although it was discussed in the previous chapter; the presentation that follows will center primarily on "hazard" applications.) While SARA was similar in many respects to the original law, it also contained new approaches to the program's operation. The 1986 Superfund legislation did the following:[11]

1. It reauthorized the program for 5 more years and increased the size of the cleanup fund from $1.6 to $8.5 billion.
2. It set specific cleanup goals and standards and stressed the achievement of permanent remedies.
3. It expanded the involvement of states and citizens in decision making.
4. It provided for new enforcement authorities and responsibilities.
5. It increased the focus on human health problems caused by hazardous waste sites.

The law is more specific than the original statute with regard to such things as remedies to be used at Superfund sites, public participation, and accomplishment of cleanup activities. The most important part of SARA with respect to public participation is Title III. Title III addresses the important issues regarding community awareness and participation in the event of an accidental chemical release and is an important part of SARA that addresses hazardous materials releases; it is subtitled the Emergency Planning and Community Right-to-Know Act of 1986. Title III establishes requirements for emergency planning, hazardous emissions reporting, emergency notification, and "community right to know." The objectives of Title III are to improve local chemical emergency response capabilities, primarily through improved emergency planning and notification, and to provide citizens and local governments with access to information about chemicals in their localities. Title III has four major sections that aid in the development of contingency plans. They are as follows:

1. Emergency Planning (Sections 301–303)
2. Emergency Notification (Section 304)
3. Community Right-to-Know Reporting Requirements (Sections 311 and 312)
4. Toxic Chemicals Release Reporting—Emissions Inventory (Section 313)

Title III has also developed time frames for the implementation of the Emergency Planning and Community Right-to-Know Act of 1986. Details on these are provided in the two sections that follow.

8.3 NEED FOR EMERGENCY RESPONSE PLANNING

Emergencies have occurred in the past and will continue to occur in the future. A few of the many commonsense reasons to plan ahead for emergencies are as follows:[12]

1. Emergencies will happen; it is only a question of time.
2. When emergencies occur, the minimization of loss and the protection of people, property, and the environment can be achieved through the proper implementation of an appropriate emergency response plan.
3. Minimizing the losses caused by an emergency requires planned procedures, understood responsibility, designated authority, accepted accountability, and trained, experienced people. A fully implemented plan can do this.
4. If an emergency occurs, it may be too late to plan. Lack of preplanning can turn an emergency into a disaster.

A particularly timely reason to plan ahead is to ease the "chemophobia" or fear of chemicals, which is so prevalent in society today. So much of the recent attention to emergency planning and newly promulgated laws are a reaction to the tragedy at Bhopal. Either a total lack of information or misinformation is the probable cause of chemophobia. Fire is hazardous, and yet it is used regularly at home. Most adults have understood the hazard associated with fire since the time of the caveman. By the same token, hazardous chemicals, necessary and useful in our technological society, are not something to be afraid of. Chemicals need to be carefully used and their hazards understood by the general public. An emergency plan that is well designed, understood by the individuals responsible for action, and understood by the public can ease the concern over emergencies and reduce chemophobia. People will react during an emergency; *how* they react can be somewhat controlled through education. The likely behavior during an emergency when ignorance is pervasive is panic.

An emergency plan can minimize loss by helping to ensure the proper response in an emergency. Accidents become crises when subsequent events and the actions of people and organizations with a stake in the outcome combine in unpredictable ways to threaten the social structures involved.[13] The wrong response can turn an accident into a disaster as easily as no response. One example is a chemical fire that is doused with water, causing the fire to emit toxic fumes; the same fire would be better left to burn itself out. Another example is the evacuation of people from a building into the path of a toxic vapor cloud; they might well be safer staying indoors with closed windows. Still another example is the members of a rescue team becoming victims because they were not wearing proper breathing protection. The proper response to an emergency requires an understanding of the hazards. A plan can provide the right people with the information needed to respond properly during an emergency.

Other than the above-mentioned commonsense reasons to plan, there are legal reasons. Recognizing the need for better preparation to deal with chemical emergencies, Congress enacted the aforementioned Superfund Amendments and

Reauthorization Act (SARA) of 1986; this act was discussed in detail in the last section. One part of SARA is a free-standing act called Title III, the Emergency Planning and Community Right-to-know Act of 1986. This act requires federal, state, and local governments to work together with industry in developing emergency plans and "community right-to-know" reporting on hazardous chemicals. These new requirements build on the EPA's Chemical Emergency Preparedness Program and numerous state and local programs that are aimed at helping communities deal with potential chemical emergencies.[14]

Most larger industries have long had emergency plans designed for onsite personnel. The protection of people, property, and thus profits has made emergency plans and prevention methods common in industry. Onsite emergency plans are often a requirement by insurance companies. Expansion of these existing industry plans to include all significant hazards and all people in the community is a way to minimize the effort required for emergency planning.

Planning Committee

Emergency planning should grow out of a team process coordinated by a team leader. The team may be the best vehicle for gathering people with various kinds of expertise into the planning process, thus producing a more accurate and complete plan. The team approach also encourages a planning process that will reflect a consensus of the entire community. Some individual communities and/or areas that include several communities had already formed advisory councils before the SARA requirements. These councils can serve as an excellent resource for the planning team.[15]

When selecting the members of a team that will bear overall responsibility for emergency planning, the following considerations are important:

1. The members of the group must have the ability, commitment, authority, and resources to get the job done.
2. The group must possess, or have ready access to, a wide range of expertise relating to the community, its industrial facilities, transportation systems, and the mechanics of emergency response and response planning.
3. The members of the group must agree on their purpose and be able to work cooperatively with one another.
4. The group must be representative of all elements of the community, with a substantial interest in reducing the risk posed by emergencies.

While many individuals have an interest in reducing the risks posed by hazards, their differing economic, political, and social perspectives may cause them to favor different means of promoting safety. For example, people who live near an industrial facility with hazardous materials are likely to be greatly concerned about avoiding any threat to their lives. They are likely to be less concerned about the costs of developing accident prevention and response measures than some of the other team members. Others in the community, those representing industry or the budgeting group, for example, are likely to be more sensitive to the costs involved. They may be more anxious to avoid expenditures for unnecessary, elaborate prevention and response measures, Also, industry facility managers, although concerned with reducing risks posed by hazards, may be reluctant, for proprietary reasons, to disclose materials and process information beyond what is required by law. These differences can be balanced by a well-coordinated team, which is responsive to the needs of its community.

Among the agencies and organizations with emergency response responsibility, there may be differing views about the role they should play in case of an incident. The local fire department, an emergency management agency, and public health agency are all likely to have some responsibilities during an emergency. However, each of these organizations might envision a very different set of actions for their respective agencies at the emergency site. The plan will serve to detail the actions of each response group during an emergency.

In organizing the community to address the problems associated with emergency planning, it is important to bear in mind that all affected parties have a legitimate interest in the choices among planning alternatives. Therefore, strong efforts should be made to ensure that all groups with an interest in the plan are included in the planning process. The need for control of the committee during the planning process, as well as for control during the plan implementation, is amplified by the number of different groups involved in the community. Each of these groups has a right to participate in the planning, and a well-structured, organized planning committee should serve all the community groups.

By law, the planning committee must include the following:[16]

1. Elected and state officials
2. Civil defense personnel
3. First-aid personnel
4. Local environmental personnel
5. Transportation personnel
6. Owners and operators of facilities subject to the SARA
7. Law enforcement personnel
8. Firefighting personnel
9. Public health personnel
10. Hospital personnel
11. Broadcast and print media
12. Community groups

There are other individuals who could also serve the community well and should be a part of the committee, such as[17]

1. Technical professionals
2. City planners

3. Academic and university researchers
4. Local volunteer help organizations

The local government has a great share of the responsibility for emergency response within its community. The official who has the power to order evacuation, fund fire and emergency units, and educate the public is a key person to emergency planning and the response effort. For example, the entire plan will fail if a timely evacuation is necessary and not ordered on time. Although politics should be disassociated from technical decisions, emergency planning is a place where such linkage is inevitable. Distasteful options that require political courage are often necessary (e.g., having to evacuate a section of town where there is some doubt about the necessity of evacuation), but the consequence of not evacuating might be deadly. A public official can build public support for future candidacy by using the issue of chemical safety as a bandwagon, but mistakes in handling emergencies are measured by a strong instrument—the election—where a failed emergency plan can be fatal to a political career. Politics is a social feedback device that, when used properly, can aid the government leader in making correct decisions. A political career can also be destroyed by an error in reading the social feedback. Developing an effective plan can save the elected official hours of media criticism after a crisis because the plan can provide the details of events organized by someone on the team as they occur. Because of the power an elected official has locally, that person is likely to take the leadership place on the committee.

The management or control of the committee during its planning, and especially during implementation of the plan, is essential. As discussed earlier, the committee will be made up of different individuals with different priorities and the emergency plan will be generated by them. The different groups will have their own legitimate interests. Each interest will have to be weighed against its value to the plan. To have a respect for the interests of all of the individuals, as well as a respect for their contributions, is an essential attribute of the committee leader. The committee leader is likely to be chosen for several reasons, but among these should be the following:

1. The degree of respect held for the person by groups and individuals with an interest in the emergency plan.
2. The availability of time and resources this person will have to serve the committee.
3. The person's history of working relationships with concerned community agencies and organizations.
4. The person's management skills and communications skills.
5. The person's existing responsibility and background experience related to emergency planning, prevention, and response.

Personal considerations, as well as institutional ones, should be weighed when selecting a committee leader. If a person being evaluated for the position of committee leader has all the right resources to address the emergency planning and implementation, but is unable to interact with local officials, then someone else may be a better choice. Since the committee leader must manage this large group of people with different priorities and expertise, the choice of the leader is critical to the success of the committee.[18]

Notification for Public and Regulatory Officials

Notifying the public of an emergency is a task that must be accomplished with caution. People will react in different ways. Many will simply not know what to do, others will not take the warning seriously, and still others will panic. Proper training in each community can help minimize any panic and condition the public to make the right response.

Methods of communicating the emergency will differ for each community, depending upon its size and resources. Some techniques for notifying the public are:

1. The sounding of fire department alarms in different ways to indicate certain kinds of emergencies
2. Chain phone calls (this method usually works well in small towns)
3. Police cars or volunteer teams with loudspeakers

Once the emergency is communicated, an appropriate response by the public must be evoked. For this response to occur, an accepted plan that people know and understand must be put into effect. Panic may occur, and this plan should be flexible enough to include the appropriate countermeasures. An emergency can quickly become a total disaster if there is a panicked public.

The reporting of information to the emergency coordinator must be carefully screened. A suspected "crank call" should be checked out before an alarm is sounded. An obvious risk in not taking immediate action is that the plan will not be implemented in time. Therefore, if a crank call cannot be verified as such, a response must begin. In this case, local police should be dispatched to the scene of a reputed emergency quickly to verify the report firsthand.

The media (e.g., news, radio, and television) can be a major resource for communication. One job of the emergency coordinator is to prepare the information that is to be reported to the media. The emergency plan should include a procedure to pass along information to the media promptly and accurately.

Besides notifying the response team and the community about what procedures to follow in an emergency, there are also requirements to report certain types of emergencies to governmental agencies. For example, state and federal laws require the reporting of hazardous releases and nuclear power plant problems. There are also more specific requirements under SARA Title III for reporting chemical releases. Facilities where a listed hazardous substance is produced, stored, or used must immediately notify the local emergency planning committee and the State Emergency Response Commission if there is a release of a substance specifically listed in SARA. These substances include (a) approximately 400 extremely hazardous chemicals on the list prepared by the Chemical Emergency Preparedness Program and (b) chemicals subject to reportable quantities requirements of the original Superfund.[14] The initial notification can be made by telephone, by radio, or in person. Emergency notification requirements involving transportation incidents can be satisfied by dialing 911 or calling the operator. The emergency planning committee should provide a means for information on transportation accidents to be reported quickly to the coordinator.

SARA requires that the notification of an industrial emergency include:

1. The name of the chemical released
2. Whether it is known to be acutely toxic
3. An estimate of the quantity of the chemical released into the environment
4. The time and duration of the release
5. Where the chemical was released (e.g., air, water, land)
6. Known health risks and necessary medical attention
7. Proper precautions, such as evacuation
8. The name and telephone number of the contact person at the plant or facility where the release occurred

A written follow-up emergency notice is required as soon as is practical after the release. The notice should include:

1. An update of the initial information
2. Additional information on response actions already taken, known or anticipated health risks, and advice on medical attention

Reporting and written notices have been required by law since October 1986.

Plan Implementation

Once an emergency plan has been developed, its successful implementation can be ensured only through constant review and revision. A number of ongoing steps that should be taken are

1. Inventory checks of equipment, personnel, hazards, and population densities on a routine basis
2. Auditing of the emergency procedure
3. Training on a routine basis
4. Practice drills

The coordinator must ensure that the emergency equipment is always in a state of readiness. The siting of the control center and the location of its equipment is also the coordinator's responsibility. A main control center and an alternate one should be provided. The location should be chosen carefully. The following items should be present at the control center:

1. Copies of the current emergency plan
2. Maps and diagrams of the area
3. Names and addresses of key functional personnel involved in the plan
4. Means to initiate alarm signals in the event of a power outage
5. Communication equipment (e.g., phones, radio, TV, and two-way radios)
6. Emergency lights
7. Evacuation routes mapped out on the area map
8. Self-contained breathing equipment for possible use by the control center crew
9. Cots plus other miscellaneous furniture

Inspection of emergency equipment such as fire trucks, police cars, medical vehicles, personal safety equipment, and alarms should be performed routinely.

The plan should be audited on a regular basis, at least annually, to ensure that it is current. Items to be updated include the list of potential hazards and emergency procedures (adapted to any newly developed technology, e.g., nanotechnology). A guideline for auditing the emergency response plan is available from the Chemical Manufacturers Association.[18]

More extensive information regarding calculational procedures can be found in the literature.[19]

8.4 EMERGENCY PLANNING[21]

Successful emergency planning begins with a thorough understanding of the event or potential disaster being planned for. The impacts on public health and the environment must also be estimated. Some of the types of emergencies that should be included in the plan are:[20]

1. Earthquakes
2. Explosions
3. Fires
4. Floods.
5. Hazardous chemical leaks—gas or liquid
6. Power or utility failures
7. Radiation incidents
8. Tornadoes or hurricanes
9. Transportation accidents

To estimate the impact on the public or the environment, the affected area or emergency zone must be studied. A hazardous gas leak, fire, or explosion may cause a toxic cloud to spread over a great distance. An estimate of the minimum affected area and thus the area to be evacuated should be performed, based on an atmospheric dispersion model. There are various models that can be used. While the more sophisticated models produce more realistic results, the models that are simpler and faster to use may provide adequate data for planning purposes.[21]

In formulating the plan, some general assumptions may be made:

1. Organizations do a good job when they have specific assignments.
2. The various resources will need coordination.
3. Most resources that are necessary are likely to be already available in the community (in plants or city departments).
4. People react more rationally when they have been apprised of the situation.
5. Coordination is basically a social process, not a legal one.
6. Disorganization and reorganization are common in a large group.
7. Flexibility and adaptability are basic requirements for a coordinated team.

The objective of the plan should be to prepare a procedure to make maximum use of the combined resources of the community in order to accomplish the following:

1. Safeguard people during emergencies.
2. Minimize damage to property and the environment.
3. Initially contain and ultimately bring the incident under control.
4. Effect the rescue and treatment of casualties.
5. Provide authoritative information to the news media (who will transmit it to the public).
6. Secure the safe rehabilitation of the affected area.
7. Preserve relevant records and equipment for the subsequent inquiry into the cause and circumstances.

The key components that should be contained in the emergency action plan include:[14]

1. Emergency actions other than evacuation
2. Escape procedures when necessary
3. Escape routes clearly marked on a site map and perhaps also on the roads
4. A method for accounting for people after evacuation
5. Rescue and medical duties
6. Reporting emergencies to the proper regulatory agencies
7. Notification of the public by an alarm system
8. Contact and coordination person responsibilities

Successful emergency planning will be enhanced with improved and more sophisticated measurement methods. These methods can help reduce nanoapplication hazards. Drexel University, for example, is developing piezoelectric microcantilever arrays for rapid and in situ quantification of multiple pathogens in water systems. This would better enable countries to respond to terrorist threats while ensuring the safety and quality of water supplies.

8.5 HAZARDS SURVEY[21]

To characterize the types and extents of potential disasters, a survey of hazards or foreseeable threats in the community must be performed and evaluated. An appropriate plan cannot be developed without information on the types of events that are possible or on the potentially affected areas. A plan for a city with a river, for example, may not be applicable to a desert city on a seismic fault. An inventory of the community assets, hazard sources, and risks must be taken prior to the actual plan writing.

Duplication can be an enemy of cost efficiency. Some emergency plans may already exist in the community, and these should be used wherever possible. Community groups that may have developed such plans are:

1. Civil defense
2. Fire department
3. Red Cross
4. Public health
5. Local industry council

Such existing plans should be studied, and their applicability to the proposed community plan should be evaluated.

The resources of the community should be listed and then compared to the needs of the plan. Local government departments, such as transportation, water, power, and sewer, may have such resources. Some examples of these are:

1. Trucks
2. Equipment (e.g., backhoes, flatbeds)
3. Laboratory services (e.g., water department)
4. Fire vehicles
5. Police vehicles
6. Emergency suits
7. Breathing apparatus
8. Gas masks
9. Number of trained emergency people
10. Number of volunteer personnel (e.g., Red Cross)

11. Buses or cars
12. Communications equipment (e.g., hand radios)
13. Local TV and radio stations
14. Ambulances
15. Trained medical technicians and first-aid personnel
16. Stockpiles of medicines
17. Burn treatment equipment
18. Fallout shelters

The potential sources of hazards should be listed for risk assessment. SARA requires certain industries to provide information to the planning committee. The committee should gather the information about small as well as large industries in order to evaluate the significant risks. The information regarding chemicals required by SARA to be provided includes:

1. The chemical name
2. The quantity stored over a period of time
3. The type of chemical hazard (e.g., toxic, flammable, ignitable, corrosive)
4. Chemical properties and characteristics (e.g., liquid at certain temperatures, gas at certain pressures, reacts violently with water)
5. Storage description and storage location on the site
6. Safeguards or prevention measures associated with the hazardous chemical storage or handling design, such as dikes, isolation of incompatible substances, and fire-resistant equipment
7. Control features for accident prevention such as temperature and pressure controllers and fail-safe design devices, if included in the handling design
8. Recycle control loops intended for accident prevention
9. Emergency shutdown features

The planning committee should designate hazard sources on a community map. This information probably already exists and can be obtained from local groups such as the transportation department, environmental agency, city planning department, community groups, and industry. Some of the data to locate on the community map are

1. Industry and other chemical locations
2. Wastewater and water treatment plants that have chlorine stored
3. Potable and surface water
4. Drainage and runoff
5. Population location and density in different areas
6. Transportation routes for children

7. Commuter routes
8. Truck transport roads
9. Railroad lines, yards, and crossings
10. Major highways, noting merges and downhill curves
11. Hospitals and nursing homes
12. Fallout shelters

The potential for a natural disaster based on the history and knowledge of the region and earth structure should be indicated in the plan. Items such as seismic fault zones, floodplains, hurricane potentials, and winter storm potentials should be noted.

The risk inventory, or risk evaluation, is the next part of the hazard survey. It is not practical to expect the plan to cover every potential accident. The hazards should be evaluated, and then the plan should be focused on the most significant ones. This risk assessment stage requires the technical expertise of many people to compare the pieces of data and determine the relevance of each.

In performing the risk evaluation, much data have to be evaluated. Among the important factors to be considered are the following:

1. The routes of transport of hazardous substances should be reviewed to determine where a release could occur.
2. Industry sites are not the only sources of hazards. The proximity of hazards to people and other sensitive environmental receptors should be examined.
3. The toxicology of different exposure levels should be reviewed.

Once the significant risks are listed, the hazard survey is complete and the plan can then be developed.

8.6 TRAINING OF PERSONNEL[2]

The education of the public regarding the real hazards in their community and how to respond to an emergency is critical to the public support of the emergency plan. Public opinion surveys show, however, that Americans are not yet prepared to deal either with the information provided by SARA Title III or with the reality of hazards in their community. Most people understand that hazards exist but do not realize that the potential hazards are all around them in their own communities. The common perception is that hazards exist elsewhere, as do the resulting emergencies.[22,23] The education of the populace about the true hazards associated with routine discharges from plants in the neighborhood and preparing that populace for emergencies is a real challenge to the community committee. People must be taught how to react to an emergency. This includes how to recognize and report an incident, how to react to alarms, and what other action to take. A possible result of SARA Title III may be, initially, a fear on the part of the public of industrial discharges. News stories, based on hazardous chemical inventory, accidental releases, or annual emissions reports of questionable accuracy or taken out of context, can be misleading. It is hoped that this can be put into perspective through training programs.

The personnel at an industrial plant, particularly the operators, are trained in the operation of the plant. These people are critical to proper emergency response. They must be taught to recognize abnormalities in operations and to report them immediately. Plant operators should also be taught how to respond to various types of accidents. Emergency squads at plants can also be trained to contain the emergency until outside help arrives, or, if possible, to terminate the emergency. Shutdown and evacuation procedures are especially important when training plant personnel.

Training is important for the emergency teams to ensure that their roles are clearly understood and that accidents can be reacted to safely and properly without delay. The emergency teams include the police, firefighters, medical people, and volunteers who will be required to take action during an emergency. These persons must be knowledgeable about the potential hazards. For example, specific antidotes for different types of medical problems must be known by medical personnel. The whole emergency team must also be taught the use of personal protective equipment.

Local governmental officials also need training. Because these officials have the power to order an evacuation, they must be aware of under what circumstances such action is necessary. The timing of an evacuation is critical; this must be understood by these people prior to the emergency itself. Local officials also control the use of city equipment and therefore must be knowledgeable as to what is needed for an appropriate response to a given emergency.

Media personnel must also be involved in the training program. The public gets information through the media; it is important that the information the public receives is accurate. If the information is incorrect or distorted, an emergency can easily cause panic. For this reason, it is important that the media people be somewhat knowledgeable about the potential hazards and the details of emergency responses.

Training for emergencies should be routinely conducted under the following circumstances:

1. When a new member is added to the group
2. When someone has a new responsibility within the community
3. When new equipment or materials are to be used in emergency response
4. When emergency procedures are revised
5. When a practice drill shows inadequacies in performance of duties
6. At least once annually

Any training program should address the following questions:

1. How are potential hazards recognized? (This can be done by periodic review of hazards and accident prevention measures.)

2. What are the necessary precautions to be taken when responding to an emergency (e.g., personal protective equipment)?
3. What are the evacuation routes?
4. To whom should a hazard be reported?
5. What actions should be taken in order to respond properly to special alarms or signals?

It is important for emergency procedures to be carried out as planned. This requires training on a regular basis so that people understand and remember how to react. The best plan on paper is likely to fail if the persons involved are reading it for the first time as the emergency is occurring. People must be trained before an emergency happens.

8.7 HAZARD RISK ASSESSMENT EVALUATION PROCESS[1,2]

Risk evaluation of accidents serves a dual purpose. It estimates the probability that an accident will occur and also assesses the severity of the consequences of an accident. Consequences may include damage to the surrounding environment, financial loss, or injury to life. This chapter is concerned primarily with the methods used to identify hazards and the causes and consequences of accidents. Issues dealing with health risks have been explored in the previous chapter. Risk assessment of accidents provides an effective way to help ensure either that a mishap does not occur or reduces the likelihood of an accident. The result of the risk assessment also allows concerned parties to take precautions to prevent an accident before it happens.

Regarding definitions, the first thing an individual needs to know is what exactly is an accident. An accident is an unexpected event that has undesirable consequences.[24] The causes of accidents have to be identified in order to help prevent accidents from occurring. Any situation or characteristic of a system, plant, or process that has the potential to cause damage to life, property, or the environment is considered a hazard. A hazard can also be defined as any characteristic that has the potential to cause an accident. The severity of a hazard plays a large part in the potential amount of damage a hazard can cause if an accident occurs.

Risk is the probability that human injury, damage to property, damage to the environment, or financial loss will occur. An acceptable risk is a risk whose probability is unlikely to occur during the lifetime of the plant or process. An acceptable risk can also be defined as an accident that has a high probability of occurring, but with negligible consequences. Risks can be ranked qualitatively in categories of high, medium, and low. Risk can also be ranked quantitatively as the annual number of fatalities per million affected individuals. This is normally denoted as a number times one millionth that is, for example, 3×10^{-6}; this representation indicates that on the average, 3 (workers) will die every year for every million individuals.

Another quantitative approach that has become popular in industry is the fatal accident rate (FAR) concept. This determines or estimates the number of fatalities over the lifetime of 1000 workers. The lifetime of a worker is defined as 10^5 hours, which is based on a 40-h work week for 50 years. A reasonable FAR for a chemical plant is 3.0, with 4.0 usually taken as a maximum. A FAR of 3.0 means that there are 3 deaths for every 1000 workers over a 50-year period. Interestingly, the FAR for an individual at home is approximately 3.0.

There are several steps in evaluating the risk of an accident (see Fig. 8.1). These are detailed below if the system in question is a chemical plant.

1. A brief description of the equipment and chemicals used in the plant is needed.

2. Any hazard in the system has to be identified. Hazards that may occur in a chemical plant include fire, toxic vapor release, slippage, corrosion, explosions, rupture of a pressurized vessel, and runaway reactions.

3. The event or series of events that will initiate an accident has to be identified. An event could be a failure to follow correct safety procedures, improperly repaired equipment, or failure of a safety mechanism.

4. The probability that the accident will occur has to be determined. For example, if a chemical plant has a given life, what is the probability that the temperature in a reactor will exceed the specified temperature range? The probability can be ranked from low to high. A low probability means that it is unlikely for the event to occur in the life of the plant. A medium probability suggests that there is a possibility that the event will occur. A high probability means that the event will probably occur during the life of the plant.

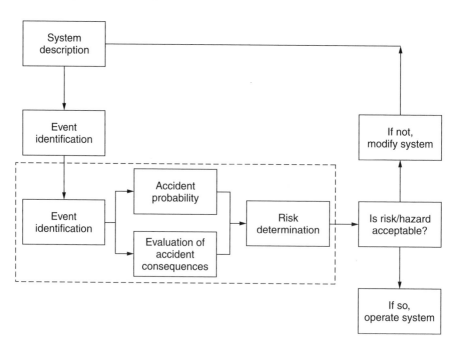

Figure 8.1   Hazard risk assessment flowchart.

5. The severity of the consequences of the accident must be determined. This will be described later in detail.

6. If the probability of the accident and the severity of its consequences are low, then the risk is usually deemed acceptable and the plant should be allowed to operate. If the probability of occurrence is too high or the damage to the surroundings is too great, then the risk is usually unacceptable and the system needs to be modified to minimize these effects.

The heart of the hazard risk assessment algorithm provided is enclosed in the dashed box (see Fig. 8.1). The overall algorithm allows for reevaluation of the process if the risk is deemed unacceptable (the process is repeated starting with either step 1 or 2).

Hazard/Events Identification

Hazard or event identification provides information on situations or chemicals and their releases that can potentially harm the environment, life, or property. Information that is required to identify hazards includes chemical identities, quantities and location of chemicals in question, chemical properties (such as boiling points), ignition temperatures, and toxicity to humans. There are several methods used to identify hazards. The methods that will be discussed are the process checklist and the hazard and operability study (HAZOP).

A process checklist evaluates equipment, materials, and safety procedures.[24] A checklist is composed of a series of questions prepared by an engineer or scientist who knows the procedure being evaluated. It compares what is in the actual plant to a set of safety and company standards.

Some questions that may be on a typical checklist are:

1. Was the equipment designed with a safety factor?
2. Does the spacing of the equipment allow for ease of maintenance?
3. Are there pressure relief valves on the equipment in question?
4. How toxic are the materials that are being used in the process, and is there adequate ventilation?
5. Will any of the materials cause corrosion to the pipe(s)/reactor(s)/system?
6. What precautions are necessary for flammable materials?
7. Is there an alternate exit in case of fire?
8. If there is a power failure, what fail-safe procedure(s) does the process contain?
9. What hazard is created if any piece of equipment malfunctions?
10. Who is first contacted in the event of an accident?

These questions and others are answered and analyzed. Changes are then made to reduce the risk of an accident. Process checklists are updated and audited at regular intervals.

A hazard and operability study is a systematic approach to recognizing and identifying possible hazards that may cause failure of a piece of equipment.[6,24] This method utilizes a team of diverse professional backgrounds to detect and minimize hazards in a plant. The process in question is divided into smaller processes (subprocesses). Guide words are used to relay the degree of deviation from the subprocesses' intended operation. The guide words can be found in Table 8.1. The causes and consequences of the deviation from the process are determined. If there are any recommendations for revision, they are recorded and a report is made.

A summary of the basic steps of a HAZOP study is[6,24]

1. Define objectives.
2. Define plant limits.
3. Appoint and train a team.
4. Obtain complete preparative work (i.e., flow diagrams, sequence of events, etc.).
5. Conduct examination meetings that select subprocesses, agree on intentions of subprocesses, state and record intentions, use guide words to find deviations from the intended purpose, determine the cause and consequences of deviation, and recommend revisions.
6. Issue meeting reports.
7. Follow-up on revisions.

There are other methods of hazard identification. A "what-if" analysis presents certain questions about a particular hazard and then tries to find the possible consequences of that hazard. The human error analysis identifies potential human errors that will lead to an accident. These two procedures can be used in conjunction with the two previously described methods.

Causes of Accidents

The primary causes of accidents are mechanical failure, operational failure (human error), unknown or miscellaneous, process upset, and design error. Figure 8.2 gives the relative number of accidents that have occurred in the petrochemical field.[25]

TABLE 8.1   Guide Words Used to Relay Degree of Deviation from Intended Subprocess Operation

Word	Subprocess Operation
No	No part of intended function is accomplished
Less	Quantitative decrease in intended activity
More	Quantitative increase in intended activity
Part of	The intention is achieved to a certain percent
As well as	The intention is accomplished along with side effects
Reverse	The opposite of the intention is achieved
Other than	A different activity replaces the intended activity

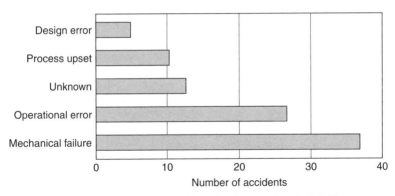

Figure 8.2 Causes of accidents in the petrochemical field.

There are three steps that normally lead to an accident:

1. Initiation
2. Propagation
3. Termination

The path that an accident takes through these three steps can be determined by means of a fault tree analysis. Generally, a fault tree may be viewed as a diagram that shows the path that a specific accident takes. The first thing needed to construct a fault tree is the definition of the initial event. The initial event is a hazard or action that will cause the process to deviate from normal operation. The next step is to define the existing conditions needed to be present in order for the accident to occur. The propagation event (e.g., the mechanical failure of equipment related to the accident) is discussed. Any other equipment or components that need to be studied must be defined. This includes safety equipment that will bring about the termination of the accident. Finally, the normal state of the system in question is determined. The termination of an accident is the event that brings the system back to its normal operation. An example of an accident would be the failure of a thermometer in a reactor. The temperature in the reactor could rise and a runaway reaction might take place. Stopping the flow to the reactor and/or cooling the contents of the reactor could terminate the accident.

The reader should view a fault tree as a technique of graphical notation used to analyze complex systems. Its objective is to spotlight conditions that cause a system to fail. Fault tree analysis also attempts to describe how and why an accident or any other undesirable event has occurred. It is also used to describe how and why a potential accident or other undesirable event could take place. Thus, fault tree analysis finds wide application in hazard analysis and risk assessment of process and plant systems.

Fault tree analysis seeks to relate the occurrence of an undesired event, the "top event," to one or more antecedent events, called "basic events." The top event may be, and usually is, related to the basic events via certain intermediate events. A fault tree diagram exhibits the causal chain linking the basic events to the intermediate events and the latter to the top event. In this chain, the logical connection between events is indicated by so-called *logic gates*. The principal logic gates are the AND gate, symbolized on the fault tree by AND, and the OR gate symbolized by OR.

As a simple example of a fault tree, consider a water pumping system consisting of two pumps, A and B, where A is the pump ordinarily operating and B is a standby unit that automatically takes over if A fails. Flow of water through the pump is regulated by a control valve in both cases. Suppose that the top event is no water flow, resulting from the following basic events: failure of a pump A *and* failure of pump B, *or* failure of the control valve. The fault tree diagram for this system is shown in Figure 8.3. Numerous illustrative examples are available in the literature.[6,7,10,20,24,25,26,28]

Consequences of Accidents

Consequences of accidents can be classified qualitatively by the degree of severity. A quantitative assessment is beyond the scope of the text; however, information is available in the literature.[7,26,27] Factors that help to determine the degree of severity are the concentration in which the hazard is released, length of time that a person within the environment is exposed to a hazard, and the toxicity of the hazard. The worst-case consequence or scenario is defined as a conservatively high estimate of the most severe accident identified.[24] On this basis one can rank the consequences

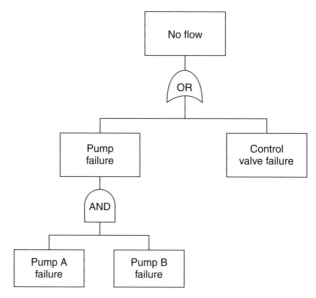

Figure 8.3   Fault tree diagram for a water pumping system consisting of two pumps (A and B).

of accidents into low, medium, and high degrees of severity. A low degree of severity means that the hazard is nearly negligible, and the injury to persons, property, or the environment is observed only after an extended period of time. The degree of severity is considered to be medium when the accident is serious, but not catastrophic, the toxicity of the chemical released is great, or the concentration of a less toxic chemical is large enough to cause injury or death to persons and damage to the environment unless immediate action is taken. There is a high degree of severity when the accident is catastrophic or the concentrations and toxicity of a hazard are large enough to cause injury or death to many persons, and there is long-term damage to the surrounding environment. Figure 8.4 provides a graphical qualitative representation of the severity of consequences.[28]

Event trees are diagrams that evaluate the consequences of a specific hazard. The safety measures and the procedures designed to deal with the event are presented. The consequences of each specific event that led to the accident are also presented. An event tree is drawn (sequence of events that led up to the accident). The accident is described. This allows the path of the accident to be traced. It shows what could be done along the way to prevent the accident. It also shows other possible outcomes that could have arisen had a single event in the sequence been changed. Thus, an event tree provides a diagrammatic representation of event sequences that begin with a so-called initiating event and terminate in one or more undesirable consequences. In contrast to a fault tree, which works backward from an undesirable consequence to possible causes, an event tree works forward from the initiating event to possible undesirable consequences. The initiating event may be equipment failure, human error, power failure, or some other event that has the potential for adversely affecting an ongoing process.

The following illustration of event tree analysis is based on one reported by Lees.[29] Consider a situation in which the probability of an external power outage

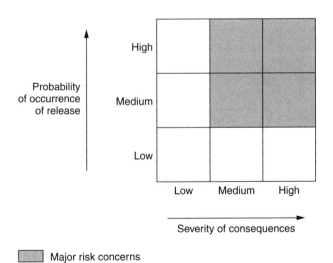

Figure 8.4 Qualitative probability–consequence analysis.

in any given year is 0.1. A backup generator is available, and the probability that it will fail to start on any given occasion is 0.02. If the generator starts, the probability that it will not supply sufficient power for the duration of the external power outage is 0.001. An emergency battery power supply is available; the probability that it will be inadequate is 0.01.

Figure 8.5 shows the event tree with the initiating event, external power outage, denoted by I. Labels for the other events on the event tree are also indicated. The event sequences I S $\bar{G}$ $\bar{B}$ and and I $\bar{S}$ $\bar{B}$ terminate in the failure of emergency power supply. Applying the applicable multiplication theorem,[6,7,25] one obtains

$$P(\text{I S } \bar{G} \bar{B}) = (0.1)(0.98)(0.001)(0.01) = 0.098 \times 10^{-5}$$

$$P(\text{I } \bar{S} \bar{B}) = (0.02)(0.01) = 2.0 \times 10^{-5}$$

Therefore the probability of emergency power supply failure in any given year is 2.098×10^{-5}, the sum of these two probabilities.

Additional calculational details are available in the literature.[6,7,10,20,24,25,26,28]

Cause–Consequence Analysis

Cause–consequence risk evaluation combines event tree and fault tree analysis to relate specific accident consequences to causes.[6] The process of cause–consequence

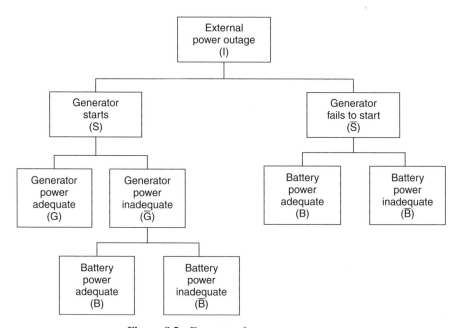

Figure 8.5 Event tree for a power outage.

evaluation usually proceeds as follows:

1. Select an event to be evaluated.
2. Describe the safety systems(s)/procedure(s)/factors(s) that interfere with the path of the accident.
3. Perform an event tree analysis to find the path(s) an accident may follow.
4. Perform a fault tree analysis to determine the safety function that failed.
5. Rank the results on a basis of severity of consequences.

As its name implies, cause–consequence analysis allows one to see how the possible causes of an accident and the possible consequences that result from that event interact with each other.

8.8 SUMMARY

1. Accidents are a fact of life, whether they are a careless mishap at home, an unavoidable collision on the freeway, or a miscalculation at a chemical plant.

2. The Superfund Amendments and Reauthorization Act (SARA) of 1986 renewed the national commitment to correcting problems arising from previous mismanagement of hazardous wastes.

3. An emergency plan can minimize loss by helping to ensure the proper response in an emergency. Emergency planning should grow out of a team process coordinated by a team leader. The team may be the best vehicle for gathering people with various kinds of expertise into the planning process, thus producing a more accurate and complete plan.

4. Successful emergency planning begins with a thorough understanding of the event or potential disaster being planned for. The impacts on public health and the environment must also be estimated.

5. In order to characterize the types and extents of potential disasters, a survey of hazards or foreseeable threats in the community must be performed and evaluated.

6. The education of the public regarding the real hazards in their community and how to respond to an emergency are critical to the public support of the emergency plan. Public opinion surveys show, however, that Americans are not yet prepared to deal either with the information provided by SARA Title III or with the reality of hazards in their community.

7. The hazard risk assessment evaluation process defines the equipment, hazards, and events leading to an accident. It determines the probability that an accident will occur. The severity and acceptability of the risk are also evaluated.

8. Notifying the public of an emergency is a task which must be accomplished with caution. People will react in different ways. Many will simply not know what to do, others will not take the warning seriously, and still others will panic.

Proper training in each community can help minimize any panic and condition the public to make the right response.

9. Once an emergency plan has been developed, its successful implementation can be ensured only through constant review and revision.

REFERENCES

1. M. K. Theodore and L. Theodore, *Major Environmental Issues Facing the 21st Century*, Theodore Tutorials, East Williston, NY, 1996.

2. Adapted from G. Burke, B Singh, and L. Theodore, *Handbook of Environmental Management and Technology*, 2nd ed., Wiley, Hoboken, NJ, 2000.

3. L. Theodore et al., *Accident and Emergency Management Student Manual*. U.S. EPA APTI, EPA, Washington, DC, contributing Author: J. O'Byrne, 1988.

4. J. Cornell, *The Great International Disaster Book*, 3rd ed., Charles Scribner's, New York, 1976.

5. *Catastrophe! When Man Loses Control*, prepared by the editors of the *Encyclopaedia Britannica*, Bantam Books, NY, 1979.

6. L. Theodore et al., *Accident and Emergency Management*, Wiley, Hoboken, NJ, 1989.

7. A. Flynn and L. Theodore, *Health, Safety, and Accident Management in the Chemical Process Industries*, Marcel Dekker, New York, 2002.

8. S. Adcock, "It's Official: Spark Led to Flight 800 Explosion," *Newsday*, August 24, 2000.

9. A. Kim, "What Is the Deal with the Year 2000 Bug? (Y2k—Demystified)," *Weeno.com*, 1999.

10. L. Theodore et al., *Accident and Emergency Management Student Manual*. U.S. EPA APTI, EPA, Washington, DC, contributing Author: L. Girardi Schoen, 1988.

11. "SARA: A First Year in Review," *Waste Age*, February 1988.

12. M. Krikorian, *Disaster and Emergency Planning*, Institute Press, Loganvile, GA, 1982.

13. W. Beranek et al., "Getting Involved in Community Right-to-Know," *C & E News*, **62**, (October 26, 1987).

14. "Other Statutory Authorities. Title III: Emergency Planning and Community Right-to-Know," *EPA J.*, **13**(1), 1987.

15. National Response Team. *Hazardous Materials Emergency Planning Guide*, March 1987.

16. R. H. Schulze, *Superfund Amendments and Reauthorization Act of 1986 (SARA Title III)*, Trinity Consultants Inc., May 1987.

17. J. T. O'Reilly, *Emergency Response to Chemical Accidents: Planning and Coordinating Solutions*, McGraw-Hill, New York, 1987.

18. P. Shrivastava, *Bhopal: Anatomy of a Crisis*, Ballinger, Cambridge, MA, 1987.

19. CMA, *Community Awareness & Emergency Response*, Program Handbook, Washington, DC, April 1985.

20. L. Theodore and Y. C. McGuinn, *Health, Safety and Accident Management: Industrial Applications*, A Theodore Tutorial, East Williston, NY, 1992.

21. E. J. Michael, O. W. Bell, and J. W. Wilson, *Emergency Planning Considerations for Specialty Chemical Plants*, Stone and Webster Engineering Corporation, Boston, MA, 1986.

22. U.S. EPA, *Title III Fact Sheet: Emergency Planning Community Right-to-Know*, 1987.

23. "CMA, Title III: The Right to Know. The Need to Plan," *Chemocology*, March 1987.

24. *AIChE, Guidelines for Hazard Evaluation Procedures*, prepared by Batelle Columbus Division for the Center for Chemical Process Safety of the American Institute of Chemical Engineers, New York, 1985.

25. L. Theodore and J. Reynolds, *Health, Safety and Accident Prevention, Industrial Applications*, Theodore Tutorial, East Williston, NY, 1996.

26. J. Crowl and J. Louvar, *Chemical Safety Fundamentals with Applications*, Prentice-Hall, Englewood Cliffs, NJ, 1990.

27. R. Ballandi, *Hazardous Waste Site Remediation, The Engineer's Perspective*, Van Nostrand Reinhold, New York, 1988.

28. Author Unknown, *Technical Guidance For Hazard Analysis*, EPA, FEMA, USDOT, Washington, DC, 1978.

29. P. Lees, *Loss Prevention in the Process Industries*, Vol. I, Butterworths, Boston, MA, 1980.

CHAPTER 9

ETHICAL CONSIDERATIONS[1]

9.1 INTRODUCTION

In 1854, President Franklin Pierce petitioned Chief Seattle—the leader of the Coastal Salish Indians of the Pacific Northwest—to sell his tribe's land to the United States. In his response to President Pierce and the white European immigrants' pursuit to own and "subdue" the earth, Chief Seattle penned thoughts as environmentally pensive and poignant as any uttered in the more than 150 years since: "Continue to contaminate your bed and you will one day lay (sic) in your own waste."[2]

His message fell on the deaf ears of the U.S. government and public. Cries for respect for the earth such as his remained few and far between for the next century. In the wake of events such as the Industrial Revolution, World Wars I and II, and the Cold War, a concern for the environment played little if any part in influencing either public policy or private endeavors.

More than 150 years later, however, Chief Seattle's words echo in every Superfund site, landfill, and oil spill. Public opinion has swung to the green side and a new ethic has evolved: an environmental ethic. As can readily be seen, however, the recent movement toward environmentalism has not created new moral codes. Instead, it has changed the emphasis and expanded the concept of the "common good" that lies at the heart of determining if an action is ethical.

The ethical behavior of engineers is more important today than at any time in the history of the profession. Engineers' ability to direct and control the technologies they master has never been stronger. In the wrong hands, the scientific advances and technologies of today's engineer could become the worst form of corruption,

Nanotechnology: Environmental Implications and Solutions, L. Theodore and R. G. Kunz.
ISBN 0-471-69976-4   Copyright © 2005 John Wiley & Sons, Inc.

manipulation, and exploitation. Engineers, however, *are* bound by a code of ethics that carry certain obligations associated with the profession. Some of these obligations include those that lead them to

1. Support one's professional society.
2. Guard privileged information.
3. Accept responsibility for one's actions.
4. Employ proper use of authority.
5. Maintain one's expertise in a state-of-the-art world.
6. Build and maintain public confidence.
7. Avoid improper gift exchange.
8. Practice conservation of resources and pollution prevention.
9. Avoid conflict of interest.
10. Apply equal opportunity employment.
11. Practice health, safety, and accident prevention.
12. Maintain honesty in dealing with employers and clients.

There are many codes of ethics that have appeared in the literature. The preamble for one of these codes is provided below.[3]

Engineers in general, in the pursuit of their profession, affect the quality of life for all people in our society. Therefore, an Engineer, in humility and with the need for Divine guidance, shall participate in none but honest enterprises. When needed, skill and knowledge shall be given without reservation for the public good. In the performance of duty and in fidelity to the profession, Engineers shall give utmost.

With regard to nanotechnology, a significant number of ethical concerns have arisen about the advisability of and the need for tampering with the traditional, or natural, order. These will continue as nanotechnology activities increase in the future. Numerous scenarios and situations will arise may disturb the conscience of the engineer and scientist. Several case studies, concerned with current, conventional technology and drawn from the work of Wilcox and Theodore,[4] are presented in the five sections that follow. There are two case studies in each section, and each is presented in the format provided by Wilcox and Theodore.[4]

9.2 AIR POLLUTION

Fact Pattern (1)

Captain Smith is a nuclear engineer for the Air Force. He is a graduate of the Air Force Academy and has always been a "model airman" by anyone's standards. Captain Smith has just been assigned his dream job, a top-level position at a former nuclear testing facility in the Southwest. It is his responsibility to measure ground and atmospheric radioactive and nanoparticle contamination levels for the outlying area (within approximately a hundred-mile radius).

The Nuclear Regulatory Commission (NRC) has strict limits on the levels of radio-active dust that can be present in the atmosphere; Captain Smith must routinely report his findings to them. Because of the end of the cold war as well as the enforce-ment of United Nations test ban treaties, the base has not performed any live nuclear tests for almost two decades. Captain Smith has never recorded contamination levels above the "acceptable" range. However, the base has been operational since the early 1950s, so out of curiosity Captain Smith decides to look through the old records. What he discovers is that, as recently as a decade ago, the base was releasing as much as three times the amount of radioactive dust permitted by the NRC into the atmosphere. He also finds that a large civilian community 150 miles outside the base has experienced a great number of health problems that are characteristic of exposure to radioactive fallout. Believing these findings must involve an oversight or a mistake, Captain Smith decides to bring the records he has found before his commanding officer.

"Some of these claims are over thirty years old, Captain!" General Brown blasts. "The NRC has given us a clean bill of health. Whose side are you on anyway? It is your duty only to check *current* contamination levels, not to go snooping through documents that were printed before you were born."

"I thought that this matter might have been kept from your attention, and I felt it was my duty as the head nuclear engineer to report it to you directly."

"I appreciate your concern, Captain Smith. But do you think that if those claims had been at all relevant, we would've just ignored them? Of course not. Now, forget you even saw the reports and return to your assigned duties."

"Yes, thank you for your time and consideration, sir," says Captain Smith reluctantly.

"Very well, Captain, Keep up the good work. Remember, what we do here is for the safety and well-being of all Americans. Dismissed."

Questions for Discussion

1. What are the facts in this case and how do they challenge Captain Smith's sense of ethics?
2. What is the environmental problem that Captain Smith is facing?
3. What could be the possible consequences if Captain Smith does not report what he has found?
4. What could be the possible consequences if Captain Smith reports his findings to the NRC or to civilian authorities?
5. Is General Brown truly interested in the welfare of the community, or just his own? If so, why his urgency to dismiss the reports?
6. What do you think Captain Smith's final action will be?

Fact Pattern (2)

Mary is an engineer in the air group in her company. Her job is to test for Volatile Organic Compound (VOC) emissions and odors from hazardous waste sites, waste-water pollution plants, and sewer treatment plants. On a routine trip to one of the waste-water treatment facilities, for the purpose of taking odor surveys around the plant and on the rooftop, Mary is asked by her boss, John, to photograph some parts of the plant and the surrounding area. The photos will be included in a presentation and a report to show all of the progress made within the last few years.

As Mary is taking pictures on the rooftop, she notices that the view is beautiful. She begins to photograph some high-rises recently built right outside the fenceline of the treatment plant.

In her excitement, Mary suddenly realizes that these buildings might have the potential to cause major problems for the plant. When the film is developed and Mary sees the pictures, she recognizes the potential problem: the stack VOC emissions. Six years ago, when the city agreed to a permit allowing the wastewater treatment plant a VOC emissions level from the stacks, it was based on the fact that the stacks were so high, there was no danger or harm to the people living or working in the area.

Many decides to test the stacks for VOC emissions and also to run an odor survey with the people living in the new buildings adjacent to the plant. She finds that the VOC emissions are still in compliance with the permit of six years ago; however, they would probably be too high and out of compliance based on today's regulations. Mary also discovers that the people living in the buildings have been complaining about sewage and fecal odors coming from the plant. Mary decides to act on these discoveries.

She says to John, "We have a problem with the stacks at the wastewater treatment plant I recently photographed. The stacks are now level with the apartments of some of the people who are living adjacent to the site."

John replies, "So, what's the problem?"

Mary continues, "The problem is that the VOC emissions would now unquestionably be out of compliance if the city were to run a test. Also, there are very bad odors coming from the stacks that the people have been complaining about."

"Hold it. Are the emissions still in compliance with the old regulations?" John asks.

"Well, yes, but . . . ," replies Mary.

"But *nothing*. Until and unless the city runs another test on that plant, we have nothing to worry about," retorts John.

"But what about the complaints from the neighbors?" asks Mary.

"We'll just use a masking agent in that area to cover up the odor, and that will do. There is no sense making a mountain out of a molehill. We are not going to start spending all this money for a problem that no one knows about. Just forget it and go back to work. I'll handle it," says John.

Questions for Discussion

1. What are the facts in this case?
2. What are the possible courses of action for Mary?
3. Mary tests the emissions and takes odor surveys to evaluate a potential problem. Should this really be her concern?
4. What are the risks if Mary remains silent and lets John "handle it"?
5. What can happen if Mary decides to act?
6. What final action do you think Mary will take?

9.3 WATER POLLUTION

Fact Pattern (1)

Julie is a second-year chemist in a relatively new but competitive nano-chemical company. She has been the assistant to senior chemist Mark, who is up for promotion this year. The two have just discovered a new chemical product, SweetSmell, which could help to reduce odor at wastewater treatment plants and landfills. Mark is ecstatic because this is the big break he has been waiting for. His company will benefit substantially from this new product, and he undoubtedly will get the promotion he has wanted for so long.

Julie too is very happy with their discovery. However, as a young, eager, and relatively new employees, she feels obliged to investigate their new product further. To her surprise, she finds the chemical could have an adverse impact on humans if it gets into their drinking water supply.

"Mark, I have bad news. We cannot introduce SweetSmell at the next meeting. I have discovered it can have harmful effects on humans."

"You do not know what you are talking about, Julie. SweetSmell is not harmful to anything or anyone."

"But Mark, I can prove it," proclaims Julie.

"Listen to me. This is my big break, and I will not have anyone ruin it. Stay quiet, or I will have you fired by tomorrow morning."

Questions for Discussions

1. What are the facts in this case?
2. What is the situation Julie is faced with?
3. Are Mark's action ethical?
4. What do you think Julie will do?
5. What *should* Julie do?

Fact Pattern (2)

John works at a wastewater treatment plant in Anytown, New York. As an entry-level engineer at the plant, he monitors the operating conditions of chlorine contact basins that are presently in operation. The different parameters that John is to monitor include the flow of wastewater into the chlorine contact basins, the coliform bacteria count of the wastewater entering and leaving the tanks, and the chlorine residual of the effluent flow. The purpose of this monitoring is to ensure that the coliform bacteria count is within an acceptable level, one that conforms to the Environmental Protection Agency's maximum contaminant level (MCL). Coliform bacteria are not harmful to humans, but this type of bacterium is an indicator of the presence of other types that *will* cause harm. Thus far, John has not recorded any values for coliform bacteria that have exceeded the acceptable level, but, as an enterprising engineer, John discovers something else of interest.

Taking all of the data for the concentration of coliform bacteria from the contact basins from the past six months, John plots concentration versus time. Upon doing this, he notices a slight upward trend in the data, one that would never be noticeable when looking at the concentration of bacteria from day to day. The plant still is meeting the maximum contaminant level for coliform bacteria, but if this upward trend were to continue, in a few months the MCL would be reached. As soon as John is sure that his analysis of the coliform counts for the past few months is correct, his initial thought is to take his results to William, his supervisor. The first opportunity that John has to mention his findings to William is while his boss is alone in his office.

"William, do you have a minute?" asks John.

"Only if it's about the plans for the flow model you are working on," replies William. "The big guys upstairs want to know what kind of progress we are making."

"It's coming along well, but I have to do a bit more research on it. I wanted to talk to you about something else, though; it's about the coliform levels in the contact basin effluent."

"What about them?" asks William. "Don't tell me we've got a reading that is not in compliance! I mean, we didn't have a plant failure just now, did we?"

"No, nothing like that. It's just that I've noticed something about the data. The coliform bacteria levels have been slowly increasing over the past few months."

"How bad?"

"Well, if the increase continues at this rate, we would be out of compliance in a couple of months."

At this point, William sits back in his chair and does not say anything. He thinks hard about what John has just said. John then asks, "What are we going to do?"

"What do you mean, 'What are we going to do?'" responds William hesitantly. "We aren't going to do anything."

"Why not? I mean, I understand that we are in compliance now, but in two months or so . . . "

"William, let me explain something to you about the way things work around here. The people upstairs—most specifically, my boss, Mr. Doe—have been hitting us with work nonstop for quite a while now. I can guarantee that if we show him the results you have compiled, he will ask me to assign someone to look into this problem, and he will still expect all of the other work to get done without increasing the number of personnel. We are not out of compliance, so we don't have to tell anyone anything."

"William, I can look into it and still handle my other work," replies John. "Let me look into it."

"I can't let you do that, I need you to finish up your part of this flow model. It's more important right now," William answers.

"What happens when we start to approach coliform levels out of compliance?" asks John.

"We let them worry about it upstairs, and we tell them that we could have done something about it if we weren't so overworked. It will be their fault, not ours."

Questions for Discussion

1. What are the facts in this case?
2. Is the decision by William *not* to address the problem an unethical one?
3. Is William justified in his decision?
4. Should John approach someone else about his findings? Should he approach someone of equal or greater authority than William?
5. Should John look into this potential problem further if it does not interfere with his regular work?

9.4 SOLID WASTE POLLUTION

Fact Pattern (1)

Bob has been hired recently by a cleaning company whose contract calls for the nightly housekeeping chores in a large office building for a major international corporation. Bob is assigned to dust, vacuum, and empty the garbage in each of the offices and the common areas of two floors of this building. After about a week and a half, Bob has a pretty good routine down: how far he can reach with the extension cord to vacuum each section of carpet, how long it takes to dust each office, and where he can leave the garbage for systematic collection. He notices the amount of garbage he hauls to the basement Dumpster is quite large.

"This place certainly produces quite a bit of garbage," he thinks to himself, "and I'm doing only two of the twelve floors here."

Bob gazes over the almost-filled Dumpster, taking notice of the various things people throw out after a day's work in the office. Bags filled with boxes, magazines, paper, cans, bottles, and books are all mixed together. Bob wonders at what point this stuff gets separated in order to follow the recycling rules for the city. To satisfy his curiosity, after Bob has finished for the night, he approaches his supervisor, Jon, about the matter.

"Separate the garbage? Recycling laws?" Jon exclaims. "We don't do that here. It's not that I don't want to, but it's too much of a hassle to go and break down cardboard boxes, sort the junk mail from the good paper, and everything else."

"Aren't you afraid of what may happen if we're caught *not* doing it?" Bob asks.

"Look, kid, first of all, laws are made to be broken, right? Besides, I know of quite a few other cleaning crews in the city that don't separate their trash either, and I'm sure there are a lot more that also don't follow the city's rules. I have heard of only a handful of cleaning crews getting fined by the city for not separating the garbage."

"It doesn't worry you that there is the possibility of getting fined?" Bob continues.

"That's the beauty of it. Even if they do catch you and fine you, the fine is insignificant compared to what it would cost to hire more people in order to follow the city's rules. So you'll pay a few thousand dollars in fines. It's a lot better than paying $20,000 a year plus benefits to a new employee."

Bob is astonished at his supervisor's attitude. He can't imagine that there can be many other supervisors with Jon's attitude in this city. How could anyone *not* recycle after seeing the commercial of the crying Native American on television? Bob finds himself so caught up in the matter, he misses his stop on the train and has to walk five blocks to get home. During his walk, various ideas pop into his head about what can be done to enforce the city's recycling laws better: He could bring the matter to the attention of a police officer perhaps, write a letter to the mayor, send an E-mail to his congressman, or maybe organize a group to help enforce the rules. Then it dawns on Bob the amount of time and energy it will take on his part to do any of these things. What could one person possibly do? Bob begins to think, is it really worth it for him to worry about such things, even though he has learned all his life that recycling is something that *should* be done?

Questions for Discussion

1. What are the facts in this case?
2. What is the issue in question?
3. Is Jon the supervisor correct in his attitude toward the law?
4. Is Bob correct in his feelings?
5. What do you think Jon may finally do?
6. What would you do if you were in Bob's place?

Fact Pattern (2)

John and David have been best friends since college, where they both studied environmental engineering. After graduation, John starts his own, relatively small R&D firm, FutureTech. David proceeded on to study environmental law. Today, they both work in the New York area and remain close.

Thursday, David is at work, and his boss calls him into his office. David receives a new case that involves cracking down on a company illegally dumping solid waste into nearby waters. To David's dismay, he discovers the company is FutureTech. Despite his mixed feelings about the case, he knows it is his job to continue with the case.

The following week, David goes to FutureTech and meets with John, telling him about the case and how he is in charge of it.

"I cannot believe this," proclaims John. "After all our years of friendship, you are going to sue my company?"

"This has nothing to do with our friendship," responds David. "Your company is involved with illegal dumping, and it is my job to see to it that it stops."

"I could lose everything. I promise it will stop. Tell your boss that it actually is not my company involved in these illegal actions. I swear to you, if you go through with this, I will no longer consider you my friend."

"I cannot believe you are putting me in this situation," exclaims David. "I am only doing my job."

"Well it's your job *or* our friendship," answers John.

Questions for Discussion

1. What are the facts in this case?
2. What are the complicating circumstances in this case?

3. Are John's actions ethical?
4. What do you think David will do?
5. What should David do?

9.5   HEALTH CONCERNS

Fact Pattern (1)

Bob is an engineer managing one of the most efficient production lines in a chemical plant. Recently, Bob's supervisor, Mr. Jay, complimented Bob on the great work he has been doing. Mr. Jay even assigned Bob the task of upgrading the equipment on the production line. More than happy to take on the assignment, Bob begins to put a lot of time and effort into his new project, researching the new equipment and technology available and preparing a budget. He is working hard because a promotion might be his reward if all goes well.

One day, Bob happens to read a report on a particular nano-chemical that has just been determined to be carcinogenic. The report states that this chemical could be hazardous to humans if they are exposed to it in concentrated amounts or for extended periods of time. Bob knows that the chemical is a minor by-product on this production line he is upgrading. He is a bit concerned about the health risk, but he knows that the chemical is produced in small, low-concentration quantities. He also knows that this chemical is then used in another production line and converted to a harmless substance.

Although there is a health risk, Bob considers it insignificant. In fact, he thinks it's not even worth mentioning. If anyone has a concern, he would gladly assure them that the plant is safe. It meets all regulations pertinent to the chemical. Besides, it would cost money and time to redesign the production line, which is currently profitable. After all of the hard work Bob has put into this project, it would be a shame to shut down the line for such a negligible risk.

So, Bob does not mention the chemical report to anyone and goes ahead with the upgrading. He successfully executes the project, which results in an increase in production. Impressed by Bob's hard work and dedication, Mr. Jay rewards him with a promotion.

Terry, Bob's coworker and friends, goes to congratulate Bob on his promotion. While talking to him, she mentions the chemical report.

Terry says, "I've been meaning to talk to you about this report. Doesn't your production line make this chemical as a by-product? Aren't you concerned about the health risk?"

Bob replies, "It's only a minor by-product. The concentrations are too low to provide any serious risk. Don't worry. And please don't mention the report to anyone else. I do not want people to get excited over nothing."

Questions for Discussion

1. What are the facts in this case?
2. Do you think there is a considerable health threat involved in this production process?
3. Should Bob have kept the report a secret?

4. Should Bob have consulted anyone before making his decision?
5. How much should the issues of money and effort have weighed in Bob's decision?
6. Does Bob deserve the promotion?

Fact Pattern (2)

Eddy is an engineer working for a small environmental consulting firm. His firm has been hired by a local swimming resort to analyze the effluent water from old pipes in order to determine the amount of organics present in the water. The resort owner hopes to eliminate the organics by treating them at optimal points along the water's path. Eddy's task is to find the various concentrations of organics along the pipelines.

Knowing lead to be a very toxic metal and corrosion of old lead piping to be a primary source, Eddy also performs tests to analyze the lead concentration. The results of the tests show that the lead in the pipe is more than double the proposed EPA concentration limit. From experience, Eddy knows that the only way to eliminate the lead is to replace the pipe.

Before writing up his report, Eddy informs the owner of the resort of his findings: "I analyzed the lead content in your pipes and found that the levels are higher than the EPA regulations."

"I didn't hire you for that. I know that there's a lot of lead in those pipes, but if you think I'm replacing them, you're crazy. Do you know how much that would cost me?" replies the owner.

Eddy realizes he shouldn't have taken it upon himself to analyze the resort's pipes for lead, but he did, and now he doesn't know what to do. Lead is toxic. It has been proven to cause all kinds of biological damage. In any event, the owner isn't going to get rid of the problem, and Eddy wouldn't have known about the problem if he'd done only what he was hired to do.

Questions for Discussion

1. What are the facts in this case?
2. Should Eddy have analyzed the pipes for lead?
3. Is the owner's response to Eddy's findings logical? ethical?
4. Based on his new knowledge, is Eddy responsible to the people who use the water at the resort?
5. What responsibility does Eddy have to his employer?
6. What responsibility does Eddy have to the reputation of his firm?

9.6 HAZARD CONCERNS

Fact Pattern (1)

Bill recently has accepted a job at an environmental chemistry lab where various kinds of toxic and nontoxic chemicals, gases, solutions, and nanochemicals/nanoparticles are used. The lab is located in a closed space with limited ventilation.

After Bill starts his new job, one of his coworkers, John, has to deal with an acid spillage. His colleagues are unable to locate specific neutralizers and ventilation equipment. This incident does not cause any physical harm but creates much commotion and leads to debates and conflicts among the lab crew. Many of the department workers then organize an unofficial OSHA training session without the permission of the lab authority, but not all workers are involved in this training.

Bill, who had a little health, safety, and accident prevention instruction in his primary training, takes matters into his own hands, sorting out the accident details and planning a meeting with the lab manager. He wants to propose mandatory OSHA training and the procurement of equipment for safety.

The next day Bill talks to his friend Henry, who says, "Bill, you are taking things way too seriously. I don't think you should worry about this. Also, as a new employee, you don't even know the managers. I suggest you should close the case right here and not utter another word about it if you want to keep your job." Bill tries to present reasons for his intervention in the name of safety, but Henry denies it's a good idea.

Bill finds himself in a predicament. He has two options: One, go to the meeting and discuss the proposal; or, two, keep quiet. Both alternatives could lead to various parties being hurt physically or financially. Any lack of action could mean his coworkers and he himself will remain in jeopardy. But acting on the safety issues may put his job at risk. It can also bring a lot of trouble to the company if the governmental authorities inspect the lab. Why is it only *his* responsibility to notify them of the problems? he wonders. Should he go with the proposal or another alternative? He certainly does not have as much support on his proposal as he would like, so he frets about his next move.

Questions for Discussion

1. What are the facts in this case?
2. How can Henry's advice affect Bill's decision?
3. Should the lab just be equipped properly, or should OSHA training also be required?
4. Is Henry's point of view an ethical view?
5. What should Bill do? Why?

Fact Pattern (2)

After several men have been injured in a cogeneration facility, the plant manager decides to establish a safety committee, which consists of the operations manager and maintenance manager as well as three technicians. Their ultimate goal is to find and eliminate all the safety hazards that exist in this facility. As an incentive, the plant manager decides to reward this committee monetarily whenever a given year passes with the number of injuries being lower than the previous year's.

In the course of several years, the committee has eliminated many of the safety hazards in the facility; concomitantly, the cost of workers' compensation has decreased considerable, and the number of injured people per year has reached an all-time low. Both the plant manager and coworkers are very happy with the results generated by this committee.

Although the committee has been solely responsible for the increase in safety awareness at the facility, the plant manager receives all the recognition and praise without acknowledging the committee; he also has not provided any of the promised monetary incentive to them.

The safety committee has become very angry and resentful. "Let's forget about using the allotted safety budget for safety; let's take some of the money for ourselves. We deserve it. And anyway, we were promised some form of monetary incentive!" shouts one of the members.

"Wait! We can't do that. Our coworkers depend on us to use those funds to help eliminate the safety hazards that still exist and to provide safety awareness courses," explains the operations manager.

Questions for Discussion

1. What are the facts in this case?
2. Do you think that the operations manager's comment will affect anyone on the committee?
3. What dilemma is the committee facing?
4. What are the two things that the plant manager failed to give the safety committee?
5. How do you think most of the members in the committee feel about what is happening?
6. What do you think this committee should do?

9.7   SUMMARY

1. In 1854, Chief Seattle penned warnings as environmentally pensive and poignant as any uttered in the 150 years since: "Continue to contaminate your bed and you will one day lay in your own waste."

2. In the traditional ethical theories, established hierarchies of duties, rights, virtues, and desired consequences exist so that situations where no single course of action satisfies all of the maxims can still be resolved.

3. "Engineers in general, in the pursuit of their profession, affect the quality of life for all people in our society. Therefore, an Engineer ... shall participate in none but honest enterprises ..."

4. At present, the concept of environmentalism is *widely* held; its future is becoming *deeply* held.

5. The case study approach is ideally suited to teaching ethics.

REFERENCES

1. Adapted from M. K. Theodore and L. Theodore, *Major Environmental Issues Facing the 21st Century*, Theodore Tutorials, East Williston, NY, 1996.
2. J. Faney, and R. Armstrong, eds., *A Peace Reader: Essential Reading on War, Justice, Non-Violence & World Order*, Paulist Press, Mahwah, NJ, 1987.
3. M. W. Martin, and R. Schinzinger, *Ethics in Engineering*, McGraw-Hill, New York, 1989.
4. J. Wilcox, and L. Theodore, *Engineering and Environmental Ethics*, Wiley, Hoboken, NJ, 1998.

CHAPTER 10

FUTURE TRENDS

10.1 INTRODUCTION

In some respects, this was probably the most difficult chapter to write. Attempting to predict future trends for a technology that in a very real sense does not currently exist leaves much to be desired. It is a topic with which many engineers and scientists are not comfortable. Nonetheless, predictions can be made that are based to some extent on judgment, experience, and common sense.

There is a certain amount of jousting in Congress over risk and cost–benefit analysis. Under this promised legislation EPA would be required to conduct risk assessments and cost–benefit analyses for environmental regulations, as well as an assessment of environmental risks as compared to one's daily risks. Risk assessment is an evolving area of science, one that will shape future decision making. Comparative risk analysis provides a way to consider and rank many problems at the same time. It is a means for comparing dissimilar environmental risks in a manner that blends public opinion with scientific data and professional judgment. This helps to provide a comprehensive economic analysis analogous to the risk analysis—sort of a bridge between risk assessment and risk management. Comparative risk analysis can lead to short-term economic gains at the expense of leaving long-term environmental problems for future generations.

What follows are discussions devoted to topics that should be of interest to those working in the nanotechnology field who are concerned with environmental impacts. The first six sections are drawn from and based on future trends as they relate to Chapters 3 to 8. The chapter concludes with two topics that have recently

Nanotechnology: Environmental Implications and Solutions, L. Theodore and R. G. Kunz.
ISBN 0-471-69976-4 Copyright © 2005 John Wiley & Sons, Inc.

surfaced in the environmental management field—environmental audits and ISO 14000—plus concluding remarks.

10.2 AIR ISSUES

The authors of this work believe that the current labyrinth of environmental statutes and administrative programs is desperately in need of reform. As many acts undergo close scrutinization from Congress, the public and private sectors are making a bid to see that their interests are protected.

Today, the EPA struggles with the difficult task of implementing the 1990 Clean Air Act Amendments (CAAA). The agency is also under the gun to produce hordes of new regulations. These regulations are complex and comprehensive in nature. The states are absorbing even a larger share of the burden of implementing the CAAAs. A greater emphasis will be placed on a multimedia approach to regulations. The multimedia regulations, which have been attempted recently in the paper and pulp industry, will regulate both air emissions and wastewater discharges. EPA is also attempting a multimedia approach to inspections and enforcement actions.

The basic design of air pollution control equipment has remained relatively unchanged since first used in the early part of the twentieth century. Some modest equipment changes and new types of devices have appeared in the last 20+ years, but all have employed essentially the same capture mechanisms used in the past. One area that has recently received some attention is hybrid systems—equipment that can in some cases operate at higher efficiency more economically than conventional devices. Tighter regulations with a greater concern for environmental control by society has placed increased emphasis on the development and application of these systems. The future will unquestionably see more activity in this area because of nanotechnology.

Recent advances in this field have been involved primarily in the treatment of metals. A dry scrubber followed by a wet scrubber has been employed in the United States to improve the collection of fine particulate metals in hazardous waste incinerators; the dry scrubber captures metals that condense at the operating temperature of the unit, and the wet scrubber captures uncollected metals (particularly mercury) and dioxin/furan compounds. Another recent application in Europe involves the injection of powdered activated carbon into a flue gas stream from a hazardous waste incinerator at a location between the spray dryer (the dry scrubber) and the baghouse (or an electrostatic precipitator). The carbon mixing with the lime particulates from the dry scrubbing system and the gas stream itself adsorbs the mercury vapors and residual dioxin/furan compounds; these are separated from the gas stream by the particulate control device. More widespread use of these types of systems is anticipated in the future.

In recent years, the EPA has increased efforts to address indoor air quality (IAQ) problems through a building systems approach. The EPA hopes to bolster awareness of the importance of prevention, and encourage a whole systems perspective to resolve indoor air problems. The EPA Office of Research and Development is also conducting a multidisciplinary IAQ research program that encompasses studies of the health effects associated with indoor air pollution exposure, assessments of indoor air pollution sources and control approaches, building studies and investigation methods, risk assessments of indoor air pollutants, and a recently initiated program on biocontaminants.

Federal research on air quality issues is driven in part by the increasing attention that IAQ has attracted from journalists as well as from scientists and engineers. EPA has performed comparative studies that have consistently ranked indoor air pollution among the top five environmental risks to public health. In analyzing over 500 IAQ investigations conducted through the end of 1988, the National Institute for Occupational Safety and Health (NIOSH) categorized its findings into seven broad sources of poor indoor air quality: inadequate ventilation (53 percent), inside contamination (15 percent), outside contamination (13 percent), microbiological contamination (5 percent), building materials contamination (4 percent), and unknown sources (13 percent).[1] Since then, ventilation has been the primary focus of most EPA programs.

Requirements for clean air are still changing rapidly, and most buildings will need to be refitted with different filters to meet with these new standards and guidelines. EPA's research will continue in these and other areas to try to ensure comfortable and clean air conditions for the indoor environment.

The Air Pollution Control Technology (APCT) Verification Center (APCTVC) is part of the EPA's Environmental Technology Verification (ETV) Program and is operated as a partnership between Research Triangle Institute (RTI) and EPA. The center verifies the environmental performance of commercial-ready air pollution control technologies. Verification provides potential purchasers and permitters with an independent and credible assessment of what they are buying and permitting. Verification tests use approved protocols, and verified performance is reported in verification statements signed by EPA. RTI contracts with Midwest Research Institute, ETS, Inc., and Southwest Research Institute to perform verification tests. This program will expand in the future.

The University of Delaware is involved with a project to develop a new technology that will determine the chemical composition of airborne particles down to about 5 nm in diameter. Two problems that have surfaced are the inefficient sampling of particles and the inefficient analysis of those particles that have been sampled. In addition, the researchers have discovered that individual nanoparticles can be efficiently detected only if quantitatively converted to atomic ions.

10.3 WATER ISSUES

Advanced wastewater treatment, known as tertiary treatment, is designed to remove those constituents that may not be adequately removed by secondary treatment. This includes removal of nitrogen, phosphorus, and heavy metals.

Biological nutrient removal has received considerable attention in recent years in regard to the inorganic constituents in the wastewater. Excessive nutrients of nitrogen and phosphorus discharged to the receiving water can lead to eutrophication, causing excessive growth of aquatic plants, and indirectly depleting oxygen sources from the aquatic life and fish. There are also other beneficial reasons for biological nutrient removal that include monetary savings through reduced aeration capacity and reduced expense of chemical treatment. This area will see more activity in the future.

Part of the tertiary treatment in a municipal treatment plant is disinfection. Currently there are controversial issues as to what type of disinfection techniques should be employed. Historically, chlorine was the choice of many facilities. However, it has been found that the disinfection by-products (of chlorides and bromides) that are being formed are toxic to the water environment. New techniques that will receive attention are ozonation and ultraviolet processes. Details on effluent management, sludge management, and recent developments in industrial wastewater treatment follow.

The development of advanced treatment technologies, along with an increasing scarcity of fresh water, has led to marked changes in effluent management. Numerous strategies for purified wastewater reuse are presently being employed in ways appropriate to the particular industrial operation. The combination of scarcity of water with increasingly stringent regulations has made effluent disposal into natural receiving bodies the option of last resort.

While it is theoretically possible to reuse ("close up") all the water of many industrial processes, inevitable limits will be encountered as a result of quality control. In a paper mill, for example, a closed system will result in increasing concentrations of dissolved organic solids. Such a buildup will lead to an increase in the cost of slime control, greater downtime on the paper machines, and possible discoloration of the paper stock. The goal is to maximize reuse while maintaining the integrity of the production process.

In the design of water reuse strategies, the particular water requirements of the process must be addressed. For example, water use by hydropulpers in a paper mill does not have to be treated for suspended solids removal. However, if these solids are not removed from shower water used on the paper machines, clogging of the shower nozzles will result.

In the production of produce, such as tomatoes, protection against microbial contamination through chlorination is of greater importance than overall purification.

In the case where two or more waste streams are generated, it is often necessary that they be separated. This is all the more important when mixing can result in health hazards. In a plating plant where both cyanide streams and acidic metal rinses are generated, a lack of separation can lead to the production of toxic hydrogen cyanide.

It is often the case that only a portion of the waste flow contains the majority of the suspended solids loading. Treatment for solids removal would then only be necessary on this part of the waste stream. Such is the case in a tannery where nearly 60 percent of the flow from the beam house yields 90 percent of the suspended solids.

Where little, if any, contamination of waste streams has occurred, such as in cooling water, the segregation of these effluents can allow for direct reuse or discharge into a receiving water.

In most industrial wastewater treatment plants, containment and control of pollutional discharges from stormwater is deemed essential. Present strategies include adequate diking around process sewers. The design of holding basins for temporary storage of contaminated stormwater is based on storm frequency for the particular region. Once collected, the water is passed through the treatment plant at a controlled rate.

That portion of wastewater that has not been reused must be disposed of in the environment, where it reenters the hydrologic cycle. Thus, disposal may be considered a first step in the long-term, global process of reuse.

The most common strategy for disposal or treated wastewater effluent is release into ambient waters. Where streams or rivers of sufficient volume or flow are not available, direct discharge of treated wastewater into lakes or reservoirs may be necessary. Oceans and large lakes (such as the Great Lakes) are used by many communities for wastewater disposal because of their tremendous assimilative capacity. Discharge in these conditions is accomplished via pipes or tunnels laid on, or buried in, the floor of the water body.

In very dry regions, land application is often utilized. Where this type of treatment is successful, the effluent will seep into the ground and replenish the underground aquifers. At the same time, a portion of the wastewater will return to the hydrologic cycle through evaporation.

Of principal importance in wastewater disposal is the consideration of environmental impact. Environmental regulations, criteria, policies, and reviews protect ambient waters from negative impacts of treated wastewater discharges. One must bear this in mind when selecting discharge locations and outfall structures, as well as the level of treatment required. Both treatment and disposal of an industrial effluent must be considered when designing a wastewater system for the facility.[2]

The utilization and disposal of sludge is jointly addressed under the Clean Water Act and the Resource Conservation and Recovery Act. Both of these federal laws emphasize the need to employ environmentally sound sludge management techniques and to use sludge beneficially whenever possible. At the same time, the national requirements for improved wastewater treatment will result in the production of a greater quantity of residuals, and possibly more concentrated forms of contaminants will be present in these residuals than ever before. The permits required for effluent discharge from sewage treatment plants will be affected by, and will contain, provisions related to the sludge management techniques employed by the facilities.

Prior to utilization or disposal, sludge may be stabilized to control odors and reduce the number of disease-causing organisms. The sludge may also be dewatered to reduce the volume to be transported or to prepare it for final processing. The liquid sludge, which contains 90 to 98 percent water, can be partially dewatered by a number of processes. Vacuum filtration, pressure filtration, and centrifugation are three of the more common dewatering technologies currently being employed.

Digestion of sludge is accomplished in heated tanks, where the material can decompose aerobically and the odors can be controlled. Anaerobic sludge digestion has the added benefit of producing methane gas, which may be used by the same treatment plant as an energy source.

Until recently, all but 20 percent of generated sludge had been incinerated, landfilled, or dumped into the ocean. Currently, much more attention is being given to sludge utilization by application to land as a soil conditioner or fertilizer, and by combustion in facilities to recover energy.

Liquid digested sludge has been used successfully as a fertilizer and for restoring areas disrupted by strip mining. Under this sludge management approach, digested sludge in semiliquid form is transported to the spoiled areas. The slurry, containing nutrients from the wastes, is spread over the land to give nature a hand in returning grass, trees, and flowers to barren land. Restoration of the countryside will also help control the flow of acids that drain from mines into streams and rivers, endangering fish and other aquatic life and adding to the difficulty of reusing the water.

With the tremendous increase in recent years in the overall generation of industrial pollutants, along with the escalating costs of pollution control, increasing efforts are being made with regard to pollution prevention. The policies of the EPA now give the highest priority to source reduction, followed by recycling, with pollution control as the final alternative.

The National Pretreatment Program is intended to eliminate the discharge of pollutants into the nation's waters. The major objective of the program is to protect publicly owned treatment works (POTWs) and the environment from the hazards of various toxic and other dangerous pollutants. Water-quality-based toxic limits and monitoring requirements are becoming a common provision in the National Pollution Discharge Elimination System (NPDES) permits of POTWs. Pretreatment of toxic pollutants is necessary in order to comply with these water-quality-based toxics requirements.[3]

Industry is responding in new, innovative ways to these mandates. The Union Carbide plant at South Charleston, West Virginia, is an example of what is happening in the organic chemicals industry. Through various in-plant changes, the average waste flow of 11.1 million gallons per day (MGD) and waste load of 55,700 lb of BOD per day was reduced to a flow of 8.3 MGD and load of 37,000 lb of BOD per day. These changes were accomplished principally by replacement of scrubbers, segregation, collection and incineration of wastewater streams, and additional processing of collected tail streams.

The largest water reclamation effort in California was initiated in the early 1990s by the Metropolitan Water District of Southern California and the West Basin Municipal Water District. By the end of the decade, the project is expected to be reclaiming nearly 70,000 acres. District officials are planning to use the water for irrigation of schools and golf courses. In addition, the local Chevron and Mobil Refineries are to receive water for use in cooling towers.[4]

10.4 SOLID WASTE ISSUES

There are several challenges facing society, including methods for improving the handling of solid wastes. The first is to change the consumption habits that have been established over many years as a result of advertising pressure that has stressed increased consumption. Next, efforts must be made to reduce the quantity of materials used in the packaging of goods, and begin the process of recycling at the source. In this way, fewer materials will become part of the disposable solid wastes of a community. Thirdly, landfills must be made safer. Every effort should be made to reduce the toxicity of wastes ending up there. Also, the design of landfills must be improved to provide the safest possible locations for the long-term storage of waste materials. Finally, the development of new technologies that not only conserve natural resources but are also cost effective is necessary for the future.

There are several trends that will affect solid waste management in the future. First, the cost of solid waste management will almost certainly continue to rise as a result of increased waste generation coupled with more restrictive regulations. Next, solid waste management will be more balanced between source reduction, recycling, energy recovery, and land disposal—as opposed to today where landfills are the most predominant waste management unit. Finally, education and training will be expanded in the future to provide competent people who will be better prepared to manage solid wastes. Additional details follow.

Often it is not necessary for the producer of the waste to reprocess the material internally. Many companies, through the ingenious application of sound engineering practices, have been able to sell waste products, particularly solid waste products, as raw materials to other processors. Numerous small companies have geared their business toward the sale of reprocessed waste materials. In the evaluation of any waste disposal program, the possible reuse or sale of the waste or its components is certain to receive more attention in the future. Successful strategies will take into account both short-term waste disposal costs and long-term site treatment liabilities. The total cost of waste management consists of the disposal costs, which include taxes and fees, transportation costs, administrative costs, and present-value for liability costs. Disposal and transportation costs are usually the only ones considered. The time spent by personnel in handling the wastes onsite, in offsite approvals, in conducting analytical testing, and in filling out any state, federal, or industry association reports is costly. Long-term costs can include those to clean up misused sites. This obligation stems from the presence of substances identified as hazardous under the Comprehensive Environmental Response, Compensation, and Liability Act (CERCLA, which established Super fund), the basis for most site remediation actions.[5]

An important factor for waste management planning in the future is the rise in the costs of disposal. Costs vary depending on the type of the waste, the quantity (bulk), its containment, and its method of disposal. Landfills close to the location of the waste generation are rapidly shrinking in capacity. The trend toward incineration, led extensively by land disposal restrictions, may also increase incineration costs in the near future.

As older sites close, the newer containment sites will refuse to take wastes or at least seriously cut back levels.[6] The ideal option will be the development of successful waste minimization programs so that there is no longer the need for disposal. Unfortunately, this attractive situation will take some time. Another option is having the companies take on the responsibility of waste treatment themselves, but not every company wants, or is able, to treat its own wastes onsite.

For some companies, hiring a contractor to do the treatment is a better option than building their own treatment plant. The amount of waste generated may not even justify the amount spent for the facility. Naturally, there are additional costs for running the plant and for the personnel to run the plant. Also, if the amount the products being produced changes from time to time, then the amount of waste treatment/management would change accordingly. There could be a large disparity between the quantities generated, and the plant may not be able to comply with that range, thereby making contracting a preferred option.

Waste management issues must be combined with realistic company factors to develop practical, attainable strategies in the future. Capital planning will influence the decision of whether to invest in new production that can use existing waste-handling equipment or to add new equipment that may not be justified. The business plan of the company is essential to estimate the future generation of waste.

The solution to industrial waste problems normally do not present themselves directly, but rather, some ingenuity and practicality must be employed so that the management of these waste can be carried out safely and efficiently. The methods, strategies, and equipment to be employed in the future will vary according to the situation and the type of waste. Regarding industry and manufacturers, the major factors that determine the way wastes are dealt with will continue to be cost and necessity.

There is significant public opposition to the siting of any type of municipal solid waste management facility. In the future, the public needs to be educated and informed so that these facilities can be properly located. Most of these facilities are found in commercial and/or industrial zones, and away from residential areas.

The effects of these facilities on health and safety have not been measured at this time. However, even with proper management, wastes containing contaminated materials and dangerous chemicals are potential hazards to millions of people. The health of an entire community can be jeopardized if these wastes are temporarily but inadequately and/or improperly managed. The whole health risk assessment area needs to be addressed in the future.

The future is also certain to bring a reduced dependence on landfilling of municipal solid waste. Source reduction and recycle/reuse options will be emphasized. And, although incineration came under pressure earlier with the Clinton Administration, it too may very well gain favor if the authorities and the public are educated as to the inherent advantages of this solid waste management option.[7]

The Superfund program also came under fire with the Clinton Administration. Critics felt that Superfund has not been efficient in cleaning up the nation's hazardous waste sites. A large portion of the money in Superfund goes to "transaction costs," which are lawyer and consultant fees. Since enforcement is a very large part of the Superfund program, much money gets spent on legal considerations instead of on actual site cleanup. The future of Superfund lies in improving the existing system.

Some argue that Superfund cannot be abolished because it is still needed. Many hazardous sites that pose threats to public health and the environment still exist. Here are two ideas that may improve Superfund. The first would cause remedies for cleaning up toxic sites to be based partly on "probable future use," and the second would reduce wasteful transaction costs.[8] The idea of probable future use means that a hazardous waste site will be cleaned up to a certain level depending on what the site will be used for after the cleanup. For example, sites that will be used as parking lots will be cleaned to different specifications than will sites that are to become hospitals. The second suggestion for reducing wasteful transaction costs involves engaging a "neutral professional," such as a judge, to be an arbitrator between the potentially responsible party (PRP) and government. This idea would reduce the legal costs incurred by Superfund because it expedites a decision in payment responsibility. These two changes in the future of Superfund may improve the program. However, based on past history, the EPA continues to move toward a legally based rather than a technology-driven agency.

Regarding nuclear waste, current regulations call for each individual state or interstate compact to store and dispose of all of the low-level waste (LLW) generated within its boundaries. An interstate compact consists of a group of states that have joined together to dispose of LLW. Interstate compacts can only be formed and dissolved by Congress. (Nuclear waste management acronyms, such as LLW and so forth, have been defined previously in Section 5.5.)

Many regulatory milestones have passed, leaving most states with restricted access to Barnwell, South Carolina, the only remaining LLW disposal facility for such wastes. It is most certain that the Barnwell facility will close before most states have centralized storage capacity online. Some states, like New York, have unsuccessfully attempted to sue the federal government, arguing that it is unconstitutional to mandate that states dispose of radioactive waste within their boundaries. Unless the individual states or compacts take immediate action to site and construct a permanent disposal facility or temporary centralized storage facility, generators of waste will be forced either to store radioactive materials onsite or to stop generating radioactive wastes by ceasing all operations that utilize radioactive materials. While these two options may seem appropriate, neither of them will solve the problem of waste disposal for any extended period of time. Many radioactive waste generators, like hospitals, do not have the storage space allocated to handle on site storage for periods exceeding 1 or 2 years. Much of the waste generated at hospitals is directly related to patient care, and it is unacceptable to assume that all processes, like chemotherapy, that produce radioactive waste will be stopped.

Both the HLW and TRU programs are limping along because of public concern for the areas surrounding the proposed facilities. The WIPP facility has performed some waste emplacement, but this has been accomplished only as a research and development activity. The HLW program has met drastic public opposition because of the amount of time that the waste will remain extremely hazardous. This time period is on the order of thousands of years. Opponents of this facility are arguing that the ability to label the disposal facility properly and guard against future intruders is lacking. Many symbols, such as thorns or unhappy faces, have been proposed.

Hopefully, hospital infectious waste management plans in the future that deal with health and safety will include a contingency plan to provide for emergency situations. It is important that these measures be selected in a timely manner so that they can be implemented quickly when needed. This plan should include, but not be limited to, procedures to be used under the following circumstances:

1. Spills of liquid infections waste, cleanup procedures, protection of personnel, and disposal of spill residue

2. Rupture of plastic bags (or other loss of containment), cleanup procedures, protection of personnel, and repackaging of waste

3. Equipment failure, alternative arrangements for waste storage and treatment (e.g., offsite treatment)

Facilities that generate waste should provide employees with waste management training. This training should include an explanation of the waste management plan and assignment of roles and responsibilities for implementation of the plan. Such education is important for all employees who generate or handle wastes regardless of the employee's role (i.e., supervisor or supervised) or type of work (i.e., technical/scientific or housekeeping/maintenance).

Training programs should be implemented when:

1. The infectious waste management plans are first developed and instituted.
2. New employees are hired.
3. Waste management practices are changed.

Continuing education is also an important part of staff training, including refresher training-aids in maintaining personnel awareness of the potential hazards posed by wastes. Training also serves to reinforce waste management policies and procedures that are detailed in the waste management plan.

Many hospitals are beginning to address the issues raised above on a comprehensive basis. Developing environmental management health and safety programs does more than meet the letter of the law. They can cut costs, reduce liability, and ensure that the hospital's primary mission of delivering health care is not jeopardized by a fine or an incident that requires shutting down a facility. On average, though, most hospitals still have attained only partial compliance. That will surely change in the future.[9]

10.5 MULTIMEDIA CONCERNS AND HAZARDS

Environmental quality and natural resources are under extreme stress in many industrialized nations and in virtually every developing nation as well. Environmental pollution is closely related to population density, energy, transportation demand, and land-use patterns, as well as industrial and urban development, The main reason for environmental pollution is the increasing rate of waste generation in terms of quantity and toxicity that has exceeded society's ability to manage it properly. Another reason is that the management approach has focused on the media-specific and the end-of-pipe strategies. There is increasing reported evidence of socioeconomic and environmental benefits realized from multimedia pollution prevention.[10] The prevention of environmental pollution in this century is going to require not only enforcement of governmental regulations and controls, but also changes in manufacturing processes and products as well as in lifestyles and behavior throughout society. Education is key in achieving the vital goal of multimedia pollution prevention.

Nanotechnology has the potential to be used to develop new "green" processing technologies that minimize or eliminate the use of toxic materials and the generation of undesirable by-products and effluents. Research may involve nanotechnology related to improved industrial processes and starting material requirements, development of new chemical and industrial procedures, and materials to replace current hazardous constituents and processes, resulting in reductions in energy, materials, and waste generation.

10.6 HEALTH AND HAZARD RISK ASSESSMENT

Regarding health risk assessment, a significant amount of work is anticipated in two fields:

1. Dose–response, that is, toxicology
2. Nanoparticle pathways in the environment

Both will emerge as key future items for those involved in the nanotechnology field.

The growing concern that risk communication was becoming a major problem led to the chartering of a National Research Council committee (May 1987–June 1988) to examine the possibilities for improving social and personal choices on technological issues by improving risk communication. The National Research Council offers advice from governments, private and nonprofit sector organizations, and concerned citizens about the process of risk communication, about the content of risk messages and ways to improve risk communication.[11]

Future goals are not to make those who disseminate formal risk messages more effective by improving their credibility, understanding, and so on; it is to "improve" their techniques. "Improvement" can occur only if recipients are also enabled to solve their problems at the same time. Generally, this means obtaining relevant information for better-informed decisions.[11] Implementation of many recommendations requires organizational resources of several kinds. One of these resources in particular is time, especially during the most difficult risk communication efforts, as when emergency conditions leave no possibility for consulting with the people concerned, or to assemble the vital information that would be necessary for them.

The committee came forward with three general conclusions that may bring to light why the task of communicating risk does not seem to be working.[11]

Conclusion 1   Even great improvement in risk communication will not resolve the problems or end the controversy. Sometimes, they will tend to create problems through poor communication. There is no ready shortcut to improving the nation's risk communication efforts. The needed improvement in performance can come only incrementally and only from constant attention to many details. For example, more interaction with the audience and the intermediaries involved is necessary for a full understanding of the issue at hand.

Conclusion 2   Better risk communication should not only be about improving procedures, but about improving the content of the risk message. It would be a mistake to believe that better communication is only a matter of a better message. To enhance the credibility, to ensure accuracy, to understand the concerned citizens and their worries, and to gain the insight necessary into how messages are actually perceived, the communicator must ultimately seek procedural solutions.

Conclusion 3   Communication should be more systematically oriented to specific audiences. The concept of openness is the best policy. It is true that the most effective risk messages are those that consciously address the specific audience's concerns. Similarly, the best procedures for formulating risk messages have been those that involved open interaction with the citizen's and their needs.

In the future, the communication of risk become more effective. It should be understood that risk communication is a two-way exchange of information and opinion among individuals, groups, and organizations. The written, verbal, or visual message containing information about the risk should include advice on waste reduction or elimination, so that in the future communication efforts will no longer be necessary.

In conclusion, a problem exists in the perception of risk because the experts' and laypeople's views differ. The experts usually base their assessment on mortality rates, while the laypeople's fears are based on "outrage" factors. In order to help solve this problem, in the future, risk managers must work to make truly serious hazards more outrageous. One example is the recent campaign for the risk involved in cigarette smoke. Another effort must be made to decrease the public's concern with low to modest hazards, that is, risk managers must diminish outrage in these areas. In addition, people must be treated fairly and honestly.

Firm data on accident impacts and a quantitative and general hazard risk assessment model is thus a necessity for all nanoapplications. Indeed, for an industry that is already producing thousands of tons of nanomaterials each year—for applications ranging from cosmetics to solar cells—the importance of characterizing potential accident impacts seems evident.

For the most part, future trends will be found in hazard accident prevention, not hazard analysis. To help promote hazard accident prevention, companies should start employee training programs. These programs should be designed to alert the technical staff and employees about the hazards they are exposed to on the job. Training should also cover company safety policies and the proper procedures to follow in case an accident does occur. A major avenue to reducing risk will involve source reduction of hazardous materials. Risk education and communication, with reference to accidents, are two other areas that will need improvement.

10.7   ENVIRONMENTAL ETHICS

Although the environmental has grown and matured in recent years, its development is far from stagnant. On the contrary, changes in individual behavior, corporate policy, and governmental regulations are occurring at a dizzying pace.

Because of the Federal Sentencing Guidelines, the Defense Industry Initiative, as well as a move from compliance to a values-based approach in the marketplace, corporations have inaugurated company-wide ethics programs, hot lines, and senior line-positions responsible for ethics training and development. The Sentencing Guidelines allow for mitigation of penalties if a company has taken the initiative in developing ethics training programs and codes of conduct.

Whenever these same guidelines apply to infractions of environmental law,[12] the corporate community will undoubtedly welcome ethics integration in engineering and science programs generally, but more so in those that emphasize environmental issues. Newly hired employees, particularly those in the environmental arena, who have a strong background in ethics education will allay fears concerning integrity and responsibility. Particular attention will be given to the role of public policy in the environmental arena as well as in the formation of an environmental ethic.

Regulations instituted by federal, state, and local agencies continue to become more and more stringent. The deadlines and fines associated with these regulations encourage corporate and industrial compliance of companies (the letter of the law), but it is in the personal convictions of corporate individuals that the spirit of the law lies, and therein also the heart of a true ecological ethic.

To bolster this conviction in the heart, a new *dominant paradigm* must emerge. This is defined as the collection of norms, beliefs, values, habits, and survival rules that provide a framework of reference for members of a society. It is a mental image of social reality that guides behavior and expectations.[13]

The general trend in personal ethics is steadily "greener" and is being achieved at a sustainable pace with realistic goals.

A modern-day author suggests the following: The flap of one butterfly's wings can drastically affect the weather.[14] While this statement sounds much like what might come from a romantic ecologist, it is actually part of a mathematical theory explored by the contemporary mathematician James Gleick in his book *Chaos, Making a New Science*. The "butterfly" theory illustrates that the concept of interdependence, as Chief Seattle professed it, is emerging as more than just a purely environmental one. This embracing of the connectedness of all things joins the new respect for simplified living and the emphasis on global justice, renewable resources, and sustainable development (as opposed to unchecked technological advancement) as the new, emerging social paradigm. The concept of environmentalism is now *widely* held; its future is becoming *deeply* held.

10.8 ENVIRONMENTAL AUDITS[15]

Environmental auditing is a systematic, documented, periodic, and objective review by regulated entities of facility operations and practices related to meeting environmental requirements. Audits can be designed to accomplish any or all of the following: verify compliance with environmental requirements, evaluate the effectiveness of environmental management systems already in place, or assess risks from regulated and unregulated materials and practices.

Environmental audits evaluate, and are not a substitute for, direct compliance activities such as obtaining permits, installing controls, monitoring compliance, reporting violations, and keeping records. Environmental auditing may verify but does not include activities required by law, regulation, or permit (e.g., continuous emissions monitoring, composite correction plans at wastewater treatment plants, etc.). Audits do not in any way replace regulatory agency inspections. However, environmental audits can improve compliance by complementing conventional federal state, and local oversight.

The EPA clearly supports auditing to help ensure the adequacy of internal systems to achieve, maintain, and monitor compliance. By voluntarily implementing environmental management and auditing programs, regulated entities can identify, resolve, and avoid environmental problems.

The EPA does not dictate or interfere with the environmental management practices of private or public organizations. The EPA also does not mandate auditing (though in certain instances EPA may seek to include provisions for environmental auditing as part of settlement agreements, as noted below). Because environmental auditing systems have been widely adopted on a voluntary basis in the past, and because audit quality depends to a large degree upon genuine management commitment to the program and its objectives, auditing should remain a voluntary activity.

An organization's auditing program will evolve according to its unique structures and circumstances. Effective environmental auditing programs appear to have certain discernible elements in common with other kinds of audit programs. These elements are important to ensure project effectiveness.[16]

Environmental auditing is fast becoming an integral component of a facility's management plan by not only promoting compliance with regulatory requirements but also limiting environmental liabilities in the form of costly penalties and third-party lawsuits. Corporations have come to realize the significant benefits that result from conducting environmental audits. These benefits range from drastic reduction of fines from federal and state environmental protection agencies through implementation of their audit policies, to participation in the flow of lucrative "green" dollars through business that promote and reward other environmentally conscious entities. Consumers often seek out and patronize these businesses for their environmental policies.

Environmental auditing has developed for sound business reasons, particularly as a means of helping regulated entities manage pollution control affirmatively over time instead of reacting to crises. Auditing can result in improved facility environmental performance, help communicate and effect solutions to common environmental problems, focus facility managers' attention on current and upcoming regulatory requirements, and generate protocols and checklists that help facilities better manage themselves. Auditing also can result in better integrated management of environmental hazards, since auditors frequently identify environmental liabilities that go beyond regulatory compliance.

One of the most compelling reasons to conduct a voluntary environmental audit should be to avoid criminal prosecution. Because senior managers of regulated entities are ultimately responsible for taking all necessary steps to ensure compliance with environmental requirements, the EPA has never recommended criminal prosecution of a regulated entity based on voluntary disclosure of violations discovered through audits and disclosed to the government before an investigation was already under way. Thus, the EPA does not recommend criminal prosecution for a regulated entity that uncovers violations through environmental audits or due diligence, promptly discloses and expeditiously corrects those violations, and meets all other conditions of Section D of the policy.

Fundamentally, there are two types of environmental audits:

1. *Compliance Audit*: An independent assessment of the current status of a party's compliance with applicable statutory and regulatory requirements. This approach always entails a requirement that effective measures be taken to remedy uncovered compliance problems, and is most effective when coupled with a requirement that the root causes of noncompliance also be remedied.

2. *Management Audit*: An independent evaluation of a party's environmental compliance policies, practices, and controls. Such evaluation may encompass the need for: (a) a formal corporate environmental compliance policy, and procedures for implementation of that policy, (b) educational and training programs for employees, (c) equipment purchase, operation, and maintenance programs, (d) environmental compliance officer programs (or other organizational structures relevant to compliance), (e) budgeting and planning systems for environmental compliance, (f) monitoring, record-keeping, and reporting systems, (g) in-plant and community emergency plans, (h) internal communications and control systems, and (i) hazard identification and risk assessment.

The elements of an effective auditing program are available in the literature.[15]

10.9 ISO 14000[15]

The International Organization for Standardization is a private, nongovernmental, international standards body based in Geneva, Switzerland. ISO promotes international harmonization and development of manufacturing product, and communications standards. ISO has promulgated over 8000 internationally accepted standards for everything from paper sizes to film speeds. More than 130 countries participate in the ISO as "Participating" members or as "Observer" members. The United States is a full voting, participating member and is officially represented by the American National Standards Institute (ANSI).

The development of International Standards began with the creation of the International Electrochemical Commission (IEC) in 1906 to make standards for the electrochemical industry. Standards in other areas were developed by the International Federation of National Standardization (ISA), which was established in 1926. ISA's emphasis was on standards pertaining to the mechanical engineering industry. Because of the advent of World War II; ISA's activities ceased.

In 1946, a meeting was held in London with delegates from 25 countries to discuss the development of international standards. After the meeting, it was decided to establish an international organization to coordinate and unify industrial standards. The ISO was formed from this international organization, and started to function in 1947. The first standard published by this body was "Standard Reference Temperatures for Industrial Length Measurement."

Many people will have noticed the seeming lack of correspondence between the official title when used in full, International Organization for Standardization, and the short form, ISO. Shouldn't the acronym be IOS? That would have been the case if it were an acronym. However, ISO is a word derived from the Greek word *isos*, meaning equal. From "equal" to "standard," the line of thinking that led to the choice of "ISO" as a name of the organization is easy to follow. In addition, the name ISO is used around the world to denote the organization, thus avoiding the plethora of acronyms resulting from the translation of "International Organization for Standardization" into the different national languages of members, for example, IOS in English or OIN in French. Whatever the country, the short form of the organization's name is always ISO.[17]

Participation is ISO 14000 is becoming one of the most sought-after statuses in the more general move toward globalization of environmental management. In recent years, there has been heightened international interest in and commitment to improved environmental management practices by both the public and private sectors. This interest is reflected in the success of collaborative international efforts to address environmental problems and in the global recognition of trade-related environmental issues. The Montreal Protocol (the environmental side agreements of The North American Free Trade Agreement) and the mandates resulting from the 1992 Earth Summit of the United Nations Conference on Environment and Development in Rio de Janeiro are all indications that industry and governments all over the world are prepared to put together a plan to manage the environment effectively.

Another indication of interest in improved environmental practices is the emergence of voluntary environmental management standards developed by national standards bodies throughout the world. To address the growing need for an international consensus approach, ISO has undertaken the development of international voluntary environmental standards.[18]

Standards are documented agreements containing technical specifications or precise criteria to be used consistently as rules, guidelines, or definitions of characteristics, to ensure that materials, products, processes, and services are fit to be used for their intended purposes. For example, the format of credit cards, phone cards, and "smart cards" that have become commonplace is derived from an ISO international standard. Adhering to the standard, which defines such features as an optimal thickness (0.76 mm), means that the cards can be used worldwide. International standards thus contribute to making life simpler and to increasing the reliability and effectiveness of goods and services.[17]

The impetus toward international standards is deeply rooted in economic rewards and an expansion into a global economy. The standardization of goods and services will not only increase potential market share but allow goods and services to be available to more consumers. Different countries producing articles using the same technologies but using different sets of standards limits the amount of trade that can be done among countries. Industries that depend on exports realized that there is need for consistent standards in order to trade freely and extensively. International standards have been established for many technologies in different industries such as textiles, packaging, communications, energy production and utilization, distribution of goods, banking, and financial services. With the expansion of global trade, the importance of international standards will continue to grow. The advent of international standards makes it possible for one to purchase a video cassette recorder made in South Korea and use a videotape made in Mexico or buy a computer made in the United States and use disks made in China. International certification programs, such as those developed by the National Association of Corrosion Engineers (NACE), set standards acceptable all over the world. Industries can be assured of qualified services in a timely manner. However, there are still some areas that need standardization; for example, a videotape that is taped in the United Kingdom cannot be played on a VCR in the United States because the systems are not standardized.

10.10 SUMMARY

This chapter has attempted to predict the future regarding environmental activities, with some reference to the nanotechnology field. It is safe to say that the authors' comments have only scratched the surface of this new and dynamic technology. Only time will complete the puzzle as it presently exists.

REFERENCES

1. J. Cox and C. Miro, "EPA, DOE and NIOSH Address IAQ Problems," *ASHRAE Journal*, **10** (July, 1993).
2. Metcalf & Eddy, Inc., *Wastewater Engineering: Treatment, Disposal, and Reuse*, 3rd ed., McGraw-Hill, New York, 1990.
3. J. Kishwara, "The National Pretreatment Program, Clearwaters," *Journal of the New York Water Pollution Control Association*, **8** (Spring 1990).
4. *U.S. Water News*, 1 and 13, September 1991.
5. S. Rice, "Waste Management: The Long View," *Chemical Technology*, 543–546, Sept 1991.
6. S. Brown, "Washing Your Hands of Waste Water Disposal," *Process Engineering*, 34–35, July 1991.
7. L. Theodore, personal lecture notes, 1994.
8. J. Cushman, "Not so Superfund," *New York Times*, National Section, February 7, 1994.

9. K. Lundy, "It's Enough to Make You Sick," *Resources*, February 1994.

10. *Chemecology*, **17**(1) (1988) and **19**(2) (1990).

11. National Research Council, Committee on Risk Perception and Communication, *Improving Risk Communication*, National Academy Press, Washington, DC, 1989.

12. Presentation by N. Cartusciello, Chief, Environmental Crimes Section, U.S. Department of Justice, May 4, 1994.

13. I. Barbour, *Ethics in an Age of Technology*, Harper, San Francisco, 1993.

14. J. Gleick, *Chaos, Making a New Science*, Viking, New York, 1987.

15. G. Burke, B. Singh, and L. Theodore, *Handbook of Environmental Management and Technology*, Wiley, Hoboken, NJ, 2000.

16. EPA, "Restatement of Polices Related to Enforcement of Auditing," FRL-5021-5, July 28, 1994.

17. Introduction to ISO: *www.isoch/infoe/intro*.

18. EPA, *EPA Standards Network Factsheet; ISO 14000. International Environmental Management Standards*, Office of Research and Development, EPA/626/F-97/004, April 1998.

NAME INDEX

Nanotechnology: Environmental Implications and Solutions, L. Theodore and R. G. Kunz
ISBN 0-471-69976-4   Copyright © 2005 John Wiley & Sons, Inc.

SUBJECT INDEX